Y0-BTA-984

# THE CONCEPT OF NUMBER:
## From Quaternions to Monads and Topological Fields

## MATHEMATICS AND ITS APPLICATIONS

*Series Editor:* G. M. BELL, Professor of Mathematics, King's College London (KQC), University of London

## NUMERICAL ANALYSIS, STATISTICS AND OPERATIONAL RESEARCH

*Editor:* B. W. CONOLLY, Professor of Mathematics (Operational Research), Queen Mary College, University of London

Mathematics and its applications are now awe-inspiring in their scope, variety and depth. Not only is there rapid growth in pure mathematics and its applications to the traditional fields of the physical sciences, engineering and statistics, but new fields of application are emerging in biology, ecology and social organization. The user of mathematics must assimilate subtle new techniques and also learn to handle the great power of the computer efficiently and economically.

The need for clear, concise and authoritative texts is thus greater than ever and our series will endeavour to supply this need. It aims to be comprehensive and yet flexible. Works surveying recent research will introduce new areas and up-to-date mathematical methods. Undergraduate texts on established topics will stimulate student interest by including applications relevant at the present day. The series will also include selected volumes of lecture notes which will enable certain important topics to be presented earlier than would otherwise be possible.

In all these ways it is hoped to render a valuable service to those who learn, teach, develop and use mathematics.

## Mathematics and its Applications

*Series Editor:* G. M. BELL, Professor of Mathematics, King's College London (KQC), University of London

Anderson, I. — **Combinatorial Designs**
Artmann, B. — **Concept of Number: From Quaternions to Monads and Topological Fields**
Arczewski, K. & Pietrucha, J. — **Mathematical Modelling in Discrete Mechanical Systems**
Arczewski, K. and Pietrucha, J. — **Mathematical Modelling in Continuous Mechanical Systems**
Bainov, D.D. & Konstantinov, M. — **The Averaging Method and its Applications**
Baker, A.C. & Porteous, H.L. — **Linear Algebra and Differential Equations**
Balcerzyk, S. & Jösefiak, T. — **Commutative Noetherian and Krull Rings**
Baldock, G.R. & Bridgeman, T. — **Mathematical Theory of Wave Motion**
Ball, M.A. — **Mathematics in the Social and Life Sciences: Theories, Models and Methods**
de Barra, G. — **Measure Theory and Integration**
Bartak, J., Herrmann, L., Lovicar, V. & Vejvoda, D. — **Partial Differential Equations of Evolution**
Bell, G.M. and Lavis, D.A. — **Co-operative Phenomena in Lattice Models, Vols. I & II**
Berkshire, F.H. — **Mountain and Lee Waves**
Berry, J.S., Burghes, D.N., Huntley, I.D., James, D.J.G. & Moscardini, A.O. — **Mathematical Modelling Courses**
Berry, J.S., Burghes, D.N., Huntley, I.D., James, D.J.G. & Moscardini, A.O. — **Mathematical Methodology, Models and Micros**
Berry, J.S., Burghes, D.N., Huntley, I.D., James, D.J.G. & Moscardini, A.O. — **Teaching and Applying Mathematical Modelling**
Blum, W. — **Applications and Modelling in Learning and Teaching Mathematics**
Brown, R. — **Topology**
Burghes, D.N. & Borrie, M. — **Modelling with Differential Equations**
Burghes, D.N. & Downs, A.M. — **Modern Introduction to Classical Mechanics and Control**
Burghes, D.N. & Graham, A. — **Introduction to Control Theory, including Optimal Control**
Burghes, D.N., Huntley, I. & McDonald, J. — **Applying Mathematics**
Burghes, D.N. & Wood, A.D. — **Mathematical Models in the Social, Management and Life Sciences**
Butkovskiy, A.G. — **Green's Functions and Transfer Functions Handbook**
Cartwright, M. — **Fourier Methods: Applications in Mathematics, Engineering and Science**
Cerny, I. — **Complex Domain Analysis**
Chorlton, F. — **Textbook of Dynamics, 2nd Edition**
Chorlton, F. — **Vector and Tensor Methods**
Cohen, D.E. — **Computability and Logic**
Cordier, J.-M. & Porter, T. — **Shape Theory: Categorical Methods of Approximation**
Crapper, G.D. — **Introduction to Water Waves**
Cross, M. & Moscardini, A.O. — **Learning the Art of Mathematical Modelling**
Cullen, M.R. — **Linear Models in Biology**
Dunning-Davies, J. — **Mathematical Methods for Mathematicians, Physical Scientists and Engineers**
Eason, G., Coles, C.W. & Gettinby, G. — **Mathematics and Statistics for the Biosciences**
El Jai, A. & Pritchard, A.J. — **Sensors and Controls in the Analysis of Distributed Systems**
Exton, H. — **Multiple Hypergeometric Functions and Applications**

*Series continued at back of book*

# THE CONCEPT OF NUMBER:
## From Quaternions to Monads and Topological Fields

BENNO ARTMANN, Dr.rer.nat.
Professor of Mathematics
Darmstadt Institute of Technology
Darmstadt, Federal Republic of Germany

Translated with additional exercises and material by
H. B. GRIFFITHS, M.Sc., Ph.D.
Professor of Pure Mathematics
University of Southampton

**ELLIS HORWOOD LIMITED**
Publishers · Chichester

Halsted Press: a division of
**JOHN WILEY & SONS**
New York · Chichester · Brisbane · Toronto

This English edition first published in 1988 by
**ELLIS HORWOOD LIMITED**
Market Cross House, Cooper Street,
Chichester, West Sussex, PO19 1EB, England
*The publisher's colophon is reproduced from James Gillison's drawing of the ancient Market Cross, Chichester.*

**Distributors:**
*Australia and New Zealand:*
JACARANDA WILEY LIMITED
GPO Box 859, Brisbane, Queensland 4001, Australia
*Canada:*
JOHN WILEY & SONS CANADA LIMITED
22 Worcester Road, Rexdale, Ontario, Canada
*Europe and Africa:*
JOHN WILEY & SONS LIMITED
Baffins Lane, Chichester, West Sussex, England
*North and South America and the rest of the world:*
Halsted Press: a division of
JOHN WILEY & SONS
605 Third Avenue, New York, NY 10158, USA

This English edition is translated from the original German edition *Der Zahlbeginff,* published in 1983 by Vandenhoeck & Rupprecht, Göttigen, © the copyright holders.



**British Library Cataloguing in Publication Data**
Artmann, B. (Benno)
The concept of number.
1. Number systems.
I. Title II. Series III. Der Zahlbeginff *English*
513.′5

**Library of Congress CIP data available**

ISBN 0–85312–749–2 (Ellis Horwood Limited)
ISBN 0–470–21323–X (Halsted Press)

Printed in Great Britain by Unwin Bros., Woking

# Table of Contents

# TRANSLATOR'S FOREWORD TO THE ENGLISH EDITION

It is perhaps unusual for a translator to begin by explaining what his translation is about, but there is an important motivation underlying Professor Artmann's text that is not apparent in the list of Contents, and which, once summarised, might draw the reader on. It is sufficiently important to make me feel that it should be made available to those who read English and not German.

Firstly, in many countries, the usual undergraduate mathematical curriculum has to make room for some discussion of the number-systems in the chain

$$\mathbb{N} \subseteq \mathbb{Z} \subseteq \mathbb{Q} \subseteq \mathbb{R} \subseteq \mathbb{C} \qquad (*)$$

before rushing on to other matters for which they might form a basis. Sometimes there is a later, optional, course that goes into greater detail, especially concerning the constructions of $\mathbb{R}$ from $\mathbb{Q}$, following Cantor and/or Dedekind. These constructions have a flavour of Analysis, and can be given a topological setting. The other inclusions in the chain (*) are normally dealt with in an 'algebraic' spirit, starting with $\mathbb{N}$ as given; in a later course, $\mathbb{N}$ may well be derived from Peano's axioms and related to Cantor's cardinal arithmetic.

But why do we usually stop with $\mathbb{C}$ ? Historically, once the mystery had been removed from $\mathbb{C}$, through its geometrical model (the Argand diagram), the way was open for Hamilton to ask for even larger systems, such as his quaternions **H** with all their rich 4-dimensional structure.

After a historical development that led to the modern structure theory of algebras (not treated in this book), there came one of those striking theorems that put bounds on the unfettered imagination – the theorem of Frobenius that the only real finite dimensional vector spaces, that can carry the structure of a skew-field, are $\mathbb{R}$, $\mathbb{C}$ or **H**. This was followed by others of the same type, as well as related questions on the coordinatization of various geometries. Artmann treats Frobenius's theorem in some detail, before giving also a proof of the related theorem of Pontrjagin, on the topological characterisation of connected, locally compact skew fields. On the way, he surveys such geometrical matters as the related work concerning vector fields on spheres, and that of Moufang on projective planes.

The arguments used in the proofs are often attractive in themselves, and show how to use, in a striking way, the concepts that undergraduates hear of in courses of 'basic algebra', but which are rarely pushed far enough to be interesting. Serious theorems like those of Frobenius and Pontrjagin display the need for inventing the concepts in the first place.

Recent applications in approximation theory and iterations have caused the old topic of continued fractions to appear again. Artmann, too, also uses continued fractions in his treatment of $\mathbb{R}$ to illustrate Eudoxus's approach to the classical Greek notions concerning the question of regarding the continuum as a number-line. He uses infinite decimals as well, and raises another question, because many of his own students are intending to teach. Thus: how do you explain (at school, to the person in the street,...) what real numbers are? If you start with decimals, say, you will have an easy lead-in, but what about multiplication? Similarly, you might start with continued fractions, and run into technical difficulties. What are the relative advantages/disadvantages of the Cantor and Dedekind approaches? Here is material extremely relevant to all who teach mathematics, at whatever (and especially University) level.

Following Weierstrass, Cantor and Dedekind, the nineteenth century provided a successful model for the intuitive notion of 'The Continuum', in which Analysis could be pursued in its now classical form. But that intuitive notion had for centuries included feelings about the infinitesimally small and the infinitely large. These feelings are not expressible in the Weierstrass model of the continuum, but even after more than a century of dogmatic teaching of that model, the feelings persist, and are to be found quite widely in the psychology of late-adolescent students, as studies by Dr. J. Mamona and others have shown. (For details, see item [48] in the list of References; it also includes a thoughtful discussion of the word 'intuitive', which mathematicians use frequently but without definition). Nowadays, however, different models of the continuum are available, within non-standard Analysis; these *can* model the feelings mentioned above. Artmann gives a technical introduction in his Chapter 7, and an argument to dissuade enthusiasts from rushing to teach non-standard Analysis to beginners.

The structure of the book is clearly defined by the list of contents, so we shall not enumerate them again at this point. We might also regard the text as an unfolding of that immensely important part of our mathematical heritage which originated from questions like 'What is a good number system?'. To comprehend it, one needs the technical apparatus commonly mentioned in undergraduate courses, — some language of sets, functions and logic, some ideas of algebraic structure (homomorphisms, residue classes, matrices, eigenvalues) and, from Analysis, the idea of a limit. As a help I have supplied an Appendix on Notation and Terminology as a revision that I would expect British students to need on passing into their Final year after a memory-erasing summer vacation. Artmann usually revises such notions when he needs them, and defines most things — rings, fields, neighbourhoods, continuity — from scratch. The rusty reader should just read on, and turn to his favourite text of algebra, analysis, or topology when courage fails. One also, of course, needs to be interested in the basic questions being raised, and to have the patience to receive fairly lengthy answers in a thoughtful manner. Incidentally, Professor Artmann shows exemplary patience himself by dividing long proofs into 'Preparation', 'Steps 1,2,...', and even 'substeps'.

Because the German language allows highly convoluted sentence structure, direct translation into simple, international English is not always possible — at least without using word patterns of the multi-adjectival, German-American "defense-related" type that nowadays so sadly infect our Engineering colleagues. Consequently, I have sometimes made small alterations, while trying to retain Artmann's spirit and informal style; sometimes too I have had to allow for cultural differences. For that reason also, I have occasionally substituted a bibliographical reference by one in English; in the text, I give the title of each German references in English, to indicate to the reader the type of writing that nowadays can be found in German. Texts originating in Britain are traditionally expected to contain exercises, and I have added some new ones, or formalised some suggestions that Artmann leaves to the reader. My colleagues Drs. David Singerman and Christopher Thompson have each at one time or another given a third-year option on Number Systems in Southampton University; and Dr. Singerman contributed exercises on quaternions, while Dr. Thompson supplied not only exercises, but the additional text in Chapter 7, on axioms for $^*\mathbb{R}$. To them I am very grateful.

I also wish to thank my typist, Mrs. Lawson, for so skilfully preparing camera-ready copy through various messy stages, starting from rough tape-recordings. Some of these tapes were made by Rachel Ford, to whom thanks are also due.

H.B. Griffiths
Southampton

# EXTRACT FROM THE AUTHOR'S FOREWORD TO THE GERMAN EDITION

This book consists of lectures that I have given, in Darmstadt since 1974, under various titles, for beginners or students of the third semester.

I have tried to concentrate on those themes concerning numbers, which seem to me to be the most important on essential and historical grounds, and which in my opinion ought to belong to any wide-ranging, general education of every mathematics student. The content of that, so it occurred to me, should be to use the study of the number systems to reveal Mathematics as a visible unity, an aspect that is just what is lost from the vision of third-semester students. In order to open up the subject of interest from all sides, tools of knowledge are used from Analysis, Algebra, Geometry and Topology. Therefore also I have preferred geometrical approaches and renounced generality, if thereby the underlying ideas of a proof became more intuitive (e.g. in Liouville's Theorem on transcendental numbers). After the essential unity of Mathematics, I have emphasised its historical continuity. In that respect, numerous indicators enrich the text, from simple citations of names and dates, to explicit quotations (Hamel, Hamilton, etc.). So also, in order to treat the conjectured earliest Greek concept of real number, as worked out by O. Becker [4], I have referred without proofs to some facts about Continued Fractions; and I have given a further development in Sections 2E and 3C. I confess gladly, that the working out of this theme, as well as of others, in this text has given me great pleasure.

Taking into account my own pleasure (and that of my students) I have above all considered the significance and implications of each theorem and concept, as a criterion for inclusion in the text. This is reflected below in the strong and oft–repeated emphasis on the completeness of the real numbers, and in the detailed treatment of the Fundamental Theorem of Algebra, with the related theorems of Frobenius and Pontrjagin.

The basis for the present version of the text was the notes of a 4–hour/week course (with exercises) in the Winter Semester 81/82. As in those lectures, so also in the book, we assume a certain experience in doing mathematics of the sort acquired in basic studies. ...

Benno Artmann.
Darmstadt.
December 1982

# 1

# The complete ordered field $\mathbb{R}$

INTRODUCTION.

The system $\mathbb{R}$ of real numbers has a property, known as that of 'Completeness', which is fundamental for the whole of Mathematical Analysis. On that account we will carefully clarify this concept and discuss its consequences. Now, besides having algebraic structure, $\mathbb{R}$ has *order–theoretic* aspects, i.e. those relating to inequalities. We shall need some preliminary propositions concerning these aspects and we shall collect them together in the first three paragraphs with many examples. In Section 1D there is a main theorem that allows us to characterise, and so recognise, $\mathbb{R}$ as a complete ordered field. In Section 1E will follow the most important variations of the concept of completeness used in Analysis. One consequence is a characterization due to Cantor, of the "chain" $(\mathbb{Q}, \geqslant)$ from which one can easily deduce the uncountability of $\mathbb{R}$. Just from these brief considerations, the role of the completeness property is made especially clear.

To establish notation, the number systems $\mathbb{N}, \mathbb{Z}, \mathbb{Q}, \mathbb{R}, \mathbb{C}$ are respectively the natural numbers $\{1, 2, \ldots\}$, the integers $\{0, \pm 1, \pm 2, \ldots\}$, the rational, the real and the complex numbers. Also we refer to the system Dez which has as elements all decimal numbers of the form $a \times 10^{-i}$ with $a \in \mathbb{Z}$ and $i \in \mathbb{N}$.

## 1A THE CONCEPT OF A FIELD: ALGEBRAIC PRELIMINARIES.

The concepts "ring" and "field" relate to calculations in systems with two operations. We note the definitions here, but for further details see for example S. Lang: *Algebraic Structures* [42]. Since an abelian group is a set $G$ together with an operation $+$ of addition, we often refer to 'the group $(G, +)$'. Similarly, the notation $(R, +, \cdot)$ is used for rings to indicate a set with operations $+$, and $\cdot$ that satisfy the following definition.

**Definition**. A system $(R, +, \cdot)$ is called a **Ring** whenever it possesses the following properties:
(a) $(R, +)$ is an abelian group (with neutral element $0$);
(b) $(R, \cdot)$ is associative and has a neutral element $1 \neq 0$;
(c) the distributive laws hold:

$$a \cdot (b + c) = ab + ac$$

$$(a + b) \cdot c = ac + bc \ ,$$

(where, as is customary, we often omit the "dot" and write $ab$ for $a \cdot b$ — but sometimes include the dot as a "spacer").

Many authors omit from the definition of a ring the existence of the neutral element $1$. Although the assumption that $1 \neq 0$ is usual, it is not necessary; but we use the above definition because we do not strive for maximum generality.

*EXAMPLES.* Of the systems mentioned in the Introduction, $(\mathbb{N},+,\cdot)$ is not a ring, but the following are all rings:

$$(\mathbb{Z},+,\cdot),\ (\mathrm{Dez},+,\cdot),\ (\mathbb{Q},+,\cdot),\ (\mathbb{R},+,\cdot),\ (\mathbb{C},+,\cdot)$$

From a given ring $R$ and a given $n \in \mathbb{N}$ we can form the ring $(R^{n \times n}, +, \cdot)$ of $n \times n$ – Matrices with entries taken from R. (This is known as the *Full* Matrix Ring because we include all possible Matrices.)

**Exercise 1**.
Show that in every ring $(-1)\cdot(-1) = 1$. Explain the difference between this equation and the equation $-(-1) = 1$. What role does the first equation play in a construction of $\mathbb{Z}$ from $\mathbb{N}$? [See Griffiths–Howson [24], p.250].

**Definition**. Let $(K, +, \cdot)$ be a ring, and let $K_0 = K\backslash\{0\}$. Then this ring is said to be a **Field**, when $(K_0, \cdot)$ is also an abelian group, i.e. if the following rules hold:
(a) $1 \neq 0$ (for, the neutral element of multiplication must lie in $K_0$). Also for every $a \neq 0$ there exists $a^{-1}$ with $a^{-1} \cdot a = 1$.
(b) $ab = ba$ for all $a, b \in K$ (Commutativity of multiplication).

*EXAMPLES.* $(\mathbb{Z},+,\cdot)$ and $(\mathrm{Dez},+,\cdot)$ are not fields, but $(\mathbb{Q},+,\cdot)$, $(\mathbb{R},+,\cdot)$ and $(\mathbb{C},+,\cdot)$ are fields. For brevity, we often say "K is a field" when the operations $(+, \cdot)$ are understood.

<u>Exercise 2</u>.

(a) Show that for each $n \geq 2$ the *full* Matrix ring $(R^{n \times n}, +, \cdot)$ is never a field.

(b) Let $K$ be a field and $r \in K$. Define a new set

$$K(r) = \{ \begin{bmatrix} a & rb \\ b & a \end{bmatrix} \mid a, b \in K\} .$$

What are the conditions on $r$ for which $K(r)$ is also a field? What role is played in $K(r)$ by the matrices

$$A = \begin{bmatrix} 0 & r \\ 1 & 0 \end{bmatrix} \text{ and } \begin{bmatrix} 0 & -r \\ -1 & 0 \end{bmatrix} ?$$

(Investigate $A^2$, etc).

One can form the *finite* fields $\mathbb{Z}_p$ from $\mathbb{Z}$ (or $\mathbb{N}$) by "calculation modulo p", if p is a prime number. (See Appendix or Griffiths-Hilton [23] Chapter 10) Using the process explained in the exercise, we can for example derive a field with 9 elements, starting from $\mathbb{Z}_3$.

## 1B ORDER RELATIONS, COMPLETENESS.

The idea of thinking of the system of real numbers as lying on the Number line makes it plausible that numbers can be ordered. We make this intuition precise by means of the following concepts.

<u>Definition</u>. The pair $(M, \subseteq)$, consisting of a set M and a relation in M, is said to be an **Ordered** set, whenever $\subseteq$ is an **Order relation**. Thus we require the following rules to hold:

(a) $\subseteq$ is transitive, i.e. $(x \subseteq y$ and $y \subseteq z) \Rightarrow x \subseteq z$

(b) $\subseteq$ is reflexive, i.e. $x \subseteq x$ for all $x$ in $M$

(c) $\subseteq$ is identitive, i.e. $(x \subseteq y$ and $y \subseteq x) \Rightarrow x = y$.

Here, $\subseteq$ is simply a symbol standing for the order relation: it need have nothing to do with the usual inclusion relation between sets.

*EXAMPLES*. $(\mathbb{N}, \leq)$, $(\mathbb{Z}, \leq)$, (Dez, $\leq$), $(\mathbb{Q}, \leq)$, $(\mathbb{R}, \leq)$ are all ordered sets. $(\mathbb{N}, \mid)$ with the divisibility relation (meaning x divides y) is an ordered set; however, this is not true of $(\mathbb{Z}, \mid)$, as here condition (c) is not satisfied (why not?) For each set M the "power set" $\mathfrak{P}M$, consisting of all subsets of M, becomes an ordered set $(\mathfrak{P}M, \subseteq)$ with the usual inclusion relation between subsets of M.

The Number line also displays a further property of the order relation between numbers.

<u>Definition</u>. An ordered set $(M, \subseteq)$ is said to be **linearly** ordered or a **chain** provided the extra condition holds:

(d) $\subseteq$ is linear, that is, for all $x, y$ in $M : x \subseteq y$ or $y \subseteq x$.

Often in an ordered set, we write

$$x \subset y \quad \text{iff} \quad x \subseteq y \quad \text{and} \quad x \neq y$$

and then refer to a "strict" or "total" order. Thus (d) allows exactly three mutually exclusive possibilities: either $x \subset y$ or $x = y$ or $y \subset x$ — the 'law of *Tri*chotomy'.

*EXAMPLES.* $(\mathbb{N}, |)$ is not a chain, nor is $(\mathfrak{P}M, \subseteq)$ whenever $M$ has more than one element; but $(\mathbb{N}, \trianglelefteq)$, $(\mathbb{Z}, \trianglelefteq)$, $(\text{Dez}, \trianglelefteq)$, $(\mathbb{Q}, \trianglelefteq)$ and $(\mathbb{R}, \trianglelefteq)$ are chains. Perhaps surprisingly, we can define a linear order on the field $\mathbb{C}$ of complex numbers and indeed in different ways, provided we recall two facts about complex numbers:

1. Every complex number can be written in the form $z = r \cdot e^{i\rho}$ with $r = |z|$ and $\rho = \arg z$. Now define $z \leqslant w$, whenever: either $|z| < |w|$, or $|z| = |w|$ and $\arg z \leqslant \arg w$. (Here, $\leqslant$ is the usual inequality relation between real numbers). In particular, in this ordering, $0$ becomes the first element of the ordered set $(\mathbb{C}, \leqslant)$.

2. Instead of splitting $z$ multiplicatively, as previously, let us split it additively and write $z = x + iy$. If also $w = u + iv$, then we can define a new order relation and write $z \leqslant w$ whenever either $x < u$, or $x = u$ and $y \leqslant v$. This is the so-called **lexicographic** order of the coordinate plane, which we shall denote in later examples by $(\mathbb{R} \times \mathbb{R}, \leqslant_{lex})$. Correspondingly we write $(\mathbb{Z} \times \mathbb{Z}, \leqslant_{lex}))$, etc.

Exercise 3.

(a) Show that with each of these order relations, $\mathbb{C}$ is linearly ordered, but that the second has no "first" element. In each case, draw a sketch to show what $x \leqslant y$ means.

(b) Define an order-relation on the plane by: $(x,y) \leqslant (u,v)$ provided $x \trianglelefteq u$ and $y \trianglelefteq v$ in $\mathbb{R}$. Show that it is not a linear order.

(c) Let $(K, \leqslant)$, $(L, \leqslant)$ be chains and $f : K \to L$ a mapping. We say that $f$ *preserves order* if $x \leqslant y$ in $K$ implies $f(x) \leqslant f(y)$ in $L$. Prove that if $f$ is also a bijection, then $f^{-1} : L \to K$ preserves order. Show that this is false without the linearity of the order (compare the lexicographic and vector orders on the plane, with $f =$ identity).

By confining ourselves to the idea of an ordered set we have isolated one structural component of the Number system. This raises the question: how far can we already — just by confining ourselves to this one structural component, — characterise the differences between the various domains of numbers? In order to make the distinctions that are appropriate for answering the question, we shall need the following further definitions:

Definition.

(a) Two elements $x, y$ of a chain $(K, \leqslant)$ are said to be **neighbouring** if* $x < y$ and there is no $z$ with $x < z < y$.

---

* No serious ambiguity arises if from now on we use the same symbol $<$ rather than $\subset$, whether or not the order relation is $\subseteq$, $\trianglelefteq$, or $\leqslant$.

(b) A chain is said to be **dense** (in itself), if the following holds: given any two elements $x, y$ with $x < y$ then there is some $z$ with $x < z < y$. In other words: A chain is dense, if it contains no neighbouring elements.

*EXAMPLES.* $(\mathbb{N}, \trianglelefteq)$ and $(\mathbb{Z}, \trianglelefteq)$ are not dense, but $(\mathrm{Dez}, \trianglelefteq)$, $(\mathbb{Q}, \trianglelefteq)$ and $(\mathbb{R}, \trianglelefteq)$ are dense.

The distinguishing feature of $(\mathbb{R}, \trianglelefteq)$ is picked out by the property described in the following definition.

<u>Definition</u>. <u>Completeness</u>.

(a) Let $(K, \leqslant)$ be a chain and $M \subseteq K$ a subset. An element $b \in K$ is called an **upper bound** of $M$, whenever

$$\text{for all } m \in M : m \leqslant b .$$

If an upper bound of $M$ exists, then $M$ is said to be **bounded above**.

(b) An element $s \in K$ is said to be the **least** upper bound of $M$ in
case: $s$ is an upper bound of $M$
and: for all $b \in K$: ($b$ is an upper bound $\Rightarrow s \leqslant b$).

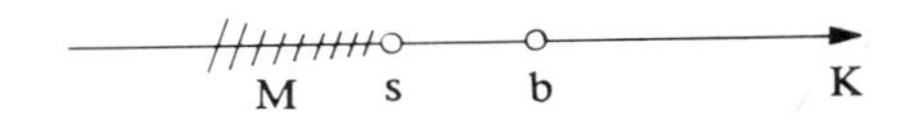


Fig. 1

Note: if a least upper bound of $M$ exists, then it is <u>unique</u>. For if $s$ and $t$ are both least upper bounds, then it follows from the last condition that both $s \leqslant t$ and $t \leqslant s$, so $s = t$.

(c) A chain $(K, \leqslant)$ is said to be **complete**, if every non-empty subset which is bounded above has a least upper bound in $K$.

Common notations for the least upper bound of $M$ are:

$$\text{lub}(M) = \sup M = \text{Supremum of } M.$$

<u>Exercise 4.</u>

(a) Formulate for yourself the terms "lower bound", "greatest lower bound" (= glb or inf) for a subset $L$ of a chain $(K, \leqslant)$.

(b) Show: If $(K, \leqslant)$ is complete, then every non empty subset $L$ which is bounded below has a greatest lower bound.

(c) Suppose $f : (K, \leqslant) \to (L, \leqslant)$ is an order preserving mapping. Which of the properties *neighbouring, dense, lub, complete* are preserved by $f$? Are all preserved if $f$ is a bijection?

*Examples and Clarifications of Completeness.*

*EXAMPLE.* (Dez, $\trianglelefteq$). In order to formulate this example, we shall use our knowlege of the embedding of Dez in $\mathbb{Q}$. Consider the subset M of Dez, defined by the equation

$$M = \{x \in \text{Dez} \mid 9x < 7\} .$$

M is not empty ($0 \in M$) and it is bounded above (e.g. by $b = 10$). In Dez however, M has no least upper bound: the rational number 7/9 does not belong to Dez, and all elements of M are smaller than 7/9 (in $\mathbb{Q}$!). Therefore 7/9 is certainly an upper bound of M in $\mathbb{Q}$, but not in Dez, because there we cannot express 7/9 in the proper decimal form. If now $b \in$ Dez and $7/9 < b$, then b is certainly not the least upper bound of M, for there is then some $c \in$ Dez with $7/9 < c < b$. (Consider the last decimal place $\neq 0$ of b.) On the other hand, if $d \in$ Dez with $d < 7/9$, then d is certainly not an upper bound of M because then there would be an $f \in$ Dez with $d < f < 7/9$. Therefore, M has no least upper bound in Dez, which means: (Dez, $\trianglelefteq$) *is not complete.*

*EXAMPLE.* Let $L \subseteq \mathbb{Q}$ be given by

$$L = \{x \in \mathbb{Q} \mid x^2 < 5\} = \{x \in \mathbb{Q} \mid -\sqrt{5} < x < \sqrt{5}\}$$

There is a well-known argument by which you should be able to show that L has no least upper bound in ($\mathbb{Q}$, $\trianglelefteq$), so ($\mathbb{Q}$, $\trianglelefteq$) is not complete. Observe how $\sqrt{5}$ here plays a role analogous to 7/9 in the previous example.

One should not however, be too heavily swayed by this example. We have already seen in Exercise 2 that one can easily create roots with the help of matrices. Because of that, the kernel of the problem can be expressed in the following way:

Suppose we would like to define a measure of length of a curve by means of successive estimates of piece-wise approximations, then we would obtain a non-empty set of approximate values which we expect intuitively to be bounded above. If we want to ascribe a length to the curve, then the most natural choice is the least upper bound of these values of length. (It is important that these arguments be possible for curves without more refined definitions.) Corresponding remarks hold for the areas of surfaces with the elementary example of the circular disc. (See, for example, Griffiths–Hilton: [23], Chap.15).

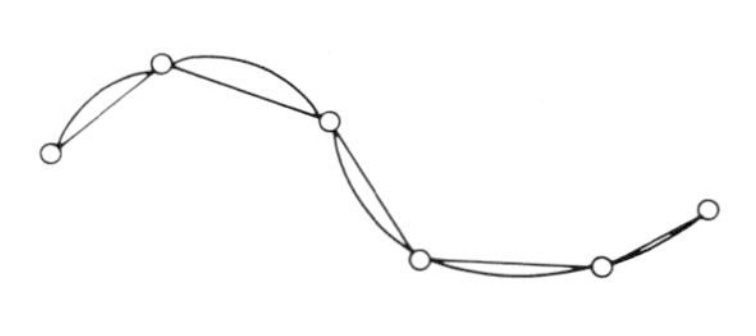

Fig.2

*EXAMPLE.* What can we say about the system $(\mathbb{Z}, \trianglelefteq)$ concerning completeness? Let $M \subseteq \mathbb{Z}$ be a non-empty subset which is bounded above. If we decrease this upper bound to try to find a *least* upper bound then we shall eventually meet a "first element" of M. Let this be S. Then

$$\text{for all } x \in M : x \leqslant s,$$

so s is an upper bound. Further: s is smaller than any other upper bound of M. Therefore s is the least upper bound of M and hence: $(\mathbb{Z}, \trianglelefteq)$ *is complete.* This might make us feel that completeness is not enough for our goal of characterising $(\mathbb{R}, \trianglelefteq)$. However, what we are looking for is a combination of "dense" and "complete".

Of the usual systems of numbers, only $(\mathbb{N}, \trianglelefteq)$, $(\mathbb{Z}, \trianglelefteq)$ and $(\mathbb{R}, \trianglelefteq)$ are complete. As $(\mathbb{Z}, \trianglelefteq)$ is not dense, the only complete chain among these is $(\mathbb{R}, \trianglelefteq)$, and this observation brings us more closely to being able to characterise $\mathbb{R}$. With the help of the decimal form of real numbers, we can at least make the completeness of $\mathbb{R}$ plausible. (By doing that, we do not ask what a decimal fraction properly is, or how one can add decimal expansions and so on. We shall come back later on to problems of that kind.) To make plausible the existence of a lub of a given non-empty subset M of $\mathbb{R}$, we now use a familiar process of using successive approximations.

As M is non-empty and bounded, there is some $z \in \mathbb{Z}$ such that

$$z \text{ is not an upper bound of } M$$

and

$$z + 1 \text{ is an upper bound of } M$$

(The existence of z will be taken for granted during this demonstration of plausibility). Now we refine our estimate through sub-division by tenths. For we have:

There exists $d_1 \in \{0,1,\ldots,8,9\}$ such that

$$z + \frac{d_1}{10} \text{ is not an upper bound of } M$$

and

$$z + \frac{d_1+1}{10} \text{ is an upper bound of } M$$

In a corresponding way we repeat this process further and there will result a decimal expansion

$$s = z + (0.d_1d_2d_3 \ldots)$$

We now maintain that s is the lub of M. For:

(a) s is an upper bound of M (otherwise some $d_i$ would have been chosen that was too small).

(b) s is the least upper bound (otherwise some $d_i$ would have been chosen too large).

Indeed, one can describe the Completeness with the help of the decimal representation as follows: *Every* infinite decimal expansion displays a genuine real number. This will be stated more precisely and there will be more about it in Section 1E and Chapter 2.

*EXAMPLE.* Is the lexicographically ordered plane (see p.4) dense, and is it complete? — It is clearly dense, as one can easily verify. As to completeness: let M be the set given by

$$M = \{(x, y) \in \mathbb{R} \times \mathbb{R} \mid x < 1\} .$$

Then M is not empty because $(0, 0) \in M$, and further, $(2, 0)$ is an example of an upper bound of M. Is there a lub of M? Certainly if $a < 1$ then any point of the form $(a, b)$ is not an upper bound. On the other hand, $(1, b)$ is an upper bound but it is certainly not the smallest because $(1, b - 1)$ is also an upper bound and $(1, b - 1) < (1, b)$. Therefore, there is no smallest upper bound of M and hence $(\mathbb{R} \times \mathbb{R}, \leqslant_{lex})$ *is not complete.* However, certain intervals in $(\mathbb{R} \times \mathbb{R}, \leqslant_{lex})$ are complete — for example the set between $(1, 0)$ and $(1, 1)$.

Survey on the Properties of Order.

So far, we have seen:
- $(\mathbb{Z}, \trianglelefteq)$ is not dense, but complete.
- $(\mathrm{Dez}, \trianglelefteq)$ is dense, but not complete.
- $(\mathbb{Q}, \trianglelefteq)$ is dense, but not complete.
- $(\mathbb{R}, \trianglelefteq)$ is dense and complete.
- $(\mathbb{R} \times \mathbb{R}, \leqslant_{lex})$ is dense, but not complete.

**Exercise 5.**
Investigate which of the properties, dense or complete, are possessed by the lexicographically ordered sets $(\mathbb{Z} \times \mathbb{Z}, \leqslant)$, $(\mathrm{Dez} \times \mathrm{Dez}, \leqslant)$, and $(\mathbb{Q} \times \mathbb{Q}, \leqslant)$.

It would be interesting to be able to distinguish between the above-mentioned chains which are not dense or not complete, by finding Order-theoretic characteristics that discriminate between them. We shall eventually do that (in particular, we shall decide the question whether $(\mathrm{Dez}, \trianglelefteq)$ and $(\mathbb{Q}, \trianglelefteq)$ are different.)

## 1C ORDERED GROUPS AND FIELDS.

So far we have been looking at the algebraic, and the order-theoretic, structures of the number system separately, but now we study the ways in which these two structures operate together.

**Definition.** Let $(G, *)$ be an abelian group and $(G, \leqslant)$ a chain. The (mixed) structure $(G, *, \leqslant)$ is said to be an **ordered** (abelian) **group**, whenever the following "Monotone law" holds which tells us how the order relation is tied to the group operation:

$$(\text{For all } x, y, z \in G)\ x < y \text{ implies: } x * z < y * z .$$

The Monotone law is therefore the essential characteristic of an ordered group. Without it we would only be able to study the group $(G, *)$ or the chain $(G, \leqslant)$ on their own and we would be in the same position as in the previous two sections.

*EXAMPLES.*

(a) $(\mathbb{Z}, +, \trianglelefteq)$, $(\text{Dez}, +, \trianglelefteq)$, $(\mathbb{Q}, +, \trianglelefteq)$ and $(\mathbb{R}, +, \trianglelefteq)$ are ordered groups.

(b) Question : What is the situation with respect to $(\mathbb{R} \times \mathbb{R}, +, \leqslant_{lex})$?

(c) Multiplicative groups of the Number domains: Let $\mathbb{Q}^{\times} = \mathbb{Q} \setminus \{0\}$, $\mathbb{R}^{\times} = \mathbb{R} \setminus \{0\}$, and $\mathbb{Q}^{+} = \{x \in \mathbb{Q} \mid 0 < x\}$, $\mathbb{R}^{+} = \{x \in \mathbb{R} \mid 0 < x\}$. Then $(\mathbb{Q}^{\times}, \cdot)$, $(\mathbb{R}^{\times}, \trianglelefteq)$, $(\mathbb{Q}^{+}, \cdot)$, $(\mathbb{R}^{+}, \cdot)$ are abelian groups. $(\mathbb{Q}^{\times}, \cdot, \trianglelefteq)$ and $(\mathbb{R}^{\times} \cdot \trianglelefteq)$ are <u>not</u> ordered groups (for negative $z$ the Monotone law does not hold!), but $(\mathbb{Q}^{+}, \cdot \trianglelefteq)$ and $(\mathbb{R}^{+}, \cdot, \trianglelefteq)$ are ordered groups.

<u>The Archimedean Property</u>.

Two ordered groups may differ either through their group structure, or through their Order structure. These differences could be distinguished by the methods of the previous sections, but they are not typical for the mixed structure called an "ordered group". The most important "mixed phenomenon" is the *Archimedean property*, so called because Archimedes discussed it around 240 B.C., relative to $(\mathbb{R}, +, \trianglelefteq)$; however he indicated that it had been used by earlier Mathematicians. For the formulation we shall need a more convenient way of writing expressions like $x * x$, $x * x * x$ etc., in a group $(G, *)$. Thus we write

$$1{*}x = x$$

$$2{*}x = x * x$$

$$3{*}x = x * x * x \quad \text{etc.}$$

$$n{*}x = x * \ldots * x \qquad (\text{with } n \text{ factors})$$

If we use additive notation (for example in the usual additive groups of numbers), then $n{*}x$ becomes familiar, for

$$n{*}x = x + x + \ldots + x = n \cdot x ;$$

but on the other hand, if we write the groups multiplicatively, $n{*}x$ becomes:

$$n{*}x = x \cdot x \cdot \ldots \cdot x = x^n .$$

<u>Definition</u>. The ordered group $(G, *, \leqslant)$ with neutral element $e$ is said to be **Archimedean ordered** whenever:

$$\text{If } e < x < y \text{, there exists } n \in \mathbb{N} \text{ such that } y < n{*}x .$$

Written additively, this condition becomes:

If $0 < x < y$, there exists $n$ with $y < nx$.

Written multiplicatively:

If $1 < x < y$, there exists $n$ with $y < x^n$.

*EXAMPLES.*

(a) The ordered additive groups of the number system are all Archimedean as are also the ordered multiplicative groups of positive numbers.

(b) The lexicographically ordered group $(\mathbb{R} \times \mathbb{R}, +, \leqslant)$ is not Archimedean. To see this let $x = (0, 3)$ and $y = (5, 0)$. Then $(0, 0) < (0, 3) < (5, 0)$. Further $n*x = (0, 3n)$ and because of the lexicographical ordering, we have for each $n : (0, 3n) < (5, 0)$. We therefore cannot express $y = (5, 0)$ as a multiple of $x = (0, 3)$, so the group is not Archimedean.

(c) In (b) we confined ourselves to integer coordinates, so we actually showed that the lexicographically ordered group $(\mathbb{Z} \times \mathbb{Z}, +, \leqslant_{lex})$ is not Archimedean. We can however, define another linear order $\subseteq$ so that $(\mathbb{Z} \times \mathbb{Z}, +, \subseteq)$ becomes an Archimedean ordered group. To do this, we use the mapping

$$(i, j) \to 2^i 5^j$$

which defines an isomorphic embedding of $(\mathbb{Z} \times \mathbb{Z}, +)$ in $(\mathbb{Q}^+, \cdot)$. (Here we assume that the reader knows about properties of injective homomorphisms). This mapping allows us to use the ordering of $\mathbb{Q}^+$ and carry it back to $\mathbb{Z} \times \mathbb{Z}$, by writing

$$(i, j) \subseteq (k, m) \Leftrightarrow_{def} : \quad 2^i 5^j \leqslant 2^k 5^m \text{ (in } \mathbb{Q}) .$$

<u>Exercise 6</u>.

(a) A mapping $h$ between ordered groups $G, H$ is a *homomorphism* if it is a group homomorphism and preserves order. Investigate whether $h$ preserves an Archimedean order.

(b) Show that with Example (c) above, all the appropriate properties of an Archimedean order are satisfied. In particular, is $\mathbb{Z} \times \mathbb{Z}$ dense when taken with this order relation?

We have thus turned one and the same group, namely $(\mathbb{Z} \times \mathbb{Z}, +)$, into an ordered group in two different ways. Furthermore, instead of using the primes 2 and 5, we could have used any two primes $p$ and $q$ (with $p \neq q$) and in each case, we would have obtained a dense Archimedean order on $(\mathbb{Z} \times \mathbb{Z}, +)$.

Next, in order to get closer to our goal — the real numbers — we consider ordered fields.

<u>**Definition**</u>. Let $(K, +, \cdot)$ be a field and $(K, \leqslant)$ a chain. Then $(K, +, \cdot, \leqslant)$ is said to be an **ordered field**, provided the following rules hold:

(a) $(K, +, \leqslant)$ is an ordered group (i.e. it satisfies the Monotone law with respect to addition)

(b) multiplication and order are connected in the sense of the following Monotone law for multiplication:

$$\text{for all } a, b, c \in K : a < b \Rightarrow \begin{cases} ac < bc & \text{iff } 0 < c \\ bc < ac & \text{iff } c < 0. \end{cases}$$

The Monotone conditions for addition and multiplication are nothing other than the usual rules for calculating with inequalities. (If we forget the possibility of the inverse of multiplication, we obtain at once the definition of an **ordered ring** $(R, +, \cdot, \leqslant)$.)

An ordered field $(K, +, \cdot, \leqslant)$ is said to be *Archimedean–ordered* whenever its additive group $(K, +, \leqslant)$ is itself Archimedean–ordered. That is, the order is Archimedean whenever we can say that to every $x, y$ with $0 < x < y$ there exists $n$ with $y < nx$. The same holds for rings in general.

*EXAMPLES*. $(\mathbb{Q}, +, \cdot, \trianglelefteq)$ and $(\mathbb{R}, +, \cdot, 6 \trianglelefteq)$ are ordered fields but $(\mathbb{Z}, +, \cdot, \trianglelefteq)$ and $(\text{Dez}, +, \cdot, \trianglelefteq)$ are ordered rings.

Of the complex numbers, we have up to now only touched on their order properties. There were good grounds for this, as our next considerations should indicate.

<u>LEMMA</u>. *In an ordered ring all squares are* $\geqslant 0$.

<u>Proof</u>. If $x = 0$, then $x^2 = 0 \geqslant 0$. If $0 < x$, then $0 = 0 \cdot x < x \cdot x = x^2$ by (b). In case $x < 0$, then $0 = 0 \cdot x < x \cdot x = x^2$ by (b). □

In particular, therefore, we always have $1 = 1^2 > 0$ in any field, (and in a ring whenever $1 \neq 0$.)

<u>**Exercise 7**</u>.
Show that in any ordered field:

(a) $x > 0 \Leftrightarrow x^{-1} > 0$, and

(b) if $x > 0$, then $(x < 1 \Leftrightarrow 1 < x^{-1})$.

Now consider the complex numbers; here $0 < i^2 = -1$. If it were possible to give them an ordering that satisfies the monotone laws, then we would have both $0 < i^2 = -1$ and $0 < 1^2 = 1$. However, with the Monotone property of addition, we would then be forced to have

$$0 < -1 \Rightarrow 1 = 0 + 1 < -1 + 1 = 0 ,$$

therefore $1 < 0$ and $0 < 1$; but this cannot hold in any chain. (Why?)

*Therefore it is impossible to define an order relation that will turn* $(\mathbb{C}, +, \cdot)$ *into an ordered field.* At first sight that is not particularly strange, because we always represent $\mathbb{C}$ by the coordinate plane and not by means of a single line. On the other hand, we have seen that at least the group $(\mathbb{C}, +)$, can be ordered even though it is nothing other than $(\mathbb{R} \times \mathbb{R}, +)$; but it cannot then be represented by the plane in any intuitive picture.

*EXAMPLES.* Fields with Different Orderings.
(1) Consider the field $K = K(5)$ of Exercise 2. In it, we have the matrices

$$I = \begin{bmatrix} 1 & 0 \\ 0 & 1 \end{bmatrix}, \quad A = \begin{bmatrix} 0 & 5 \\ 1 & 0 \end{bmatrix},$$

and each element of $K(5)$ can be written $aI + bA$ for some $a, b \in \mathbb{Q}$. The matrices of the form $aI$ form a sub-field of $K$ that is isomorphic to $\mathbb{Q}$; hence the ordering of $\mathbb{Q}$ can be carried over to this set of matrices. The matrices $A$, and $B = -A$ have the property that $A^2 = B^2 = 5I$, so they can both be considered as square-roots of 5, because they are algebraically equivalent solutions of the equation $X^2 = 5I$. Let us now order $K$ differently by using an isomorphic embedding of $K$ in $\mathbb{R}$. We can do this using two different mappings — namely

$$\varphi : aI + bA \to a + b\sqrt{5}, \qquad \text{with } \varphi(A) = \sqrt{5}$$

and

$$\psi : aI + bA \to a - b\sqrt{5}, \qquad \text{with } \psi(A) = -\sqrt{5}.$$

(We leave the reader to verify that $\varphi$ and $\psi$ preserve addition and multiplication). Thus the image $\varphi(K)$ of $K$ is a sub-field of $\mathbb{R}$, usually denoted by $\mathbb{Q}(\sqrt{5})$.) From these we can derive two different orderings of $K$; using $\varphi$ we define $X \leqslant_\varphi Y$ in $(K, \leqslant_\varphi)$ to mean $\varphi(X) \leqslant \varphi(Y)$ in $(\mathbb{R}, \leqslant)$, and similarly we define $X \leqslant_\psi Y$. Then, if $0$ denotes the zero matrix, we have in particular, $-A = B <_\varphi 0 <_\varphi A$ and $A <_\psi 0 <_\psi B$. Since we have defined these two orderings by embeddings in $\mathbb{R}$ they are Archimedean. (Remark: one can see this without using $\mathbb{R}$; for a description with calculations, see the last chapter of B. Artmann: *Introduction to Modern Algebra* [2].)

(2) Let $\mathbb{Q}[t]$ denote the polynomial ring in $t$ over $\mathbb{Q}$ with elements

$$p(t) = a_n t^n + a_{n-1} t^{n-1} + \ldots + a_o \qquad \text{with } a_i \in \mathbb{Q}$$

which we consider as functions $p : \mathbb{Q} \to \mathbb{Q}$. (Compare Griffiths-Hilton, [23] Chap.22). The quotient field of this ring is the field $\mathbb{Q}(t)$ of all rational functions with elements of the form $p/q$, where $q$ is not the zero polynomial. If we could define an ordering on $\mathbb{Q}[t]$, then we could easily extend it to this quotient field by the same kind of process by which we pass from $\mathbb{Z}$ to $\mathbb{Q}$. (A rational function $r = p/q \in \mathbb{Q}(t)$ is declared to be positive exactly when $pq > 0$ in $\mathbb{Q}[t]$. This is 'invariant' relative to cancellation i.e., $p/q$ is positive iff $pr/qr$ is positive for any non-zero $r \in \mathbb{Q}[t]$.) We now give two orderings of $\mathbb{Q}(t)$.

First Ordering of $\mathbb{Q}(t)$.

Here we need the important fact that the real number $\pi$ is not a zero of a rational polynomial, since $\pi$ is 'trancendental'. By means of the mapping we induce an isomorphic embedding $\varphi : \mathbb{Q}[t]: \to \mathbb{R}$ that sends the polynomial $\mathbb{Q}(t)$ to the number $p(\pi)$. Thus we obtain an Archimedean ordering of $\mathbb{Q}[t]$, and extend it to $\mathbb{Q}(t)$ by the process described earlier.

Second Ordering of $\mathbb{Q}(t)$.

Here we first consider the polynomial ring $\mathbb{Q}[t]$. In this ring we declare the polynomial

$$p = a_n t^n + \dots + a_o$$

to be positive, whenever $a_n > 0$. This expresses the geometrical property that the curve represented by $p(t)$ is eventually above the t-axis. As $p(t)$ has only finitely many zeros, then $p(t)$ must from a certain point on, always be above or always be below the t-axis. Then for any two polynomials f, g, we write:

$$f < g, \quad \text{whenever} \quad g - f \quad \text{is positive.}$$

The properties of this linear ordering for $(\mathbb{Q}[t], +, \cdot, \leqslant)$ can be verified by routine calculations. The ordering however, is not Archimedean. For example, let:

$$f(t) = t \quad \text{and} \quad g(t) = t^2$$

so $0 < f < g$. However, for every $n \in \mathbb{N}$ we have

$$nf < g,$$

that is to say: g is a "whole class" bigger than f. In this sense, one can think of $t^2$ as "infinitely large" relative to t and correspondingly, t is "infinitely large" relative to the number 1. (See Fig.3)

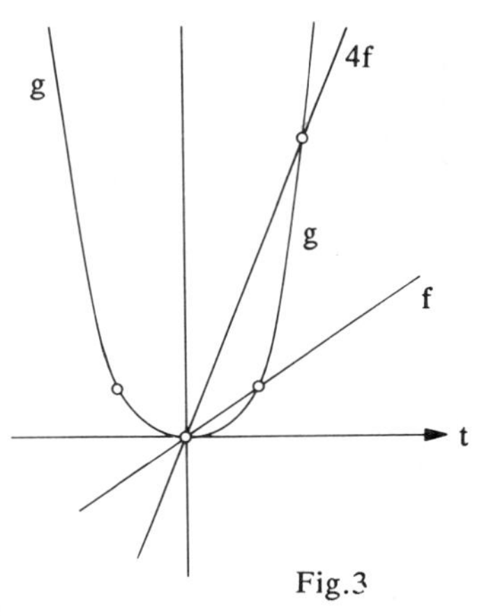

Fig.3

*REMARK*. **Positivity Domains.** Instead of considering the relation $\leqslant$ in a ring R we could instead, distinguish the "positive elements" of R and then define $\leqslant$ in terms of them. As this process is often advantageous, we shall consider it briefly here.

If $(R, +, \cdot, \leqslant)$ is an ordered ring, then we put

$$P = \{x \in R \mid 0 < x\}$$

and name $P$ the *positivity domain* of $R$. $P$ has the following properties:

(Pi) $x, y \in P$ implies: $x + y \in P$ and $xy \in P$ and $x^{-1} \in P$

(Pii) $0 \notin P$ and $1 \in P$ (when these exist)

(Piii) $\forall x \neq 0$ : either $x \in P$ or $-x \in P$ .

Conversely, if there is a subset $P$ of $R$ which displays these properties, then one can define

$$x < y \text{ iff: } y - x \in P ,$$

and we derive in this manner a linear ordering in $R$. The two processes are therefore equivalent. We use positivity domains to prove the following theorem.

Orderings of $\mathbb{R}$.

THEOREM. *Except for the usual ordering, there is no other order relation* $\leqslant$, *with which* $(\mathbb{R}, +, \cdot, \leqslant)$ *becomes an ordered field.*

*PROOF*. We have already seen that in an ordered field, all squares $\neq 0$ must be positive. Moreover, given $a \neq 0$ only one of the two elements $a, -a$ can be positive. But in $(\mathbb{R}, +, \cdot)$, only one of the two elements $a$ or $-a$ is a square. Consequently to find a positivity domain $P$ in $\mathbb{R}$, there is no other choice but to take $P$ to be the set of squares in $\mathbb{R}$, and this just gives the usual ordering. □

Note: In $\mathbb{Q}(\sqrt{5})$ and $\mathbb{Q}(t)$ there are *'relatively few'* squares. Consequently, several different orderings are possible here. But in $\mathbb{C}$ there are *too many* squares for an ordering. Among the real numbers there are just enough squares for one and only one ordering to be possible.

Automorphisms of $\mathbb{R}$.

The field $\mathbb{Q}(\sqrt{5})$ was described in the example on p.12. It possesses the automorphism $\alpha : a + b\sqrt{5} \to a - b\sqrt{5}$, which is analogous to $z \to \bar{z}$ for the complex numbers. Automorphisms of fields often come up in algebra, but we now show that in $\mathbb{R}$, nothing much can be gained along that line.

Definition. Let $(K, +, \cdot)$ be a field. We call $\alpha : K \to K$ an ***automorphism*** of $K$ iff $\alpha$ is bijective and for all $x, y \in K$,

$$\alpha(x + y) = \alpha(x) + \alpha(y)$$

$$\alpha(xy) = \alpha(x)\,\alpha(y) \quad .$$

It is easy to prove that $\alpha(0) = 0$ and $\alpha(1) = 1$. The identity mapping of $K$ is clearly an automorphism.

THEOREM. *Apart from the identity mapping,* $\mathbb{R}$ *has no automorphisms.*

*PROOF*. From $\alpha(1) = 1$ follows $\alpha(2) = \alpha(1+1) = \alpha(1) + \alpha(1) = 2$ etc. This type of argument then shows that $\alpha$ leaves fixed every $r \in \mathbb{Q}$. Further, $\alpha$ maps all squares again into squares, since $\alpha(x^2) = (\alpha(x))^2$. Consequently, $\alpha$ respects the ordering of $\mathbb{R}$.

Moreover, if $y \in \mathbb{R}$, and $\alpha(y) < y$, then we could find $q \in \mathbb{Q}$ with $\alpha(y) < q < y$; but $\alpha$ preserves order so $q = \alpha(q) < \alpha(y) < q$, a contradiction. Hence $y \leqslant \alpha(y)$, and similarly we can show that for all $y \in \mathbb{R}$, $y \geqslant \alpha(y)$. This leaves only $\alpha(y) = y$ for all $y \in \mathbb{R}$, so $\alpha$ is the identity mapping of $\mathbb{R}$. □

*NOTE.* These findings have an interesting application in geometry. They can be used in the proof of the theorem that every collineation (i.e. mapping that preserves lines) of the real affine plane can be described by a matrix.

We have seen that $(\mathbb{C}, +, \cdot)$ is order-theoretically uninteresting, so we turn our attention to the "other end" of the number system, namely to $\mathbb{Z}$ and $\mathbb{Q}$, – but always with an eye on our goal, the characterisation of $\mathbb{R}$.

THEOREM. *Each ordered ring* $(R, +, \cdot, \leqslant)$ *with* $1 \neq 0$ *contains (a copy of)* $(\mathbb{Z}, +, \cdot, \trianglelefteq)$.

*PROOF.* Since $0 \neq 1$ then $0 < 1$. Hence $0 + 1 < 1 + 1$. So $1 < 2$ etc., and $n < n + 1$ for all $n \in \mathbb{N}$. Therefore all natural numbers lie in R. Also $-1 < 0$, so $0 < 1$ implies $0 + (-1) < 1 + (-1) = 0$ and one derives correspondingly $-1 > -2 > -3$ etc. Hence we have found a copy of $\mathbb{Z}$ in $\mathbb{R}$. □

As a contrast we have the following example:

*CONTRAST EXAMPLE.* In the finite field $\mathbb{Z}_5$ of residues modulo 5, we have $1+1+1+1+1 = 0$ (modulo 5), so $\mathbb{Z}$ does not lie in $\mathbb{Z}_5$. (From this it follows that $(\mathbb{Z}_5, +, \cdot)$ cannot be ordered.) One should also keep in mind, that the 1 in R can be, e.g. the $n \times n$ unit matrix, so the "whole numbers" in R are then written differently from their usual notation in $\mathbb{Z}$.

THEOREM. *Every ordered field* $(K, +, \cdot, \leqslant)$ *contains (a copy of)* $(\mathbb{Q}, + \cdot, \trianglelefteq)$.

*PROOF.* In K we have $0 \neq 1$, and as K is in particular an ordered ring, then $(\mathbb{Z}, + \cdot, \trianglelefteq)$ lies in K. As K contains, for each element $\neq 0$, a multiplicative inverse, then in particular we have $n^{-1} \in K$ for each $n \in \mathbb{N}$; and then $a \cdot n^{-1} \in K$ for each $a \in \mathbb{Z}$ and $n \in \mathbb{N}$. This gives, however, all rational numbers. (Two different rational numbers cannot become identified in K, for otherwise the same would hold for two distinct whole numbers.) – Again, we should note that the rationals in K can appear in a notation different from the usual one. □

Characterisation of $(\mathbb{Q}, +, \cdot, \leqslant)$ as an Ordered Field.

THEOREM. *Let* $(L, +, \cdot, \leqslant)$ *be an ordered field with the property:*
(*) *(A copy of)* $(L, +, \cdot, \leqslant)$ *is contained in every ordered field* $(K, +, \cdot, \leqslant)$.
*Then (apart from notation) we have* $(L, +, \cdot, \leqslant) = (\mathbb{Q}, +, \cdot, \trianglelefteq)$.

*PROOF.* As $\mathbb{Q}$ is an ordered field, it follows from (*) that $L \subseteq \mathbb{Q}$. On the other hand, we have from the previous theorem that $\mathbb{Q} \subseteq L$. Therefore $\mathbb{Q} = L$. □

*COMMENTARY.* By a **recognition** or **characterisation** of a mathematical object, we understand generally a specification of the object by means of *properties* which are independent of any construction or concrete representation. A typical example is the theorem in linear algebra, that any two n-dimensional real vector spaces are isomorphic. The combination of properties signified by the phrase "n-dimensional real vector space" is so strong, that it allows of only one mathematical object to within isomorphism – that is to say, ignoring structurally immaterial differences in notation. For the ordered field of rational numbers, we have found (*) to be a similarly effective (if also not very exciting) property.

THEOREM. *For an ordered field* $K$, *we have:* $K$ *is Archimedean* $\Leftrightarrow \mathbb{N}$ *is unbounded in* $K$. *Equivalently (contrapositive):* $K$ *is not Archimedean* $\Leftrightarrow \mathbb{N}$ *is bounded in* $K$.

*PROOF.*

(a) Assume that $\mathbb{N}$ is bounded in $K$, so there exists $b \in K$ with $n < b$ for all $n \in \mathbb{N}$. With $0 < x = 1 < b$ we see that we can never find $n \in \mathbb{N}$ with $b < nx$ so $K$ is not Archimedean.

(b) If $K$ is not Archimedean, there exist $0 < x < y \in K$ with $0 < nx < y$ for all $n \in \mathbb{N}$. Since $x^{-1} > 0$ we have $0 < n \cdot x \cdot x^{-1} = n < yx^{-1}$ for all $n \in \mathbb{N}$, so $yx^{-1}$ is an upper bound of $\mathbb{N}$.

**Exercise 8.**

(a) Show that $K$ is Archimedean iff $\forall x > 0, \exists n \in \mathbb{N} : 1/n < x$.

(b) Show that, for every ordered field $K$, the chain $(K \leqslant)$ is dense.

**Definition.** Let $(K, \leqslant)$ be a chain and $L$ a subset of $K$. $L$ is said to be **dense** in $K$ if we have: for all $x, y \in K$ with $x < y$ there exists $z \in L$ with $x < z < y$.

THEOREM. *Given any Archimedean ordered field* $K$, $\mathbb{Q}$ *is dense in* $K$.

*PROOF.* As $K$ is ordered, then $\mathbb{Q}$ is a subset of $K$. Now suppose $x < y$ in $K$. Then $0 = x - x < y - x$. Since $K$ is Archimedean, there exists $n \in \mathbb{N}$ with $1 < n(y-x)$, so that $nx + 1 < ny$. Also, there exists a natural number $r > nx$, and hence a smallest natural number $s > nx$. Then, $s \leqslant nx + 1$ and it follows that $nx < s < ny$. We now obtain $x < s/n < y$ by multiplication with $n^{-1}$, so $\mathbb{Q}$ is dense in $K$. □

## 1D RECOGNITION OF $\mathbb{R}$

Definition. An ordered field K is called **complete** if the chain $(K, \leqslant)$ is complete.

One example is $(\mathbb{R}, +, \cdot \trianglelefteq)$. Our goal is to prove the main Theorem below, which asserts that there are no other examples. First we have:

THEOREM. *A complete ordered field is Archimedean.*

*PROOF*. We shall establish the contrapositive. If K is not Archimedean, then the subset $\mathbb{N}$ of K, being non-empty, has an upper bound. But then also $b - 1$ is an upper bound, for if there were some $n \in \mathbb{N}$ with $b - 1 < n$, then we would have $b < n + 1$, and b would not be an upper bound. This shows that no upper bound of $\mathbb{N}$ can be a least upper bound, so K is not complete. □

COROLLARY. *In a complete field, the rationals lie densely.*

Isomorphism.
As a help towards the proof of the proposed theorem about $\mathbb{R}$, we use the concept of isomorphism between ordered fields.

Definition. Let $(K, +, \cdot, \leqslant)$ and $(L, +, \cdot, \leqslant)$ be ordered fields and $\rho : K \to L$ a mapping. We call $\rho$ an *isomorphism of the ordered fields* (and $(K, +, \cdot, \leqslant)$ and $(L, +, \cdot, \leqslant)$ *order-isomorphic*), if the following conditions hold:

(i) $\rho$ is bijective
(ii) $\rho(a + b) = \rho a + \rho b$ for all $a, b \in K$.
(iii) $\rho(ab) = \rho a \cdot \rho b$ $a, b \in K$
(iv) $a \leqslant b \Rightarrow \rho a \leqslant \rho b$

Isomorphism ought to be a symmetric relation, so we need to show here that also the inverse mapping $\rho^{-1} : L \to K$ has the properties (ii)-(iv). To do this, there is a standard trick: if $x, y \in L$, we put $a = \rho^{-1} x$ and $b = \rho^{-1} y$, whence $\rho a = x$ and $\rho b = y$. Then we prove (ii) thus:

$$\rho^{-1}(x+y) = \rho^{-1}(\rho a+\rho b) = \rho^{-1}(\rho(a+b))$$
$$= a + b = \rho^{-1}x + \rho^{-1}y.$$

Correspondingly for (iii).

For (iv) assume $x \leqslant y$. If now $a > b$, then by (iv) for $\rho$ we would have $\rho a = x > \rho b = y$. As this is false, we must have $a \leqslant b$; so $\rho^{-1} x \leqslant \rho^{-1} y$, for $(K, \leqslant)$ is a chain. (Warning: in arbitrary ordered sets, which are not chains, the last conclusion for $\rho^{-1}$ need not hold!) We now state the Main Theorem:

MAIN THEOREM. Recognition of $\mathbb{R}$: *Any two complete ordered fields are isomorphic.*

An alternative formulation, without using the fact that $(\mathbb{R}, +, \cdot, \trianglelefteq)$ is complete, is:

*Every ordered complete field* $(K, +, \cdot, \leqslant)$ *is isomorphic to* $(\mathbb{R}, +, \cdot, \trianglelefteq)$.

*COMMENT*. Thus a complete field $(K, +, \cdot, \leqslant)$ differs from $\mathbb{R}$ only in notation. Concerning this, we shall later deal more precisely with the question of what the real numbers actually 'are'. As completeness is also the most important property for the differential and integral calculus in $\mathbb{R}$, we shall then see also why this theory (at first) is only developed over the real numbers. One can also say: such a thing as $\mathbb{R}$ happens only once. The following proof is long, and the reader may prefer to return to it later, after taking up the story after Exercise 9 (p.22).

*Indication of the basic idea of the Proof*. We shall represent the elements of K (as of L) as least upper bounds of sets of rational numbers. If one were to allow the decimal numbers instead of the rationals (which is possible) then the identification of the corresponding least upper bounds with the (infinite) decimal expansions would be obvious; and one would have displayed the elements of K, as also those of L, as such decimal expansions. The interested reader may care to examine, whether in the following proof $\mathbb{Q}$ could everywhere be replaced by Dez.

*PROOF OF THE MAIN THEOREM*. We assume that $(K, +, \cdot, \leqslant)$ and $(L, +, \cdot, \leqslant)$ are both complete fields. For the proof we need to construct a mapping $\rho$ with the above properties (i)÷(iv). This requires many separate steps. (For the proof, see also the book [52] of Oberschelp Section 23.) The reader is forewarned that the completeness of K is only used in the third step, part (2). Everything else holds just as well for a field K that is merely Archimedean.

First Step : *Definition of* $\rho$.

(1) Preparation. $\mathbb{Q}$ lies densely in both K and L, since both are Archimedean fields. For each $x \in K$ we form the set

$$M_x = \{r \in \mathbb{Q} \mid r < x\}.$$

This set is not empty (rational number a between $x - 1$ and $x$) and has a (rational) upper bound (rational number b between $x$ and $x+1$). For the lub in K we have $x = \sup M_x$, for by definition, $x$ is an upper bound of $M_x$ and if $y < x$ then between $y$ and $x$ there is a rational number and $y$ is not an upper bound — therefore $x$ must be the least. In particular, we have for each $q \in \mathbb{Q}$ the equation $q = \sup M_q$. In what follows, we have to keep not only the sets L and K distinct, but also the suprema that we have just defined; therefore we will denote the upper bound in K by Sup and that in L by fin (for 'finis superior'). As in K we have similarly in L: for each $y \in L$ we have (cf. Fig.4)

$$y = \operatorname{fin} M_y = \operatorname{fin} \{r \in \mathbb{Q} \mid r < y\};$$

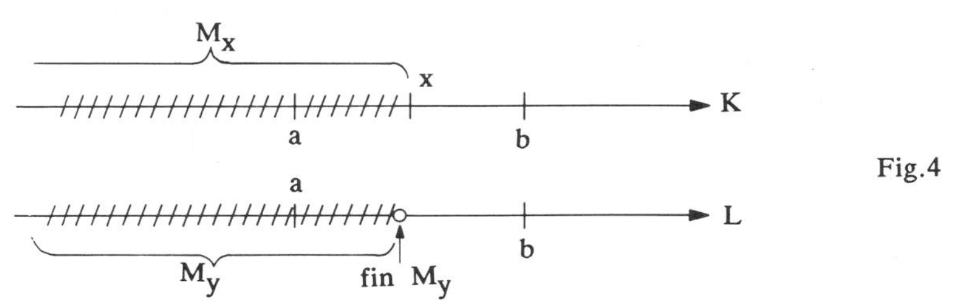


Fig.4

(2) Definition of $\rho$.

Again let $x \in K$. The uniquely determined non-empty set $M_x$ in (1) above lies also in L, since $\mathbb{Q} \subseteq L$, and is there bounded by the afore-mentioned $b \in \mathbb{Q}$ with $x < b < x+1$. Therefore, *because of the completeness of* L, there exists the (again uniquely determined) lub fin $M_x$ in L and we define a function $\rho: K \to L$ by

$$\rho x = \text{fin } M_x \in L\ .$$

Basic properties of $\rho$ are as follows:

(a) For $q \in \mathbb{Q}$, we have $\rho q = q$, since $q = \text{fin } M_q$.

(b) Instead of $M_x$ we can also choose a 'shorter set' $N_x \subseteq M_x$ provided we preserve the same lub – for example, a rational interval such as $N_x = \{r \in \mathbb{Q} \mid x - 1 < r < x\}$, as in Fig.5.

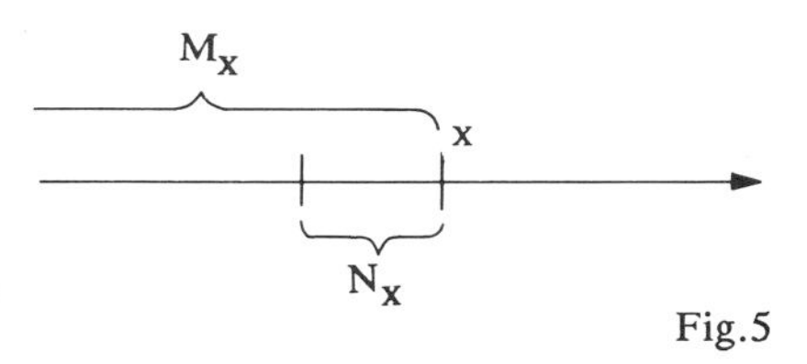


Fig.5

Second Step : *$\rho$ is bijective.*

*Sub-step (1)* : *$\rho$ is injective.* Suppose $a \neq b$ in K. We have to show $\rho a \neq \rho b$ in L. As K is linearly ordered, we can assume the notation such that $a < b$. There is a rational number r with $a < r < b$. From this follows fin $M_a < r <$ fin $M_b$, therefore $\rho a < \rho b$ and in particular, $\rho a \neq \rho b$.

*Sub-step (2)* : *$\rho$ is surjective.* Given $y \in L$ we form $M_y$ in L. Then, *because* K *is complete*, $M_y$ possesses in K an upper bound $x = \sup M_y$ (same reason as with fin $M_x$). If we form now in K the set $M_x$, then $M_x = M_y$ (both are sets of rational numbers!) and it follows that fin $M_x$ = fin $M_y = y$. Therefore $\rho x = y$ and y is a value of the function, so $\rho$ is surjective.

*REMARK*: We have also constructed the inverse mapping $\rho^{-1}$ to $\rho$, namely $\rho^{-1} y = \sup M_y$ in K.

Third Step: *Monotone property (iv).*

We have to show $a \leqslant b \Rightarrow \rho a \leqslant \rho b$, but we already did this in the proof of the injectivity.

*REMARK.* We have defined $\rho$ with the help of the order relation without mentioning the associative laws etc. for calculating in K and L. On that account it is no wonder that those properties that concern addition and multiplication are somewhat more complicated to prove.

4th Step : *Property (ii).*

$$\rho(a + b) = \rho a + \rho b \quad .$$

For this, we introduce as an auxiliary set the "sum set"

$$S_{a,b} = \{r+s \mid r, s \in \mathbb{Q} \text{ and } r < a \text{ and } s < b\} \quad .$$

*Sub–Step (1)* : $\rho(a+b) = \text{fin } S_{a,b}$ .

Because $\rho(a+b) = \text{fin } M_{a+b}$ we have to show that $M_{a+b} = S_{a,b}$. Let $t \in M_{a+b}$, so $t$ is rational and $t < a+b$ in K. Then there is a rational $r$ with $t-b < r < a$.

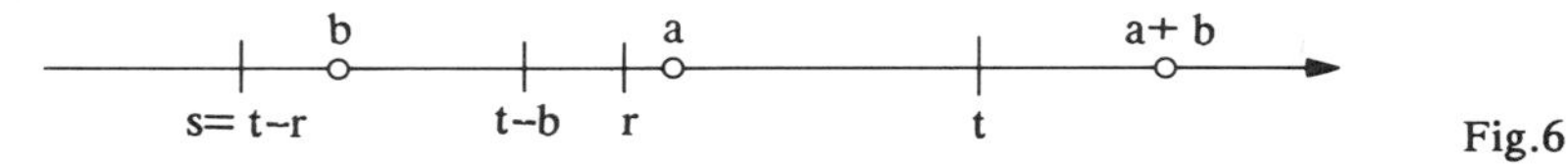


Fig.6

Since $t-b < r$ then $t-r < b$, so $t-r = s \in M_b$ and $t = r+s \in S_{a,b}$. Thus $M_{a+b} \subseteq S_{a,b}$. If now conversely, $t \in S_{a,b}$, then there exist $r < a$ and $s < b$ with $r, s \in \mathbb{Q}$ and $r+s = t$. Hence $t = r+s < a+b$ so $t \in M_{a+b}$. Therefore $M_{a+b} \subseteq S_{a,b}$, whence $M_{a+b} = S_{a,b}$, as required.

*Sub–Step (2)* : $\rho a + \rho b = \text{fin } S_{a,b}$ .

(a) $\rho a + \rho b$ is an upper bound of $S_{a,b}$; for, with $r < a$ and $s < b$ we have $r < \rho a$ and $s < \rho b$; therefore $t = r + s < \rho a + \rho b$, so each element $t$ of $S_{a,b}$ is $< \rho a + \rho b$.

(b) $\rho a + \rho b$ is the least upper bound of $S_{a,b}$. For, let $y > \rho a + \rho b$ in L. Then there exists $t \in \mathbb{Q}$ with $y < t < \rho a + \rho b$. Similarly to our previous argument, we find a rational $r$ with $t - \rho b < r < \rho a$ in L. Then also $r < a$ in K. Further let $s = t - r$, so $s$ is rational and $s < \rho b$ (since $s + r - \rho b < r$); therefore $s < b$ in K. It follows that $y < s + r = t < \rho a + \rho b$, so $t \in S_{a,b}$. Therefore y is not an upper bound of $S_{a,b}$, so the element $\rho a + \rho b$ is in fact the least upper bound.

*Sub–step (3)* : From the two previous steps, there follows

$$\rho(a+b) = \text{fin } S_{a,b} = \rho a + \rho b \quad .$$

Consequence: $\rho 0 = 0$ and $\rho(-x) = -\rho x$. (Reason: since $0 \in \mathbb{Q}$ we have $\rho 0 = 0$. Further, $\rho(-x) + \rho x = \rho(-x+x) = \rho 0 = 0$, so $\rho(-x) = -\rho x$.)

5. Fifth Step : *Property (iii).*

$$\rho(ab) = (\rho a) \cdot (\rho b)$$

*Sub–step (1)* : In order to avoid difficulties with the signs, we consider first the case $a > 0$ and $b > 0$ in K. Now let

$$N_a = \{r \in \mathbb{Q} | 0 < r < a\} \ , \ N_b = \{s \in \mathbb{Q} | 0 < s < b\} \ .$$

We use these sets in place of the previous $M_a$, $M_b$, so that only positive numbers are involved in the multiplications that follow. Recalling (b) in Step 1, we have

$$\rho a = \text{fin } N_a = \text{fin } M_a, \qquad \rho b = \text{fin } N_b = \text{fin } M_b$$

and

$$\rho(ab) = \text{fin } N_{ab} \ .$$

Just as with addition, we introduce now as an auxiliary set the 'product set'

$$P_{a,b} = \{rs | r, \ s \in \mathbb{Q} \ \text{ and } \ 0 < r < a \ \text{ and } \ 0 < s \quad b\} \ .$$

The following sub-steps (2), (3) and (4) consist only of simple modifications of items (1), (2) and (3) in sub-step 4 above, but everywhere we must substitute $a + b$ by $a.b$ and $-b$ by $b^{-1}$ etc.

*Sub-step (2)* : $\rho(ab) = \text{fin } P_{a,b}$.

Since $\rho(ab) = \text{fin } N_{ab}$ it is sufficient to show $N_{ab} = P_{ab}$.

(a) $N_{ab} \subseteq P_{a,b}$. For, let $t \in N_{ab}$, so $t$ is rational and $0 < t < ab$ in $K$.

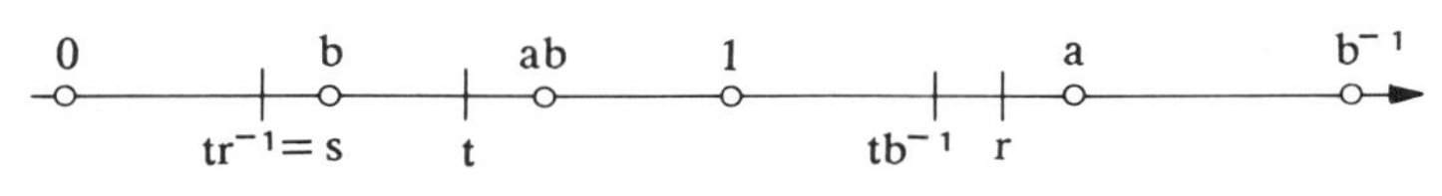


Fig.7

Then there exists a (positive) rational $r$ with $tb^{-1} < r < a$. Since $s = tr^{-1}$ it follows that $0 < s < b$. (For, $tb^{-1} = srb^{-1} < r$ so $sb^{-1} < 1$, whence $s < b$). Now $t = rs$ with $0 < r < a$ and $0 < s < b$, so $t \in P_{a,b}$.

(b) $P_{a,b} \subseteq N_{ab}$. For if $t \in P_{a,b}$ there exist rational $r, s$ with $0 < r < a$ and $0 < s < b$, and $t = rs$. Then we have $t = rs < ab$ and hence $t \in N_{ab}$.

From (a) and (b) follows $N_{ab} = P_{a,b}$.

*Sub-step (3)* : $\rho a.\rho b$ is the least upper bound of $P_{a,b}$.

(a) $\rho a.\rho b$ is an upper bound of $P_{a,b}$; for, from $r < a$ and $s < b$ follows $r < \rho a$ and $s < \rho b$. Therefore $t = rs < \rho a.\rho b$, which proves that each element of $P_{a,b}$ is smaller than $\rho a.\rho b$.

(b) $\rho a.\rho b$ is the least upper bound of $P_{a,b}$. For, let $y < \rho a.\rho b$ in $L$. In case $y \leqslant 0$ there is certainly a rational $t$ with $y < t \in P_{a,b}$, so $y$ is not an upper bound. Therefore we need only bother with positive $y$ and hence have $0 < y < \rho a.\rho b$.

There is a rational $t$ with $y < t < \rho a.\rho b$ and further a rational $r$ with $t(\rho b)^{-1} < r < \rho a$. Hence also $r < a$ in $K$. Again let $s = tr^{-1}$, so that $s > 0$ as $s$ is also rational with $s < \rho b$. (Since $t(\rho b)^{-1} = rs(\rho b)^{-1} < r$, we have $s(\rho b)^{-1} < 1$, so $s < \rho b$.) Therefore $s < b$ in $K$, and it follows that $y < t = rs < \rho a.\rho b$ and $t \in P_{a,b}$. Therefore $y$ is not an upper bound of $P_{a,b}$, and the element $\rho a.\rho b$ is in fact the least upper bound.

*Sub-step (4)* : From the two previous steps follows

$$\rho(ab) = \text{fin } P_{a,b} = \rho a.\rho b \quad \text{for} \quad a, b > 0.$$

*Sub-step (5)* : We must consider still the cases in which $a \leqslant 0$ or $b \leqslant 0$. If one of the two elements is $0$, say $a = 0$ we have

$$\rho(0b) = \rho 0 = 0 = (\rho 0)\cdot\rho b \ .$$

The remaining cases in which both elements $a, b$ are different from $0$ and at least one is $< 0$ can be led back to the result of the last sub-step, via the consequence $\rho(-x) = -\rho x$ from Step 4. With this we use the known rules of signs in multiplication ("minus times minus is plus"), which one can prove easily in any ring (see Exercise 1). These remaining cases are:
(i) $a < 0 < b$, (ii) $b < 0 < a$ and (iii) $a, b < 0$ .
*Case* (i). Since $a < 0$ we have $0 < -a$. Then

$$\begin{aligned}\rho(ab) &= \rho(-(-a)b) \\ &= -\rho((-a)b) && \text{as} \quad \rho(-x) = -\rho x \\ &= -\rho(-a)\rho b && \text{as} \quad -a > 0 \text{ and } b > 0 \\ &= -[-\rho a\rho b] && \text{as} \quad \rho(-x) = -\rho x \\ &= \rho a\rho b \ .\end{aligned}$$

Similarly, (ii) and (iii) can be dealt with.

**Exercise 9**. Complete the details.
From this we have

$$\rho(ab) = \rho a.\rho b \quad \text{for all} \quad a, b \in K$$

and the proof of the main theorem is concluded. □

Remarks. (Concerning Steps 4 and 5 of the Proof.) With Step 5 we indicated that for positive a,b one should argue wholly analogously to Step 4. This situation raises the question: in what way does the analogy hold more fully? One can scrutinise the proof carefully in order to find the essential form of the argument. Such an analysis brings to light the common basic concepts and assumptions of the two cases — here essentially that of the Archimedean ordered group. According to taste, one calls such a process "axiomatisation" or "generalisation". The more general theory will, in our case, lead us to a theorem on the isomorphism of complete ordered groups with $(\mathbb{Z}, +, \leqslant)$ or $(\mathbb{R}, +, \leqslant)$. (Compare, say Salzmann [65] I, Paragraph 7). In preparing a lecture on this material, one naturally asks oneself whether the more general theorem cannot be obtained from the earlier work. It seems more convenient to use the opportunity raised here to talk about the question of generalization. We see here a beautiful example, of the way in which the two proofs compel one to derive them from a common version, and that is what leads one to an axiomatic theory. *Axioms result from necessity, not from some arbitrary decree*, and this reason is often misunderstood.

A Consequence of the Previous Proof.

From the proof of the main theorem we can extract even more information in the form of the assertion that any two complete ordered fields are isomorphic. Already at the beginning of the proof, it was indicated that the completeness of K was used only in the proof of the surjectivity of $\rho : K \to L$. In all other steps and sub–steps it was sufficient that K was Archimedean. On that account we have also proved the following theorem.

THEOREM. *If* K *is Archimedean and* L *is complete then there is an injective mapping* $\rho : K \to L$ *which satisfies conditions (ii), (iii) and (iv) of the definition on page 17, that is to say,* $\rho$ *preserves addition, multiplication and ordering.*

An injective mapping of this type is called an **embedding**. Now we no longer need to say that K and L are equal 'apart from notation', or that K lies 'to within notation' in L; for we have shown how $\rho$ gives us a completely faithful copy of K in L, namely the image $\rho K$ of K in L. With this understanding, we derive now, without trouble, a further characterisation of $\mathbb{R}$.

Characterization of $\mathbb{R}$ as a Maximal Archimedean Field.

THEOREM. *Let* $(L, +, \cdot, \leqslant)$ *be an Archimedean ordered field with the property*

(*) $(L, +, \cdot, \leqslant)$ *contains (a copy of) every (other) Archimedean ordered field* $(K, +, \cdot, \leqslant)$.

*Then (to within isomorphism) we have* $(L, +, \cdot, \leqslant) = (\mathbb{R}, + \cdot, \trianglelefteq)$.

*PROOF*. $\mathbb{R}$ is Archimedean and therefore by (*) we have $\mathbb{R} \subseteq L$. On the other hand, $\mathbb{R}$ itself has property (*), for $\mathbb{R}$ is complete. Therefore we have $L \subseteq \mathbb{R}$, hence $L = \mathbb{R}$. □

*COMMENTARY*. A field with the property (*) is called *maximal Archimedean*. We can then formulate the theorem in the following way.

*Every maximal Archimedean field is equal (isomorphic) to* $\mathbb{R}$.

This second characterization of $\mathbb{R}$ is wholly analogous to that of $\mathbb{Q}$ in the previous paragraphs, where we learned to characterise $\mathbb{Q}$ as the *smallest* of all ordered fields. Here $\mathbb{R}$ is the largest of all *Archimedean* ordered fields. A consequence is that all Archimedean fields K lie between $\mathbb{Q}$ and $\mathbb{R}$, so $\mathbb{Q} \subseteq K \subseteq \mathbb{R}$. On the other hand, if one has such an embedding of K (as an ordered field) in $\mathbb{R}$, then also K must be Archimedean. This gives us an intuitive characterization of the concept "Archimedean": an ordered field is Archimedean precisely when one can regard it as a sub-set of the Number line (here "regard it" means "embed it isomorphically").

It is natural now to ask whether we can set aside the Archimedean property and ask simply for a maximal field, and eventually find a similar kind of characterization. Here, however, we would be disappointed; no such characterization exists. The Archimedean fields can increase only as far as $\mathbb{R}$ and then stop; but without this restriction, ordered fields can be built on each other ever further, like a tower, without coming to an end.

## 1E. ON PROPERTIES EQUIVALENT TO COMPLETION.

### The Meaning of Completion in Analysis.

The existence of the least upper bound (lub) is decisive for all processes of Analysis, as will become clear in the following results. On that account the characterizing power of this condition is especially noteworthy from the viewpoint of Analysis. As we have already said, we can gain insight into the meaning of the lub through the examples of simple measuring processes. For example, if one wishes to determine the length of a curve, one can consider approximating polygons (as in Fig.2), and calculate the length of each; then if the set of all these approximating lengths has a lub, one can define this to be the length of the curve. Correspondingly, one can proceed for surfaces, using the Riemann integral. The usual argument for the passage from $\mathbb{Q}$ to $\mathbb{R}$ with the help of the irrationality of $\sqrt{2}$ stands on weak foundations because there one can easily manage without mentioning completeness and it avoids the central problem. Just how seriously such gaps can affect the domain of definition of a function can be best understood with the simple but typical example of the differential equation $f'(x) = 1/x$ for the logarithm. One has no possibility of satisfactorily bridging together the solution curves across the gap at $x = 0$.

### The Equivalence of the Conditions "Complete" and "Maximal Archimedean".

We have seen that if either of these two conditions is assumed in $\mathbb{R}$, then we can characterise $\mathbb{R}$. If one has a complete field K, then this is isomorphic to $\mathbb{R}$, therefore K is also maximal Archimedean; conversely a maximal Archimedean field K is isomorphic to $\mathbb{R}$ and therefore, it is also complete. The following equivalent statements are not as easy to make as the previous ones, and some preparations are needed to formulate the corresponding conditions.

<u>Modulus Function and Intervals in Ordered Fields.</u>
Let $(K, +, \cdot, \leqslant)$ be an ordered field. One puts

$$|a| = \begin{cases} a & a \geqslant 0 \\ -a & a < 0 \end{cases}$$

and one thinks of $|a|$ as the distance of the element $a$ from 0.

Intervals are defined in $K$ as is usual in $\mathbb{R}$: if $a, b \in K$ and $a < b$ we denote by $\langle\!\langle a, b\rangle\!\rangle$ the closed, and by $\langle a,b\rangle$ the open interval determined by the end points $a$ and $b$.

<u>Continuity.</u>
No essential role is played by the domains of the functions we are about to consider. For simplicity therefore, we shall assume that such domains are always intervals, with infinite ones also being allowed. It will be helpful if the reader already knows the definition of continuity in the special case when $K = \mathbb{R}$, in order to convert the following symbolic definition into the already familiar literary form.

A function $f : D \to K$ (with $D \subseteq K$) is said to be continuous at $x_0 \in D$, whenever we have

(a) $$(\forall \varepsilon > 0)\ (\exists \delta > 0)\ (\forall x \in D) : |x - x_0| < \delta \Rightarrow \left|f(x) - f(x_0)\right| < \varepsilon .$$

Here $\varepsilon, \delta$ are to be taken in $K$. As usual $f$ is said to be continuous on $D$ if $f$ is continuous at each $x_0 \in D$. Often it is worth while to have a condition for continuity which is equivalent to (a) but formulated with intervals. Here it is: define first a **neighbourhood** of $x \in K$ to be an open interval $U$ that contains $x$. Then $f$ is continuous at $x_0$ precisely when

(b) *to each neighbourhood* $V$ *of* $f(x_0)$ *there is a neighbourhood* $U$ *of* $x_0$ *with* $f(U) \subseteq V$.

(In (a), it is a matter of the $\varepsilon$-neighbourhood $V$ of $f(x_0)$ and the $\delta$-neighbourhood $U$ of $x_0$).

<u>Cauchy Sequences and Further Concepts.</u>
Just as in $\mathbb{R}$, Cauchy sequences can be defined in arbitrary ordered fields. The sequence $(a_n)$ with $a_i \in K$ is said to be a **Cauchy sequence** whenever we have

$$(\forall \varepsilon > 0)\ (\exists N \in \mathbb{N}) : \text{if } n, m \geq N \text{ then } |a_m - a_n| < \varepsilon ;$$

equivalently

$$\text{if} \quad m > N \quad \text{then} \quad |a_N - a_m| < \varepsilon/2 \ .$$

Further concepts such as limit point, open and closed sets, are defined just as in the usual manner of Analysis.

Properties Equivalent to Completeness.

The following presentation arises from the article of H.G. Steiner: *Equivalent formulations of the completeness axioms for the theory of real numbers* in [69] pp.180÷201. We shall give here only the most important part of the equivalence proof; the rest one can read in Steiner. In particular, we do not go into the question of nested intervals.

MAIN THEOREM. *Each of the following properties of an ordered field* K *is equivalent to completeness.*

[MS]* Each monotone bounded sequence converges (i.e. has a limit in K).

[BW] Bolzano–Weierstrass: An infinite bounded sub–set of K has at least one limit point in K. This is the 'accumulation principle'.

[CA] Connectivity axiom: K cannot be split into two disjoint non-empty open sets.

[DS] Every Dedekind section in K is generated by some $s \in K$.

[IV] Intermediate value theorem for continuous functions (see, for example Scott and Tims [67] Ch.6).

[MX] Maximum property for continuous functions: a continuous function on a closed interval takes a maximum value.

[NB] Continuous functions on closed intervals are bounded by natural numbers.

[HB] Heine–Borel: Every covering by open sets, of a closed bounded interval of K, contains a finite sub–covering. (Thus K is locally compact.)

[A&I] K is Archimedean and fulfills the 'nested interval axiom' (the Chinese Box principle, see Armitage–Griffiths [2] p.113).

[A&CS] K is Archimedean and every Cauchy sequence has a limit in K.

---

* *Translator's note* : the initials in these abbreviations are occasionally altered from the German, to help the reader to associate them with the English terms. E.g. [MS] here was originally [MF], with F for "Folge" corresponding to our S for "sequence".

For simplicity we formulate yet another: our preferred definition of completeness.

[UB] Each non-empty subset of K that is bounded above, possesses a least upper bound.

<u>Grouping of the Listed Conditions with Respect to their Possibilities of Generalization</u>.

Although all the conditions above are equivalent to completeness, they are formulated according to specific purposes of Analysis. We can distinguish three types which however, are not quite clearly separated from each other.

<u>Metric Type</u>. Here the modulus function plays the essential role: with it we can describe the distance between two elements of K. It only occurs directly in the condition [A&CS], but it can easily be carried over to arbitrary metric spaces. Thus it makes sense to regard $\mathbb{C}$ and then $R^n$ as complete. In these generalizations, the Archimedean part of the axiom is redundant because there is no question of ordering.

<u>Topological Type</u>. These are those formulations in which essentially only open and closed sets are in question. To this type belongs also continuity, for it can be defined in (b) above with the help of open sets. In this group belongs [CA], [IV], [MX] and [HB]. In topology, these conditions play an essential role (via connectivity, and compactness).

<u>Types Belonging to Analysis</u>. These are [MS], [BW], [NB], [A&I] as well as [DS], and [UB]. The last two can be generalised to arbitrary ordered sets. Monotone sequences, and nested intervals, seldom appear outside Analysis.

*PROOF OF THE MAIN THEOREM*. Many of the parts of the proof we are about to give will be recognized by the reader from Analysis lectures or books – for example, a proof of the Intermediate Value Theorem with the help of one of the other assertions of the list. For that reason, we shall only select 'typical cases' and display the connection with topology by proving the equivalence [UB] $\Leftrightarrow$ [HB], and the connection with metric spaces through [UB] $\Leftrightarrow$ [A&CS].

*PROOF–PORTION:* [UB] $\Leftrightarrow$ [HB].

We carry the proof through only for intervals. Let $\langle a, b\rangle$ be a closed interval in K and with each $y \in \langle a, b\rangle$ let there be given a neighbourhood $U_y$ (for example an open interval). We must select *from these* neighbourhoods *finitely many*, $U_{x_1},\dots,U_{x_k}$ with the property

$$\langle a, b\rangle \subseteq U_{x_1} \cup U_{x_2} \cup \dots \cup U_{x_k} .$$

For the proof, we consider those sub-intervals $\langle a, x\rangle$ of $\langle a, b\rangle$ which possess such finite coverings and we form

$$M = \{x \in \langle a,b\rangle \mid \langle a,x\rangle \text{ has a finite covering}\}.$$

(Obviously for such coverings, only neighbourhoods $U_y$ from the original one are permitted.) To this set M we apply the upper bound principle, first carrying out the appropriate checking. First, M is not empty, since $a \in M$ with the neighbourhood $U_a$; second, M is bounded above ($x \leqslant b$ for each $x \in M$). Therefore by [UB], there exists $s = \sup M$.

We now show $s = b$. For if $s < b$, we would have in $U_s$ elements r, t with $r < s < t$ and (in case $a < s$) a finite covering $U_{x_1},\ldots,U_{x_n}$ of $\langle a, r\rangle$ which can be enlarged by $U_s$ to a finite covering of $\langle a, t\rangle$. No $s < b$ can therefore be an upper-bound of M. It follows that $s = b$. Then there is for each $r < b$ a finite covering $U_{x_1},\ldots,U_{x_n}$ of $\langle a, r\rangle$. We choose such an $r \in U_b$ and add $U_b$ to the given covering as well; so then we have

$$\langle a, b\rangle \subseteq U_{x_1} \cup \ldots \cup U_{x_n} \cup U_b ,$$

which is therefore a finite covering of the interval $\langle a, b\rangle$ with open sets from the originally given system.

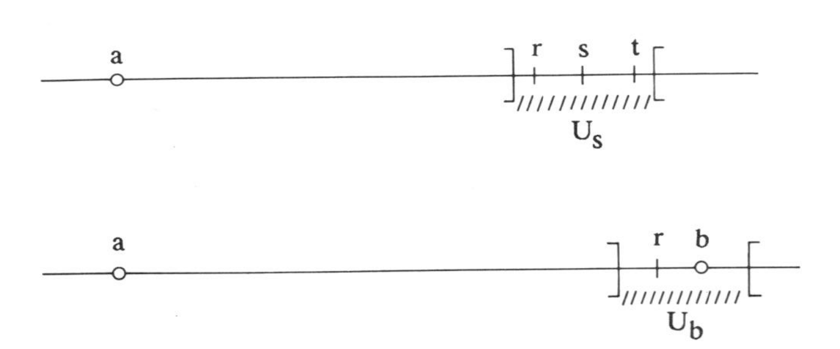


Fig.8

*PROOF–PORTION :* [HB] ⇒ [UB].

We prove this in the form: not [UB] ⇒ not [HB].
Because 'not [UB]', there is in K a non-empty set M bounded above without an upper bound. Let $a \in M$ and suppose b is an upper bound for M. First we observe that $a < b$, for otherwise b would be the least upper bound of M, which cannot be. We will find a covering of the interval $\langle a, b\rangle$ without a finite subcovering. Given $y \in \langle a, b\rangle$, we define $U_y$ in the following way, depending on the possibilities that (see Fig.9) either

(i) there exists $s \in M$ with $y \leqslant s$, or

(ii) no such $s \in M$ exists.

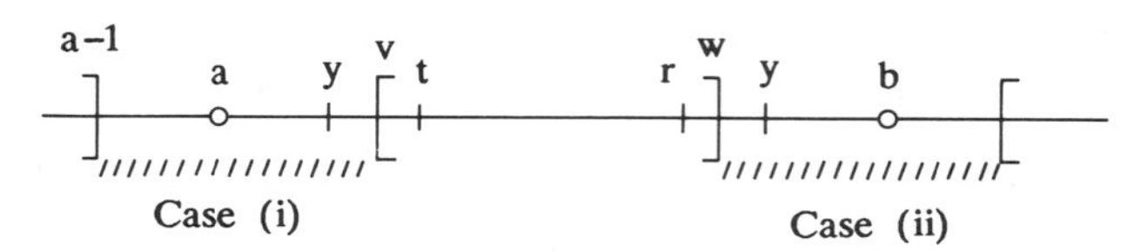


Fig. 9

Suppose (i) holds; then there also exists $t \in M$, with $y < t$, for otherwise $y$ would be the lub of $M$. We put $v = (y + t)/2$ and $U_y = \langle a - 1, v\rangle$.

If, instead, (ii) holds, there is no $s \in M$ with $y \leqslant s$. Then $y$ is an upper bound and there must still be another upper bound $r < y$. We put

$$w = (r+y)/2 \quad \text{and} \quad U_y = \langle w,\ b + 1\rangle.$$

In this way, every $y \in \langle\!\langle a, b\rangle\!\rangle$ is provided with a neighbourhood.

Because of the special choice of the $U_y$ we can reduce each finite sub–covering of $\langle\!\langle a,b\rangle\!\rangle$ to (at most) two of the neighbourhoods $U_y$. But two such neighbourhoods always leave 'a hole' free in the middle and do not cover $\langle\!\langle a, b\rangle\!\rangle$ completely. Thus 'not [HB]' holds.

**Exercise 10.**

Carry through the proof for an arbitrary closed bounded sub–set $A$ of $K$ instead of for an interval.

*PROOF–PORTION* : [UB] $\Rightarrow$ [A&CS].

We have already proved earlier that a complete field is Archimedean. Let therefore $(a_n)$ be a Cauchy sequence in $K$. We must somehow arrive at a set $M$ of which the upper bound $\sup M$ is the limit of $(a_n)$. Contemplation of Fig.10 leads us to the correct clue.

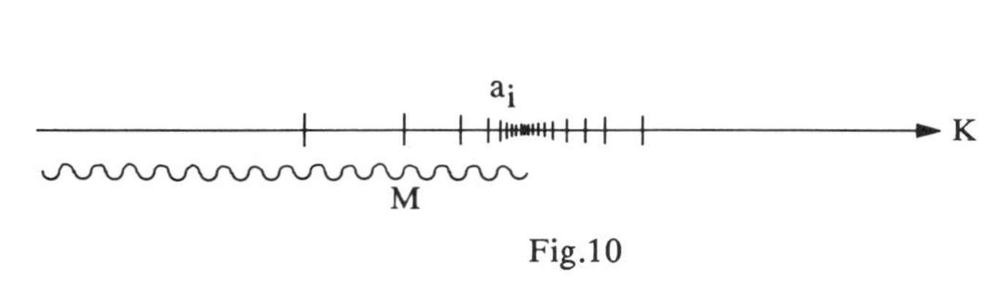


Fig.10

We should like to approach "from below" the spot at which the $a_i$ accumulate. We can manage this by defining:

$$M = \{m \in K \mid \text{there are only finitely many } i \text{ with } a_i < m\}.$$

Has $M$ a supremum? $M$ is not empty, for if $\varepsilon = 1$ we can find an $N \in \mathbb{N}$ in such a way that for almost all $i$, (that is, all but finitely many) the terms $a_i$ of the sequence lie in the interval $\langle a_N - 1, a_N + 1\rangle$. If $x < a_N$ then $x \in M$. Also for infinitely many $i$, the terms $a_i$ lie in the named interval, so $a_N + 1 = b$ is certainly an upper bound of $M$. Therefore $\sup M$ exists. We put $a = \sup M$ and assert: $\lim a_n = a = \sup M$. To establish this we must show:

$$(\forall\, \varepsilon > 0)\ (\exists\, k \in \mathbb{N}):\ \text{if}\ \ k < j\ \ \text{then}\ \ |a_j - a| < \varepsilon\ .$$

(The usual definition of Limit is also used in $K$.) Here, $\varepsilon$ is given and we seek $k$; therefore we substitute $\eta = \varepsilon/2 > 0$ as the '$\varepsilon$' in the definition of Cauchy–sequence. Because $(a_n)$ is a Cauchy sequence, we have:

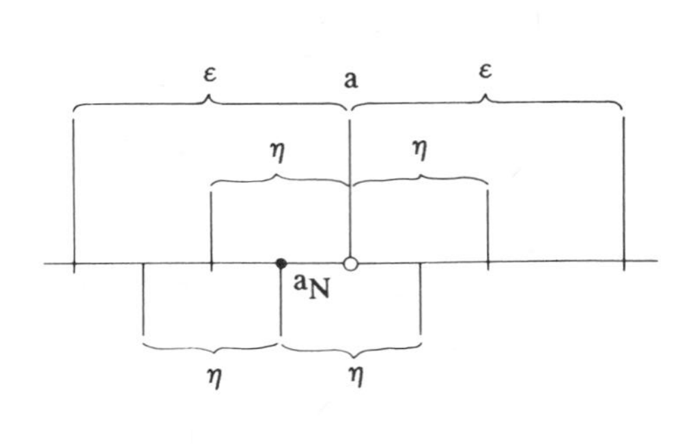


Fig. 11

There exists $n \in \mathbb{N}$ such that for almost all i, $a_i$ lies in the interval $\langle a_N - \eta, a_N + \eta\rangle$. Where does $a_N$ lie? To answer, we look at Fig.11. Thus $a_N - \eta$ belongs to M and $a_N + \eta$ is certainly an upper bound of M. Therefore we cannot have $a_N < a - \eta$, otherwise $a_N + \eta < a$; also $a + \eta < a_N$ is impossible (otherwise $a < a_N - \eta \in M$). Since $\eta = \varepsilon/2$, we now have for all $j > N$

$$a - \varepsilon < a_N - \eta < a_j < a_N + \eta < a + \varepsilon$$

so we have fulfilled the condition of the limit definition above with $k = N$.

*PROOF–PORTION :* [A&CS] $\Rightarrow$ [UB].

Because K is Archimedean, we may assume $K \subseteq \mathbb{R}$. Now consider a decimal expansion of the form

$$z + \frac{d_1}{10} + \frac{d_2}{10^2} + \ldots + \frac{d_n}{10^n} + \ldots$$

with $z \in \mathbb{Z}$ and $d_i \in \{0, 1, \ldots, 9\}$. The partial sums

$$\sum_{\nu=1}^{i} \frac{d_\nu}{10^\nu} = a_i$$

form a Cauchy sequence, and by [A&CS] this has a limit $a \in K$. Therefore *every* decimal expansion belongs to K and we have $K = \mathbb{R}$, so [UB] holds.

<u>*REMARK*</u>. It is worth pointing out here, something about our 'naïve' model of $\mathbb{R}$. In place of the last portion of the proof, we might think that we could argue as we did earlier, when we found an upper bound for a set $M \subseteq \mathbb{R}$ by means of a decimal fraction. Careful inspection shows that we simply asserted there the existence $a = \lim a_i$ of the limit of the approximating fraction and therefore, implicitly but certainly, we were working with [A&CS]. Honour therefore requires us here to point out that not everything there was done in a healthy manner. Indeed, throughout this chapter we must simply assume that [UB] or a property equivalent to it is assumed as for $\mathbb{R}$. But now remains the problem of specifying a complete ordered field in some concrete way. For a first attempt one might say: $\mathbb{R}$ "is" the number line which has all the required properties. If one is not satisfied with that, then one must construct $\mathbb{R}$

in some way, starting from other simpler mathematical objects. To this problem we shall return in the second chapter.

<u>The Intermediate Value Theorem</u>.

The following indicates how one can prove [IV] $\Rightarrow$ [UB]. For this purpose, it is useful to take the contrapositive, and prove not [UB] $\Rightarrow$ not [IV]. 'Not [UB]' means that there is in K a non-empty set M which is bounded above but has no supremum. Without loss of generality, we can assume that if $m \in M$, then for all $m' < m$, $m' \in M$; for, that has no effect on the existence of the supremum.

Now we define a function f, with graph as in Fig.12, where

$$f(x) = \begin{cases} -1, & \text{if } x \in M \\ 1, & \text{if } x \notin M \end{cases}$$

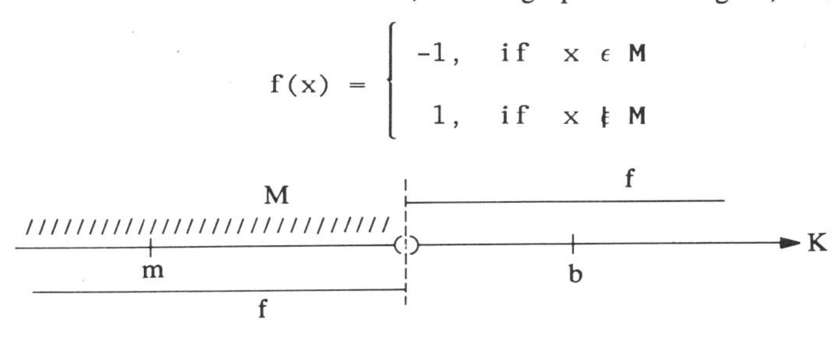


Fig.12

As M is not empty, there exists $m \in M$ with $f(m) = -1$. As M has an upper bound $b \notin M$ there exists b with $m < b$ and $f(b) = 1$. In an ordered field, we have $-1 < 0 < 1$, and f does not take the intermediate value 0 in the interval $\langle m, b \rangle$. To obtain a contradiction, it remains to show that f is continuous, so the intermediate value theorem can not hold. The continuity proof is left as an exercise, in which one must assume that M has no least upper bound.

*OBSERVATION*. If, with the set M used in the previous proof, we associate the set B of upper bounds of M, then B cannot have a greatest lower bound. We see then that M and B are non-empty, disjoint and open, and they cover K. Therefore K is not connected. This perhaps makes clear the close interrelation between the intermediate value theorem and the concept of connectivity. (Often one formulates the intermediate value theorem in the following way: *The continuous image of a connected set is again connected.*)

An important consequence of completeness is the *uncountability* of $\mathbb{R}$, wherein a set M is called **countable**, if there is a bijective mapping $\alpha : \mathbb{N} \to M$.

<u>THEOREM</u>. (Cantor) $\mathbb{R}$ ***is not countable.***

*PROOF*

Later we shall enter into the details concerning different powers of sets, but here we consider only the real interval $\langle 0, 1 \rangle$. One can easily give an injective mapping $\beta : \mathbb{N} \to \langle 0, 1 \rangle$. However, we shall show that because $\mathbb{R}$ is complete, there is no surjective mapping $\alpha : \mathbb{N} \to \langle 0, 1 \rangle$. For this purpose, we use the representation of the real numbers as decimal fractions (without recurring nines, so if necessary we must attach an infinite sequence of zeros). The proof originated with Cantor and is often called the "Second Cantor Diagonal Process". Let therefore, $\alpha : \mathbb{N} \to \langle 0, 1 \rangle$ be any mapping. We write the values of $\alpha$ successively, for example say:

$$\alpha(1) = 0.\ 1\ 4\ 1\ 5\ 9\ 2\ 6\ \ldots$$

$$\alpha(2) = 0.\ 4\ 1\ 4\ 2\ 1\ 3\ 6\ \ldots$$

$$\alpha(3) = 0.\ 7\ 3\ 2\ 0\ 5\ 0\ 8\ \ldots$$

$$\alpha(4) = 0.\ 7\ 1\ 8\ 2\ 8\ 1\ 8\ \ldots$$

$$\alpha(5) = 0.\ 2\ 3\ 6\ 0\ 6\ 8\ 0\ \ldots$$

$$\alpha(6) = 0.\ 4\ 4\ 9\ 4\ 8\ 9\ 7\ \ldots$$

this suggests that we go down the diagonal to display the digits 1, 1, 2, 2, 6, 9,... which will be important for the definition of the number $x \in \langle 0, 1 \rangle$ we are about to construct. This $x$ does not appear among the $\alpha(i)$ and to construct it we use the following trick.

The first digit $x_1$ shall be:
1, in case the first digit of $\alpha(1)$ is not equal to 1,
2, in case the first digit of $\alpha(1)$ equals 1.

For $n > 1$, the $n^{th}$ digit $x_n$ of $x$ shall be
1, in case the $n^{th}$ digit of $\alpha(n)$ is not equal to 1,
2, in case the $n^{th}$ digit of $\alpha(n)$ equals 1.

This process results in a decimal expansion $0.x_1 x_2 x_3 \ldots$ without recurring nines. Therefore because of the completeness of $\mathbb{R}$, this is a real number $x \in \langle 0, 1 \rangle$. In our example, $x$ is 0.221111...) This $x$ is certainly not a value of $\alpha$ because it differs from $\alpha(i)$ in the $i^{th}$ place of the decimal expansion. Thus $\alpha : \mathbb{N} \to \langle 0, 1 \rangle$ is not surjective, as claimed above. Even more then, there can be no surjective mapping $\mathbb{N} \to \mathbb{R}$ and therefore no bijective mapping. Hence $\mathbb{R}$ is not countable. □

Once again, only because of the completeness of $\mathbb{R}$ can we be certain that our newly constructed decimal fraction is a real number.

Exercise 11.

Make this clearer by attempting to use the same method for proving there is no surjective mapping $\gamma : \mathbb{N} \to \langle 0, 1 \rangle \cap \mathbb{Q}$, (therefore on the rational interval!) to see how things break down.

## 1F. CANTOR'S CHARACTERISATION OF $(\mathbb{Q}, \trianglelefteq)$.

When we considered the number domains as ordered sets, it remained open in which order÷theoretic respect they can be distinguished from the chains (Dez, $\trianglelefteq$), $(\mathbb{Q}, \trianglelefteq)$ and $(\mathbb{Q} \times \mathbb{Q}, \leqslant_{lex})$. A beautiful theorem of Cantor tells us that there is no difference, i.e. these chains are order÷isomorphic. (G. Cantor, *Works* pp.303÷306). More precisely, his theorem states:

THEOREM. *Let* (K, $\leqslant$) *be a chain with the properties:*

*(1)* K *is countable*

*(2)* (K, $\leqslant$) *has no greatest and no smallest element*

*(3)* (K, $\leqslant$) *is dense.*

*Then* (K, $\leqslant$) *is order–isomorphic to* $(\mathbb{Q}, \trianglelefteq)$.

This theorem is at the same time a characterization of $(\mathbb{Q}, \trianglelefteq)$ by means of the given properties, (1), (2), and (3).

Consequences.

Cantor's characterization of $(\mathbb{Q}, \trianglelefteq)$ by such simple properties is in itself interesting but the theorem has far–reaching consequences, which we review before giving the proof. For simplicity, we shall say (K, $\leqslant$) has "holes", if (K, $\leqslant$) is not complete. (The missing lubs are just such holes.)

(i) As the chains (Dez, $\trianglelefteq$) and $(\mathbb{Q} \times \mathbb{Q}, \leqslant_{lex})$ have the properties (1), (2) and (3), they are isomorphic to $(\mathbb{Q}, \leqslant)$. Recalling Cantor's theorem that the set A of real algebraic numbers is countable, we then have also an isomorphism betweem (A, $\trianglelefteq$) and $(\mathbb{Q}, \trianglelefteq)$.
*Caution*: this is not an isomorphism between ordered *fields* since it neglects the arithmetic operations.

(ii) With the help of the embedding of (Dez, $\trianglelefteq$) in $(\mathbb{Q}, \trianglelefteq)$ we have seen in the previous example, that (Dez, $\trianglelefteq$) has holes. Using the isomorphism (Dez, $\trianglelefteq$) $\approx$ $(\mathbb{Q}, \trianglelefteq)$ it follows that $(\mathbb{Q}, \trianglelefteq)$ also has holes and indeed, without the mention of irrational numbers in any form! In other words: since $(\mathbb{R}, \trianglelefteq)$ *is complete*, $\mathbb{R}$ must contain other numbers besides the rationals.

(iii) By the same considerations, we see that there must be transcendental numbers in $\mathbb{R}$; for, the chain (A, $\trianglelefteq$) of algebraic numbers, being isomorphic with (Dez, $\trianglelefteq$), also has holes.

(iv) Without mentioning algebraic numbers, we can derive, in a new way, Cantor's theorem asserting the non–countability of $\mathbb{R}$. If $\mathbb{R}$ were countable, then $(\mathbb{R} \trianglelefteq)$ would be isomorphic to (Dez, $\trianglelefteq$) and it would therefore have holes. This is impossible, *because* $(\mathbb{R}, \trianglelefteq)$ *is complete*, so $\mathbb{R}$ cannot be countable.
Here we encounter $\mathbb{R}$ without mentioning decimal fractions and we see the influence of the completeness more directly than in the proof at the end of the previous paragraphs.

*PROOF OF THE THEOREM.* For the definition of the required isomorphism $\gamma$ we give an intuitive representation of an order-isomorphism in Fig.13:

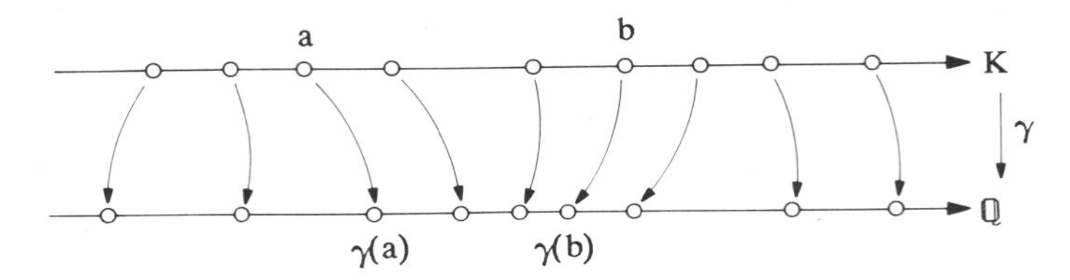


Fig. 13

If $a < b$ we shall always have $\gamma a < \gamma b$ provided the ordering arrows $a \to \gamma a$ and $b \to \gamma b$ never cross. Following this intuitive idea, we now construct $\gamma : K \to \mathbb{Q}$. As both $K$ and $\mathbb{Q}$ are countable, we choose some enumeration for each of them – say $k_1$, $k_2$, $k_3$, $k_4$,.. for $K$ and $q_1$, $q_2$, $q_3$... for $\mathbb{Q}$. With the help of these enumerations, we define $\gamma$ "recursively" (i.e. using induction).

We start the recursion by setting $\gamma k_1 = q_1$. Now suppose, say, $k_1 < k_2$; if it happens that $q_1 < q_2$, we put $\gamma k_2 = q_2$. Possibly, however, $q_2 < q_1$; so we consider those $x \in \mathbb{Q}$ with $q_1 < x$. Such elements $x$ clearly exist because $\mathbb{Q}$ has no largest element. Among them we choose that $q_n$ with the smallest index $n$ and put $\gamma k_2 = q_n$. For the further procedure, consider the sketch in Fig.14.

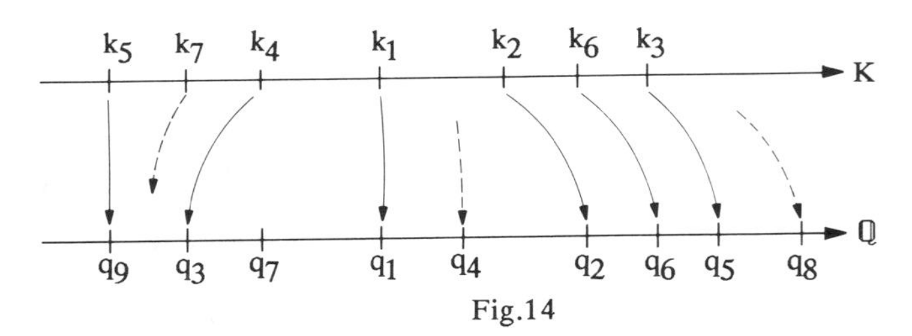


Fig.14

Suppose then that $\gamma k_1, \ldots, \gamma k_{r-1}$ have already been so constructed that the (finite) sets $(\{k_1,\ldots,k_{r-1}\}, \leqslant)$ and $(\{\gamma k_1,\ldots,\gamma k_{r-1}\}, \trianglelefteq)$ are order-isomorphic. We will then specify $\gamma k_r$ in such a way that the isomorphism will still hold for the sets enlarged by $k_r$ and $\gamma k_r$. To achieve this, we shall so choose $\gamma k_r$ that it lies, with respect to $\gamma k_1,\ldots,\gamma k_{r-1}$, in the same way as $k_r$ does to $k_1,\ldots,k_{r-1}$.

Case 1. Suppose $k_r$ is larger than $k_1,\ldots,k_{r-1}$. Then we consider in $\mathbb{Q}$ the set of all $x$ with $\gamma k_1,\ldots,\gamma k_{r-1} < x$. This set is not empty *because* $\mathbb{Q}$ *has no greatest element.* We choose among these $x$ the one with the smallest index – say $q_n$, and put $\gamma k_r = q_n$.

Case 2. Suppose $k_r$ lies in one of the intervals determined by $k_1,\ldots,k_{r-1}$, suppose $k_i < k_r < k_j$. Then we consider in $\mathbb{Q}$ the set of all $x$ with $\gamma k_i < x < \gamma k_j$. This set is not empty *because* $\mathbb{Q}$ *is dense*. From these $x$ we choose that with the smallest index – say $q_m$, and put $\gamma k_r = q_m$.

Case 3. Suppose $k_r$ is smaller than $k_1,\ldots,k_{r-1}$. We proceed analogously to Case 1, but using this time the fact that $\mathbb{Q}$ *has no smallest element*.

This recursive definition gives us, for each $x \in K$, a function value $\gamma x$. The construction ensures that $\gamma$ respects the order and is also injective. What is still lacking is a proof of the surjectivity of $\gamma$. Does every $q \in \mathbb{Q}$ really appear in this procedure as a value of $\gamma$? We answer this question affirmatively, again using induction.

Beginning of the Induction.
By construction, $q_1$ is the function-value $\gamma(k_1)$ of $k_1$. Does this happen with $q_2$? Suppose $q_1 < q_2$ but $k_2 < k_1$; then $q_2$ is not the function value of $k_2$ and possibly also not of $k_3$ and $k_4$. However, there must be a first $k_i$ that exceeds $k_1$ since K has no greatest element. Then since $q_2$ is that element $> q_1$ with smallest index, $q_2$ is (by the construction) the function value of $k_i$.

General Inductive Step. Suppose $q_1,\ldots,q_{s-1}$ have already been attained as function values – say $q_{s-1} = \gamma k_n$ with $n \geqslant s-1$. As the k's are not necessarily selected in order, it can happen that further rational numbers $q_j$ with $j > s-1$ are function values. If $q_s$ is one of them then we are finished. If $q_s$ is not, then we have three possibilities:

Case 1. $q_s$ exceeds all $\gamma k_1,\ldots,\gamma k_n$. Of these, let $\gamma k_i$ be the biggest. *Because* K *has no greatest element*, there must appear in the enumeration of K, a (first) element $k_j > k_i$. Corresponding to this first $k_j$, $q_s$ is that element of $\mathbb{Q}$ in the correct position with the smallest index; and therefore we have $\gamma k_j = q_s$.

Case 2. $q_s$ lies in an interval $\gamma k_i < q_s < \gamma k_j$, in which otherwise no further function-value lies. One argues as previously, and uses the fact that K *is dense*.

Case 3. $q_s$ is smaller than all $\gamma k_1,\ldots,\gamma k_n$. Analogously as before, but we use the assumption that K *has no smallest element*.

This proves that $\gamma : K \to \mathbb{Q}$ is surjective. Hence we have shown $\gamma$ to be an order-preserving bijection, so $\gamma^{-1}$ preserves order (see Exercise 3) and hence $\gamma$ is an order-isomorphism. Cantor's theorem is therefore established. □

Exercise 12.

(a) List the theorems proved in this Chapter, and draw a diagram to show the logical interrelations between them.

(b) Let $F$ be an ordered field. Prove that if $a, b \in F$ and $a^2 + b^2 = 0$ then $a = b = 0$. Also prove that $F$ has characteristic zero and contains a subfield (the <u>rational</u> subfield of $F$) which is isomorphic to $\mathbb{Q}$.

(c) Prove that the following statements are equivalent in any ordered field $F$ :

(i) the order of $F$ is Archimedean.

(ii) for every $a \in F, a > 0$, there exists $n \in \mathbb{N}$ such that $2^n > a$. (Here $2 = 1 + 1$, and $1$ is the multiplicative identity of $F$).

(iii) the rational subfield of $F$ is dense in $F$.

(iv) every monotonic increasing sequence of elements of $F$ which is bounded above is a Cauchy sequence.

(v) the set of dyadic rational elements of $F$ is dense in $F$. [The dyadic rational elements are those of the form $(2^n.1_F)^{-1}\ (m.1_F)$ where $m \in \mathbb{Z}$, $n \in \mathbb{N}$].

(vi) $$1 = \lim_{n\to\infty} \sum_{i=1}^{n} 1/2^i$$

(d) State four equivalent notions of completeness of an ordered field. Use any one of them to prove that every element of a complete ordered field has a cube root.

(e) Prove that every Cauchy sequence $(x_n)$ of real numbers either converges to $0$ or else is ultimately of constant sign and bounded away from $0$. Deduce that if $x_n \geqq 0$ and $(x_n{}^2)$ is Cauchy, then $(x_n)$ is Cauchy.

(f) Let $\mathbb{R}(t)$ be the ordered field of all formal rational expressions $x$ with real coefficients

$$x = \frac{a_o+a_1t+\ldots+a_mt^m}{b_o+b_1t+\ldots+b_nt^n} \qquad (b_n > 0)$$

where $x$ is positive iff $a_mb_n > 0$. Prove that if $x$ is any positive element of $\mathbb{R}(t)$ then

$$t^{-k} < x$$

for all sufficiently large $k$ in $\mathbb{N}$.
Now let $y = 1 = t^{-1}$ and define $(x_n)$ inductively by

$$x_1 = 1 + \frac{1}{2}t^{-1}, \quad x_{n+1} = \frac{1}{2}\left[x_n + \frac{y}{x_n}\right].$$

By considering $x_{n+1}^2 - y$ and $x_n - x_{n+1}$, or otherwise, prove that

(i) $$x_n > x_{nm+1} > 1 ,$$

(ii) $$0 < x_n^2 - y < t^{-2^n} ,$$

(iii) $(x_n)$ is a Cauchy sequence.

Deduce that $\mathbb{R}(t)$ is not Cauchy complete (i.e. not every Cauchy sequence converges).

(g) Assuming that $\mathbb{R}$ has the least upper bound property, prove that every Cauchy sequence of real numbers is convergent in $\mathbb{R}$.

Let $K$ denote the set of all formal Laurent series

$$\sum_{-\infty}^{\infty} r_j t^j$$

where $r_j \in \mathbb{R}$ and $r_j = 0$ for all integers $j \leqslant$ some integer $k$. Show how $K$ can be made into a non-Archimedean field (do not verify all the details, but pay attention to the existence of a multiplicative inverse of a non-zero element). Prove that $K$ is Cauchy complete, and exhibit a sequence of elements of $K$ which is increasing, bounded above but not convergent in $K$.

(Questions (b) - (g) are taken from various Final Examination papers of Southampton University.)

# 2

# Constructions of $\mathbb{R}$

## INTRODUCTION.

In Chapter 1 we have seen how $\mathbb{R}$ is characterized as a complete ordered field, and what significance completeness has for Analysis. At the same time it has become clear, that to build up Analysis, we do not need an actual *construction* of the real numbers (say as equivalence classes of Cauchy sequences from $\mathbb{Q}$). Here, when giving an introductory view of the theory, one can quietly rely on the intuitively given Number line in order to lighten the load on beginners.

One can therefore on the one hand assume that the real numbers are intuitively given and – as we have done – essentially describe them through their properties. On the other hand, one can also ask whether the reals could be obtained by a construction from simpler number–domains, for example, $\mathbb{N}$ or $\mathbb{Q}$. This latter policy has the advantage that in building up the number systems, one would need to fall back on intuition only once, namely by assuming the existence of natural numbers. Nevertheless, there are prominent mathematicians, who grant to the number line $\mathbb{R}$ of geometry a greater intuitiveness than to the natural numbers, (see for example, the essay of René Thom in the collection of M. Otte: *Mathematicians on Mathematics* [53], p.384). However one may stand regarding that question (it cannot be decided within mathematics), one can still recognize that one of the greatest mathematical achievements of the nineteenth century was the systematic building up of the number systems from $\mathbb{N}$ to $\mathbb{R}$ and $\mathbb{C}$ and beyond, and the gaining of the insights that have been won through that process. More on this theme in Section 2F!

## 2A DECIMAL REPRESENTATIONS OF THE REAL NUMBERS.

*PRELIMINARY REMARKS*.

(a) For the real numbers, one often uses the conveniently reassuring definition that "they are the infinite decimal fractions". In this section the correctness of that definition will be established.

(b) In Chapter 1 we saw that every Archimedean field $K$ lies in $\mathbb{R}$. Therefore all elements of $K$ have decimal expansions. Conversely however, one cannot expect every decimal sequence to correspond to an element from $K$. In the following, $K$ will always denote an Archimedean ordered field.

(c) We only need decimal expansions for numbers $> 0$. For handling the rest, one then needs only to use the minus sign.

(d) We write $d$ (= decem) for 10 because that is simpler, and then we simultaneously prove everything for arbitrary $d$ with $1 < d \in \mathbb{N}$; but we speak however always of decimal sequences etc. (The preference for 10 rests on biological, rather than mathematical reasons. Euler is reputed to have calculated in a system with base 8.)

**Division with Remainder in $\mathbb{N}$.**

*Suppose* $1 < d \in \mathbb{N}$ *and* $m \in \mathbb{N}$. *Then there exist uniquely determined numbers* $q, r \in \mathbb{N}$ *with*

$$m = qd + r \quad \text{and} \quad 0 \leq r < d$$

(*Proof*. By induction, see books on algebra.) □.

**Decimal Representations of Natural Numbers.**

*Let* $1 < d \in \mathbb{N}$ *and* $m \in \mathbb{N}_0$. *Then there exist uniquely determined numbers* $n, c_i \in \mathbb{N}_0$ *with* $0 \leq c_i < d$ *and*

$$m = c_o + c_1 d + \ldots + c_n d^n .$$

*PROOF*. This results from repeated application of the process of division with remainder. We have

$$m = q_1 d + c_o, \quad \text{where} \quad 0 \leq c_o < d ,$$

with uniquely determined $q_1, c_o$. Since $1 < d$, we have $q_1 < m$. The next step yields

$$q_1 = q_2 d + c_1, \quad \text{where} \quad 0 \leq c_1 < d .$$

Since $1 < d$, then $q_2 < q_1$; and we repeat the process, which breaks off since $m > q_1 > q_2 \ldots \geq 0$. All the numbers $c_i$ and $q_i$ are uniquely determined. Therefore we obtain

$$m = [\ldots (c_n d + c_{n-1})d + \ldots + c_1]d + c_o$$

$$= c_n d^n + \ldots + c_1 d + c_o \qquad \square$$

Now let K be Archimedean and $0 < x \in K$. Then there is a natural number s with $x < s$ (since the natural numbers are unbounded) and hence there is a smallest natural number $m + 1$ with $x < m + 1$. Therefore

$$x = m + \alpha, \text{ where } 0 \leqq \alpha < 1 \text{ and } m \in \mathbb{N}_o$$

For m we have obtained already a decimal representation, but for $\alpha$ we construct one in the next theorem.

LEMMA. *Let* $x \in K$ *with* $0 \leqq x < 1$ *and* $1 < d \in \mathbb{N}$. *For each* $n \in \mathbb{N}$ *there is a unique representation*

$$x = \frac{a_1}{d} + \frac{a_2}{d^2} + \ldots + \frac{a_n}{d^n} + \alpha_n ,$$

*where* $0 \leqq a_i < d$ *and* $0 \leqq \alpha_n < \frac{1}{d^n}$.

(This representation is usually expressed in the form $x = 0.a_1a_2a_3 \ldots a_n + \alpha_n$ with $0 \leqq a_i \leqq 9$.)

*PROOF*. This can be carried through simply using the previous results. Let m be the greatest whole number $\leqq d^n x$. Then $m \geqq 0$ and $d^n x = m + \alpha_n$, where $0 \leqq \alpha_n < 1$. If we expand m as above, and divide finally by $d^n$, then the proof is completed. □

Putting the previous assertions together, we derive for each $x \in K$ with $0 < x$ a decimal representation, just as one learns at school. With m playing the same role as before, one can write

$$x = m.a_1a_2a_3 \ldots$$

$$= c_nc_{n-1} \ldots c_1c_o.a_1a_2a_3 \ldots ,$$

in which the sequence of digits of the $a_i$ is determined by the previous process. This holds, as we have said previously for every Archimedean field K. It is with the converse question, whether every decimal sequence also represents a genuine real number, that we eventually concern ourselves. First we need to deal conclusively with the irritating problem of "recurring nines".

Thus, let $a_1, a_2, a_3, \ldots$ be a sequence of natural numbers with $0 \leqq a_i < d$. We call $(a_n)$ a d-sequence. Further, we say that $(a_n)$ **has no recurring nines**, provided that for each $n \in \mathbb{N}$ there exists $i > n$ with $a_i \neq d - 1$.

CLAIM. *Let* K *be complete and* $1 < d \in \mathbb{N}$. *Given a d-sequence* $(a_n)$, *there is an* $x \in K$ *such that for every* $n \in \mathbb{N}$,

$$x = \frac{a_1}{d} + \frac{a_2}{d^2} + \ldots + \frac{a_n}{d^n} + \alpha_n , \quad \text{with} \quad 0 \leqq \alpha_n \leqq 1/d^n .$$

LEMMA. *If the d–sequence* $(a_n)$ *has no recurring nines, then* $0 \leq \alpha_n < 1/d^n$.

*PROOF*. Since K is a complete field we can handle matters to within notation as if we were in $\mathbb{R}$. Therefore we may use all known facts about infinite series in $\mathbb{R}$. With the d-sequence $a_1, a_2, a_3, \ldots$ we associate the partial sums

$$s_n = \frac{a_1}{d} + \frac{a_2}{d^2} + \ldots + \frac{a_n}{d^n}$$

The set S of partial sums $s_n$ is not empty, and is bounded above – e.g. by

$$\sum_{i=1}^{\infty} (d-1)/d^i .$$

Therefore, there is an upper bound

$$x = \sum_{i=1}^{\infty} a_i/d^i$$

of S. Hence

$$x = s_n + \alpha_n \quad \text{with} \quad \alpha_n = \sum_{i=n+1}^{\infty} a_i/d^i$$

Certainly, since $a_i \leq d - 1$,

$$\alpha_n \leq \beta_n = \sum_{i=n+1}^{\infty} (d-1)/d^i$$

so, in case $(a_n)$ has no recurring nines, we have $\alpha_n < \beta_n$. Using the formula for the sum of a geometric series, we can calculate $\beta_n$, and obtain

$$\beta_n = \frac{d-1}{d^{n+1}} \sum_{i=0}^{\infty} \frac{1}{d^i} = \frac{d-1}{d^{n+1}} \cdot \frac{1}{1 - \frac{1}{d}} = \frac{1}{d^n} .$$

Therefore we derive $\alpha_n \leq 1/d^n$ without the hypothesis of the lemma and $\alpha_n < 1/d^n$ with the hypothesis. □

Recurring Nines are Superfluous.

We have seen that in the Archimedean field K, each $x > 0$ determines uniquely a decimal sequence. We will show that with this no recurring nines appear. An example suffices. Can it happen that $a_i = d - 1$ for all $i \in \mathbb{N}$? By our calculation of $\beta_n$, we must then have $x = 1/d^0 = 1$. That is, in determining the digit sequence, we would already at the first step

$$x = \frac{d-1}{d} + \alpha_1$$

have excluded the condition $0 \leq \alpha_1 < 1/d$ since $\alpha_1 = 1/d$. (Correspondingly with recurring nines that start further on.).

## 2B CONSTRUCTIONS OF ℝ WITH DECIMAL SEQUENCES.

For decimal sequences without recurring nines

$$c_n c_{n-1} \cdots c_1 c_o . a_1\ a_2\ a_3\ \cdots$$

$$y_m y_{m-1} \cdots y_o . x_1\ x_2\ x_3\ \cdots$$

we introduce the lexicographical order. First we make them both have the same length by adding, if necessary, zeros to a sequence; and then we say that the smaller sequence is that which has the first smaller letter from the alphabet $0, 1, \ldots, d - 1$. (E.g. 1234.567 ... < 1234.576 ...) Therefore, the set of decimal sequences becomes a complete chain which we denote by $(\Delta^+, \trianglelefteq)$.

The description in Section 2.A of the positive elements of a complete field K by means of decimal sequences, now lies close to the following process. Starting from ℕ one can construct a complete field K as follows.

We take the set $(\Delta^+, \trianglelefteq)$ and define in it an addition and a multiplication, so that all the arithmetic laws in $(\Delta^+, +, \cdot, \trianglelefteq)$ are satisfied that one expects of the positive elements of a complete field. Having achieved this, we can use a simple extension process (like that of ℕ in ℤ) in order to obtain a complete field $(\Delta, +, \cdot, \trianglelefteq)$.

Literature for this Process.

F.A. Behrend: "A Contribution to the Theory of Magnitudes and the Foundations of Analysis" [7], p.345÷362. Here the process indicated above is precisely carried out. Additionally, it is shown how to lead from completeness in a natural way to the elementary transcendental functions exp, ln, sin, cos.

An article that is somewhat more explicit and concentrates wholly on addition and multiplication of positive decimal sequences, (and on that account, is certainly simpler, if also not simple, to read) is that of G. Holland: "A Proposal for Introducing Real Numbers as Decimal Fractions" [33], p.87÷110.

Correspondingly, there is a didactic preparation for schools: G. Holland: "Decimal Fractions and Real Numbers" [34], p.5÷26.

Further, we should mention: W. Rautenberg: "A Short and Direct Path from the Natural Numbers to the Real Numbers with Associated Foundation for the Calculus of Fractions" [61], and his *Real Numbers in an Elementary Exposition* [62].

Evaluation of this Process.
The decisive advantage lies in the simple definition of the positive real numbers as decimal sequences. The objects themselves are directly accessible and (at least partially) given in an intuitively concrete way. Examples can be written down without difficulty. From the structural properties, the ordering is easy to define and completeness is especially easy to prove. The disadvantages of the process show themselves with the definitions of addition and multiplication and the proof of the various laws of arithmetic. Here the process becomes troublesome and often obscure. There is a trick with which one can get round the especially complicated problem of multiplication, but the previously cited advantages start to be lost in a drift towards abstraction; one uses endomorphisms of the additive structure (see Behrend). G. Holland has made convincingly clear in his articles, that this process is to be preferred as "background theory" above all others for the handling of real numbers in school. Therefore it is also important for school that the process should be based on $\mathbb{N}$ and hence rational numbers are not used. (In the Author's opinion, an approach through simply defined objects with relatively complicated laws of calculation, is always easier for beginners to understand than one which introduces abstractly defined objects with simple operations — which many mathematicians prefer. (Compare say, matrices and linear mappings.)

## 2C CONSTRUCTION OF $\mathbb{R}$ FOLLOWING DEDEKIND.

Richard Dedekind (1831-1916) took his Ph.D. under Gauss, and was Professor in Göttingen, Zürich, and Brunswick. He was the first to construct the real numbers from the rational numbers, and to indicate explicitly the decisive significance of completeness. He himself named the date of his discovery of this connection as 24th November, 1858, in Zürich. His pamphlet "Continuity and Irrational Numbers" of 1872 brings the theory together. Essential sections from that are expressed in: O. Becker: *Foundations of Mathematics in Historical Development* [4], a book which is especially a mine of information, strongly recommended to the reader.

Dedekind is one of the outstanding mathematicians of the nineteenth century; the founder, together with Galois, of algebra in the modern sense.

Dedekind Sections.
In the pamphlet mentioned above, Dedekind gives the following formulation of completeness (called by him "Continuity") of the Number line. "Separate all points of the line into two classes, in such a way that each point of the first class lies to the left of each point of the second class; then there is one and only one point which produces this partition of all points in to two classes, this cutting of the line into two pieces."

We denote this consequence by [DS]. It is equivalent to the various formulations of completeness introduced in the main theorem of Section 1.E. In particular, we shall compare it with the connectivity axiom [CA].

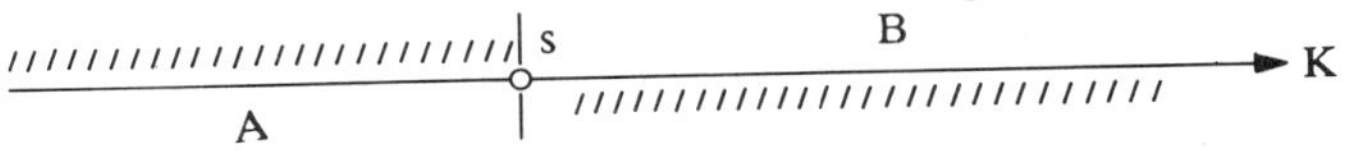

Fig.15

By [CA], $A$ and $B$ cannot both be open, if at the same time the equations $A \cap B = 0$ and $A \cup B = K$ hold. Therefore either $A = \langle -\infty, s\rangle$ and $B = \langle s, \infty\rangle$ or $s \in B$, and in any case the splitting is induced by the number $s$ (= sup $A$). The important thing is that $s$ *exists on the line*.

## Construction of $\mathbb{R}$.

After Dedekind had found this formulation of completeness, he used "cuts in $\mathbb{Q}$" for the definition of $\mathbb{R}$.

For technical reasons arising from his proof, we shall not follow Dedekind word for word, but we shall use a variant of his process. We do not deal with the whole of $\mathbb{Q}$ but only with the set

$$\mathbb{Q}^+ = \{q \in \mathbb{Q} \mid 0 < q\}$$

of positive rational numbers; instead of upper and lower classes, we use only the lower classes which we call "sections".

***Definition***. A section $A \subseteq \mathbb{Q}^+$ is a non-empty subset of $\mathbb{Q}^+$ which is bounded above and without greatest element but with the property: if $a \in A$ and $b < a$, then $b \in A$.

We denote by $\mathbf{A}^+$ the set of all sections.

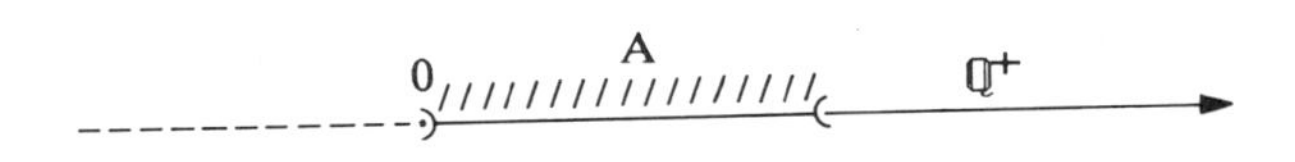


Fig.16

(Analogously with $\Delta^+$ in Section 2.B, we have decided to take just those sets which will later correspond to 'positive real numbers'. The point of this procedure is clear. In place of the least upper bound of $A$ which is missing in $\mathbb{Q}$ we shall be taking simply $A$ itself.

**Exercise 1.**

Let $R = \{x \in \mathbb{Q}^+ \mid x^2 \geq 2\}$ and let $L = \{x \in \mathbb{Q} \mid x \notin R\}$. Prove that $R$ has no smallest element and that $L$ has no largest element. Hence prove that $\mathbb{Q}$ is not Dedekind complete.

Definition of Order and Operations in $A^+$. Given $A, B \in A^+$ let

$$A + B = \{a + b \mid a \in A,\ b \in B\}$$

$$A \cdot B = \{a \cdot b \mid a \in A,\ b \in B\}$$

$$A \leqq B \Leftrightarrow A \subseteq B \,.$$

it is now relatively simple (in contrast to the decimal fractions!) to verify those properties of the structure $(A^+, +, \cdot, \leqq)$ which one needs for extending it to a complete field $(A, +, \cdot, \leqq)$. This extension follows again the standard process "from $\mathbb{N}$ to $\mathbb{Z}$". Example: the neutral element of multiplication is

$$E = \{q \in \mathbb{Q}^+ \mid q < 1) \;;$$

it is easy to verify that $A \cdot B = B \cdot A$. The least upper bound, of a non-empty subset $M$ of $A^+$ which is bounded above, is simply

$$\bigcup_{A \in M} A = S.$$

A Second Variant of this Process.

Instead of using $\mathbb{Q}^+$ and defining addition and multiplication for the sections, we could go from $\mathbb{Q}$ and similarly define the set $A$ (of all real numbers) in the form of lower classes. Doing this, one finds that multiplication makes for difficulties (the product of two lower classes is not a lower class again). On that account, we avoid that procedure, and construct first the complete ordered group $(A, +, \leqq)$. Again there is a standard process (with endomorphisms) for introducing multiplication. This process is worked through by, for example, H. Lenz: *Foundations of Elementary Mathematics* [45] (Chapter VII).

Evaluation of the Dedekind Process.

If we compare the Dedekind process with the path we took through decimal sequences, or with the Cantor process (to be dealt with below) and work through each process in all details, there is no doubt that the Dedekind process is by far the simplest. That is to say, if one puts value on every detail of the proof and its understanding, then with Dedekind we have the least trouble. For use in schools however, there are disadvantages in beginning with objects which are less easy to describe, (namely the lower sets) and the fact that we need to operate upon entire sets. For use in general work it is disadvantageous that the process needs too much familiarity with the real numbers and uses the order relation in the definition of the sections. On that account, generalizations of this path are only possible in the not so central area of order structures (the so-called Dedekind – McNeille completion). Nowadays in school, one does not give all proofs as much as previously, so one usually uses the decimal sequences there. And in High School lessons*, one stops at the Cauchy sequences which were chosen by Cantor on account of their possibilities for generalization. That's a pity!

## 2D. CANTOR'S CONSTRUCTION OF ℝ.

George Cantor, (1845÷1918, Professor in Halle) was studying Fourier series, and around 1870, he was struck by the necessity for a sharper version of the concept "real number". In many respects his conclusions are closely connected with those of his teacher Weierstrass (1815÷1897, Professor in Berlin). As Weierstrass only gave his conclusions in lectures and published no notes or book about them, the origin of the ideas cannot be established any more precisely. Even in the great work of P. Dugac: *Elements of Analysis of Karl Weierstrass* [12], pp.41÷176) the details still remain unclear.

Cantor is the founder of Set÷theory (*Mengenlehre* in his German) and here we shall go further into some of his main results in that area.

### Cantor's Fundamental Sequences.

Cantor's "fundamental sequences" (he said: fundamental series), are nothing more than the Cauchy sequences, but with *rational* terms, that we well know already. In order to emphasize this restriction to rationals, we shall retain here the designation "fundamental sequence". It then turns out that with fundamental sequences, we shall have to create non–existent limits, just as previously we had to create the least upper bounds with the sections. In so doing, the essential idea is the by–passing of the passage to the limit concept, through the definition of the fundamental (Cauchy) sequence: one grasps convergence without having to name an actual limit. However, there results a new (and for this process a typical) complication: the terms of different fundamental sequences can accumulate to the same point. If we want to create, therefore, these "accumulation points" as new "points", we must lump together all sequences which tend towards such a "point".

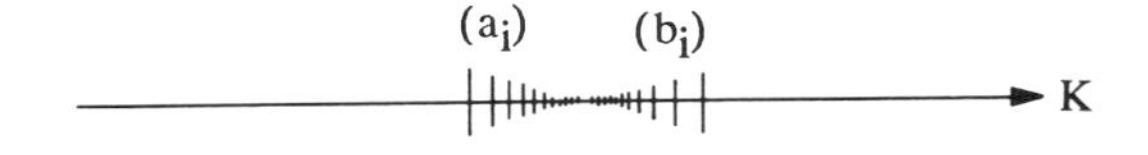


Fig.17

Definition. A *fundamental sequence* $(a_i)$ is a Cauchy sequence with each $a_i$ in $\mathbb{Q}$. If $(a_i)$ converges to zero, then the sequence is called a **null** sequence. Two fundamental sequences $(a_i)$, $(b_i)$ are said to be **equivalent** if their difference $(a_i \div b_i)$ is a null sequence.

---

* Translator's Note: The author is here referring to schools in Germany; British schools would be unlikely to mention such approaches at all.

Within the set K of equivalence classes of fundamental sequences, we can introduce addition, multiplication and order in the following way. For this, let $[(a_i)]$ be the equivalence class of $(a_i)$, etc. Then we define the following operations on equivalence classes:

$$[(a_i)] + [(b_i)] = [(a_i + b_i)]$$

$$[(a_i)] \cdot [(b_i)] = [(a_i b_i)] .$$

The order relation is defined by:

$[(a_i)] \trianglelefteq [(b_i)]$ iff $\exists\ \varepsilon > 0$ such that for almost all $i$ we have $b_i - a_i > \varepsilon$ .

Already, there is here plenty of work to attend to, in proving that these definitions are "independent of representatives of the equivalence classes" — that is, that they really are operations on the classes and not just operations on the sequences. Similarly for the order relation. It is not quite so troublesome to carry through the proof that $(K, +, \cdot, \trianglelefteq)$ is an Archimedean ordered field. However, the proof of completeness of K is considerably more difficult. (For the carrying through of the details, see say, the book of Oberschelp [52], Sections 17÷22, or Endl÷Luh [15], Chapter 1.) Using concepts from algebra, one can describe Cantor's process in the following way: the set F of fundamental sequences is turned into a ring $(F, +, \cdot)$ using operations defined component÷wise. In this ring the null sequences form an ideal N, and K is nothing more than the "residue class ring" F/N.

<u>Evaluation of the Cantor Process</u>.
The advantages lie in the great ease with which the process can be generalised in the theory of metric spaces.

(The concept of *distance*, on which the process depends, comes to us from Nature, through our perception of Space. On that account, it has a greater potential for generalisation than the concept of an ordering, which we meet only on lines.) Moreover, the construction of $(K, +, \cdot)$ is carried out in a trouble-free manner, using simple concepts of algebra. Also, one obtains in this way the whole of K and not initially only the positive numbers, as in the two previous processes. The disadvantages of the Cantor process lie in the didactic domain, because the objects $[(a_i)]$ are very abstract and many proof-steps are involved; also the proof that $(K, \trianglelefteq)$ is complete is rather difficult. Altogether, one can regard all this as an algebraic process, relative to that of Dedekind.

<u>Exercise 2</u>.

(a) Prove that an ordered field is Archimedean if and only if every sequence which is increasing and bounded above is a Cauchy sequence. Outline the construction of $\mathbb{R}$ from $\mathbb{Q}$ by the method of Cauchy sequences, and prove the existence of multiplicative inverses in $\mathbb{R}$.

(b) Let $(F, +, X, \leqslant)$ be an ordered field. Explain what is meant by saying that F is (a) Archimedean, (b) Cauchy complete, (c) Dedekind complete.

Prove that the following are equivalent:
(i) F is Archimedean,
(ii) every bounded monotonic sequence in F is Cauchy,
(iii) $1 = \lim_{n\to\infty} \sum_{i=1}^{n} 1/2^i$, where $2 = 1+1$ and 1 is the unit element of F.

Prove that if F is Archimedean and Cauchy complete then F is Dedekind complete. (Southampton University)

## 2E CONTINUED FRACTIONS.

From about 1600 until Dedekind, the real numbers had been defined as ratios of magnitudes, and more especially (in modern parlance) as ratios between length of line-segments. Ultimately, these depended on Euclid's definition – or description of – proportions (cf. Gericke [22]).

The difficulty is that if $s, t$ are two segments, then we ought to be able to represent the "ratio" $s/t$ as another segment. This problem leads to the theory sketched here. In any case there stems from Euclid a type of approximation method for determining ratios, of which the historical significance can be inferred only indirectly and arduously. We look next at this method. It is nothing other than the so-called Euclidean Algorithm (also called Exchange process), but applied to segments rather than to integers.

Fig.18

Thus with the two given segments $s, t$ suppose first that each is an integral multiple of a common unit of length $d$. Then $s$ and $t$ are called **commensurable**, and there exist $n, m \in N$, such that $s = nd$ and $t = md$. Clearly, any proposed definition of the "ratio" $s/t$ should, *in this case*, give the rational fraction $n/m$, which it is easy to represent as a segment.

Therefore, following Euclid, we now proceed in the following way. We determine successive numbers $a_i \in \mathbb{N}_0$ and lengths $r_i$ so that

(0) $s = a_0 t + r_1$ with $a_0 \in \mathbb{N}_0$ and $0 \leqq r_1 < t$

(1) $t = a_1 r_1 + r_2$ with $a_1 \in \mathbb{N}$ and $0 \leqq r_2 < r_1$

(2) $r_1 = a_2 r_2 + r_3$ with $a_2 \in \mathbb{N}$ and $0 \leqq r_3 < r_2$

(3) $r_2 = a_3 r_3 + r_4$ with $a_3 \in \mathbb{N}$ and $0 \leqq r_4 < r_3$

....and so on.

This is the *Exchange process*, with each remainder being 'exchanged' for the next.

If the initial magnitudes had been whole numbers rather than segments, then the process always breaks off after a finite number of steps because $r_i \in \mathbb{N}_0$. We would eventually obtain, say,

$$(n-1) \qquad r_{n-2} = a_{n-1}r_{r-1} + r_n$$

$$(n) \qquad r_{n-1} = a_n r_n \ ,$$

and $r_n$ is the highest common factor of $s$ and $t$. We write:

$$r_n = \text{hcf}(s,t) \ .$$

The process may also terminate in the case of segments. This occurs precisely in the case when the segments are comensurable, in the sense explained above. (See Euclid, beginning of Book 10.) We then have

$$\frac{s}{t} = \frac{nd}{md} = \frac{n}{m} \in \mathbb{Q}$$

so here we have expressed $s/t$ as expected, as a "ratio of whole numbers". If the segments are not commensurable the above Exchange process can still be carried out, but it does not terminate and yields an infinite coefficient sequence $[a_0, a_1, a_2, \ldots]$. One might guess that the Greek Mathematicians in the times before Plato, accepted as a substitute for the "ratio" of $s$ and $t$, the coefficient sequence $[a_0, a_1, a_2, \ldots]$ itself. Concerning this, there is an ancient commentary on a passage of Aristotle, in which it is said: "The following definition of proportional magnitudes is what the ancients use: magnitudes stand in proportion to one another, when they give rise to the same Exchange process." (Quoted from O. Becker, "Eudoxus-Studies" in [4] pp.311÷333. By the word "ancients" is meant the mathematicians before Eudoxus and Plato. We shall discuss the later definition that comes from Eudoxus further on. This oldest "Definition of a real number" we now see in a more modern version and follow Hardy-Wright: *Number Theory* [27], Chapter 10, Continued Fractions. Details and proofs can be found there, and we give only a survey of the material. Here we regard the real numbers as given.

For the two (positive) real numbers $s$ and $t$ let the (finite or infinite) sequence $[a_0; a_1, a_2, \ldots]$ be defined as in the Exchange process (0), (1),... above. We call this sequence the *Continued Fraction sequence* belonging to $s/t$. (Often $t$ will equal 1, and then we speak of the Continued Fraction sequence *belonging* to $s$). The description "Continued Fraction" is immediately understandable from the following example.

*EXAMPLE.* Let $[2; 1, 3, 5, 1]$ be the Continued Fraction sequence of $s/t$. This means that

$$s = 2t + r_1$$

$$t = 1r_1 + r_2$$

$$r_1 = 3r_2 + r_3$$

$$r_2 = 5_3 r + r_4$$

$$r_3 = 1r_4 + 0$$

(One speaks here of a *finite* Continued Fraction, with the obvious definition). We now calculate s/t "from the top down" in the following way:

$$\frac{s}{t} = 2 + \frac{r_1}{t}$$

$$= 2 + \frac{r_1}{1r_1 + r_2} = 2 + \frac{1}{1 + \frac{r_2}{r_1}}$$

$$= 2 + \frac{1}{1 + \frac{r_2}{3r_2 + r_3}}$$

$$= 2 + \frac{1}{1 + \frac{1}{3 + \frac{1}{5 + \frac{1}{1}}}}$$

$$= \frac{69}{23}$$

The last display, with the long bars, is usually called the Continued Fraction itself (of 69/23). Thus we often omit the word "sequence" or speak by contrast of the "coefficient sequence".

From these calculations we immediately see a certain ambiguity in the sequence: we could just as well have written [2; 1, 3, 6] and this would have given the same answer with the division process. With finite sequences therefore, we "forbid" this last 1 in front of the zero. This convention will always be observed from here on.

From the treatment of the Continued Fractions, the first significant conclusion is this: we can get approximate fractions for the value s/t from its infinite Continued Fraction if we simply break off the coefficient sequence $[a_o; a_1, a_2, \ldots]$ and calculate the resulting number from it as in the previous finite case. These approximate fractions are usually called the ***successive convergents***, and are described by the "approximating" numerators $p_i$ and denominators $q_i$. For the example [2; 1, 3, 5, ...] we obtain

$$\text{(with } r_1 \text{ put to zero)} \quad \frac{p_o}{q_o} = a_o = 2$$

$$\text{(with } r_2 \text{ put to zero)} \quad \frac{p_1}{q_1} = a_o + \frac{r_1}{1r_1} = 3 \ ,$$

$$\frac{p_2}{q_2} = a_o + \cfrac{1}{1 + \cfrac{r_2}{3r_2}} = \frac{11}{4}$$

$$\frac{p_3}{q_3} = a_o + \cfrac{1}{1 + \cfrac{1}{3 + \cfrac{r_3}{5r_3}}} = \frac{64}{23}$$

Approximating numerators and denominators can be recursively defined:

$$p_o = a_o \quad \text{and} \quad p_1 = a_1 a_o + 1 \quad \text{and} \quad p_i = a_i p_{i-1} + p_{i-2}$$

$$q_o = 1 \quad \text{and} \quad q_1 = a_1 \quad \text{and} \quad q_i = a_i q_{i-1} + q_{i-2}$$

This is so arranged that if $s/t = [a_o; a_1, a_2, \ldots, a_n]$ then $s/t = p_n/q_n$.

The coefficient sequence $s/t$ can be easily worked out with a pocket calculator when one notes the following small trick, where $[x]$ as usual means the biggest integer $\leqq x$. Just note:

$$s = a_o t + r_1 \Longrightarrow \frac{s}{t} = \left[\frac{s}{t}\right] + \frac{r_1}{t} \quad \text{with} \quad a_o = \left[\frac{s}{t}\right]$$

<u>Exercise 3</u>.

(a) Justify the correctness of the last statement.

(b) Show that at each stage (0), (1),... of the Exchange process (p.49), if $s,t \in \mathbb{Z}$, then the remainder can be expressed in the form $as + bt$, with $a, b \in \mathbb{Z}$. In particular then, hcf$(s,t)$ is of this form.

Now continue the Exchange process, so:

$$t = a_1 r_1 + r_2 \Longrightarrow \frac{t}{r_1} = \left[\frac{t}{r_1}\right] + \frac{r_2}{r_1} \quad \text{with} \quad a_1 = \left[\frac{t}{r_1}\right]$$

$$r_1 = a_2 r_2 + r_3 \Longrightarrow \frac{r_1}{r_2} = \left[\frac{r_1}{r_2}\right] + \frac{r_3}{r_2} \quad \text{with} \quad a_2 = \left[\frac{r_1}{r_2}\right]$$

....

We therefore obtain the coefficient sequence from the whole-number parts in the process

$$x_i \to [x_i] \to x_i - [x_i] \to \frac{1}{x_i - [x_i]} = x_{i+1} \ ,$$

as long as $x_i \neq [x_i]$.

For $\surd 3 = \surd 3/1$ the author's calculator produced with this procedure the sequence:

[1; 1, 2, 1, 2, 1, 2, 1, 2, 1, 2, 1, 2, 1, 2, 1, 2, 1, 1,...]

To within the calculator's accuracy, $\surd 3 = 1.7320508$, and this is reached with $p_{13}/q_{13}$.

<u>Exercise 4</u>. Use a calculator to follow through the same procedure, for the cube and fifth roots of 3, and then for e and $\pi$.

In the following theorems it is always assumed that $t = 1$ in $s/t$, so that only the expansion of the coefficient sequence belonging to s is in question. For the proof, see the cited chapter of Hardy–Wright or books on Number Theory. Within our conventions the following formulae are valid for the (infinite or finite) Continued Fraction sequences:

$$a_o \in \mathbb{N}_o \text{ and } [(a_i \in \mathbb{N} \text{ for all } i > 0) \text{ or}$$

$$(a_i \in \mathbb{N} \text{ for } 1 \leqslant i < N \text{ and } a_N > 1)] .$$

<u>THEOREM A</u>. *Each* $s \in \mathbb{R}^+$ *determines uniquely a Continued Fraction sequence*

$$\kappa(s) = [a_o; a_1, a_2, a_3, ...]$$

*of the given sort. Also*

$$s \in \mathbb{Q} \textit{ if } \kappa(s) \textit{ is finite.}$$

<u>THEOREM B</u>. *Given a Continued Fraction sequence*

$$[a_o; a_1, a_2, a_3, ...]$$

*then for the sequence of convergents* $p_i/q_i$, *we have*

$$p_i/q_i \to s \in \mathbb{R},$$

*and*

$$\kappa(s) = [a_o; a_1, a_2, a_3, ...] .$$

We have formulated Theorems A and B for positive numbers only. One can read in Hardy–Wright that for negative numbers one must allow $a_o \in \mathbb{Z}$, and for 0 the obvious sequence. The theorems are then valid as before and through them $\kappa$ defines a bijection:

$$\kappa : x \to \textit{Continued Fraction expansion of } x$$

from the real numbers onto the coefficient sequences. The inverse mapping is given by the limit of the sequence of convergents.

This theory allows us, in principle, the possibility of defining the field $\mathbb{R}$ as the set of Continued Fraction sequences. Because the elements of a coefficient sequence are whole numbers, it is thus possible to base a definition of $\mathbb{R}$ on the existence of $\mathbb{Z}$. As we have seen, such a foundation was actually looked for by the Greeks. We can also say, however, why the Greeks had to give up this task: even today, it is very complicated to define addition and multiplication of real numbers via the coefficient sequences (e.g. express $\kappa(x+y)$ in terms of $\kappa(x)$ and $\kappa(y)$; see C.J. Rieger [64]). Therefore such a strategy was unusable for $(\mathbb{R}, +, \cdot)$. Had the "ancients" said that two "ratios" are the same when their coefficient sequences agree then we would find the definition formulated, around 370 BC by Eudoxus, to be in accord with the Dedekind Section. For, we read in Euclid Book V, Definition 5, that given "magnitudes" a, b, c, d, then

$$a : b = c : d \Leftrightarrow_{\text{def}} (\forall m, n \in \mathbb{N} : nb < ma \Leftrightarrow nd < mc)$$

If we write the right hand side of this in more modern notation, we have

$$\forall m, n \in \mathbb{N} : \frac{n}{m} < \frac{a}{b} \Leftrightarrow \frac{n}{m} < \frac{c}{d}$$

One can see here the description of the "left÷hand" (lower) class of a Dedekind Section, but with a significant difference: the least upper bound is already given. The difficulty of the lub is circumvented in Eudoxus' formula, and one can clearly see that he was conscious of this problem in the following definition of "ratio of two magnitudes" (which we would now regard as "real number a : b"). Thus, Definition 3: "ratio is the same behaviour of two magnitudes relative to measurement". It has therefore been left undefined what a ratio is. The "ancients" would have had the possibility of using the Exchange process, to say the ratio "is" the coefficient sequence itself. For a more precise discussion of the Eudoxus theory of magnitudes, we may refer to the cited works of O. Becker and W. Krull in Felscher [17], Vol.II, and in particular on the extraordinarily lucid analysis that is supported on these works, by F. Beckmann: *New Viewpoints on Euclid's Book V* [5], pp.1÷144. (For a first look at Greek Mathematics, one should in any case read B.L. van der Waerden's book "*Science Awakening*" [71]). With these historical observations, one must however always be aware that the Greeks never thought of real numbers in our sense, and therefore the interpretations given here by the author are false in the sense of the historians, even though they seem "right" in the sense of mathematicians.

## 2F CLOSING REMARKS ON CHAPTER 2.

1. What are Real Numbers? (This concludes a discussion in the Introduction). Everybody expects a mathematician to be able to say briefly and concisely what the real numbers are. The mathematician will first say that he is here talking about the most important subject in mathematics, and then become embarrassed. What should he say? A real number is a Dedekind section? An equivalence class of fundamental sequences? A decimal expansion? The answer should not depend at all on his having read our initial Chapter! We have already seen previously that the Greeks also had difficulty with this question. A way out is offered by the following statement: a real number is an element of a complete ordered field. But that is also unsatisfactory, because there could be many examples of this type of field, with elements described in different ways, and the vagueness is not removed. One can now bring the Number line into play, because we can describe geometrical constructions for the addition

and multiplication of points on the Number line, as indicated in Fig.19. (See also Griffiths–Howson [24] Ch.18 for a discussion relating to school mathematics.)

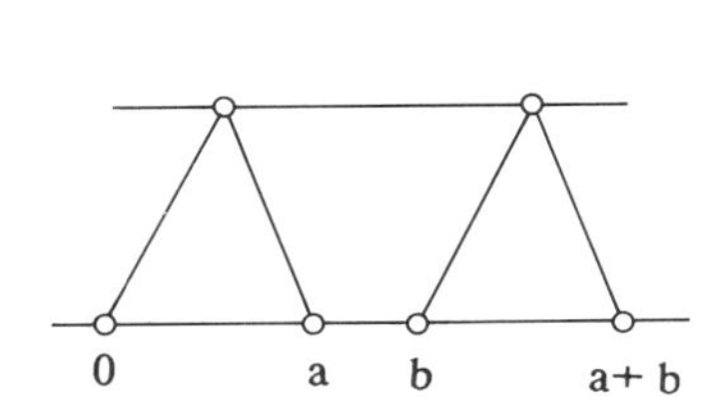


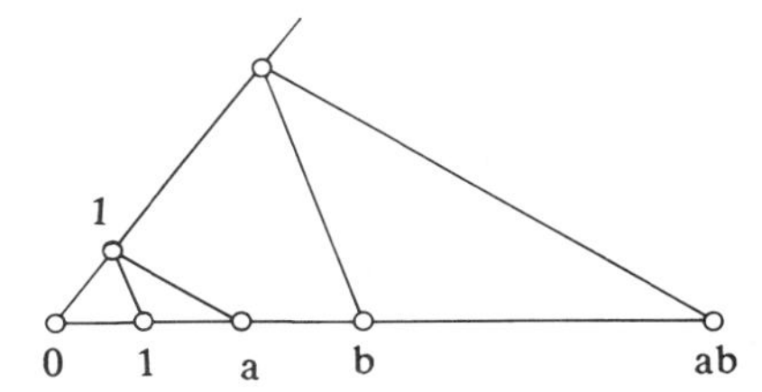


Fig. 19

In order to prove that the operations thus defined satisfy the field axioms, we need to make sure we have some geometrical axioms and deductions, to allow both ordering and completeness of the field. (See books on "Foundations of Geometry", e.g. Hilbert [32] 1899 and later editions). But then we would need some work to lead back the concept of the metric plane, say, to the concept of the real numbers, because to define the metric plane $\mathbb{R}^2$ with euclidean distance, we need to assume the existence of real numbers. The "Number line" has its undisputed justification as "intuitive background" for the concept of the real numbers, but this does not help any further with their formal definition. In addition to this, geometric ideas and formal presentations are difficult to digest.

Now, one thinks of a point of the Number line as an "indivisible entity". Can something like this really be an entire equivalence class of Cauchy sequences, and therefore an extraordinarily complicated and structured set? Dedekind mentions these problems in a letter to H. Weber (24th January, 1888, and in other places). "...It is exactly the same question, of which you say that the irrational number ought to be nothing other than the Section itself, whereas I prefer it to be created as something new (different from the Section) which corresponds to the Section and which produces the Section. We have the right to allow ourselves such a power of creation and it is more appropriate to proceed thus, on account of treating all numbers equally. The rational numbers also produce Sections but I shall certainly not regard rational numbers as identical with the Sections that they produce..." (Dedekind, *Works* Vol III p.489).

After we have been left in the lurch by Geometry, there remains nothing else for us than to admit our hopeless state. The only unique feature that is fixed is the "structure type" of the complete ordered field, of which there are many specimens. One can "give" an individual specimen only by (say) using the Number line to demonstrate it. The ambivalence would not be removed by fixing once and for all one particular method of introduction, perhaps using Dedekind Sections. For, one would need to take into consideration different possible definitions of the rational numbers; and if a convention had been chosen, one would then have to fix also what a natural number should be. So, individual objects are not really part of mathematics; we can handle only their properties and mutual inter-relationships.

"The mathematician abstracts completely from the constitution of the object and the content of its relations. His business is with the enumeration of relations and their mutual comparisons.?" (Gauss, *Works* II, p.176.) On this subject the following remarks by Felscher [17] also apply, which we take from his Volume III pp.212/213. (The chapter citations etc below relate to Felscher's book. We have said nothing on the completion of a totally ordered group; we ourselves characterized $\mathbb{R}$ as a complete ordered field).

"*A note on the set–theoretic construction.* In Chapter III the natural numbers were constructed from Set–theory, from them the integers and rational numbers and then (in Chapters IV and V) onwards to the algebraic and the real numbers. More precisely, the concern there was with a *reduction* of the different types of numbers from each to a simpler one, and thence to Set–theory: through this it is *proved* that a model of Set–theory contains copies of all these number–domains (therefore, say, complete ordered fields). That is a valuable piece of knowledge about Set–theory; as knowledge about the number domains however, it is little more than a curiosity. Indeed, neither algebraic nor order–theoretic insights into the natural numbers follow from the Set–theoretic constructions described in III.3; however, it is very significant for the theory of the real numbers that one can, with the help of Set–theory deduce the *existence* of the real numbers from that of the rational numbers, but it is also quite irrelevant in such a connection, whether one represents the real numbers as lower classes, as upper classes, or as a pair of both classes of Dedekind sections – or also as equivalence classes of Cauchy sequences. On the other hand, what is important for the real numbers is that such descriptions contain characterisations but only to within *isomorphism* – be it in connection with the Archimedean totally–ordered groups, or with the completion of commutative totally–ordered groups. Often one finds, in the Literature, the opinion that the Set–theoretic definition of the real numbers is a *construction* in the sense that these numbers were actually *produced* and their previous, possibly questionable, existence is now secure and *well–grounded.* With such a view it is easy to overlook the point that in mathematics just as in life, one never gets anything for nothing, and that the tools of construction which allow us to get to such an existence proof are at least only as strong – and therefore just as questionable – as the construction produced with their help. This methodological triviality escapes attention in the copiously luxurious technical build–up of the mathematical execution (in which mathematicians love to lose themselves) – just as in similar situations the complexity of the bureaucracy makes us forget that helpful presents from well–meaning authorities are ultimately paid by the recipients through their own taxes".

The Problem of the Continuum in the History of Mathematics.
We have already seen that the Greek mathematicians had been confronted with the problem of giving a definition of the real numbers in the form of "ratios of magnitudes". In connection with the first efforts in this direction, there can be seen already the famous paradoxes of the philosopher Zeno of Elea (450 BC: "Achilles and the Tortoise" etc.). By the time of Zeno, we can also see a mixture of Physics and Mathematics forming a theory of motion, which was for a long time conceived as part of the question as to the nature of the "Continuum".

Aristotle (around 340 BC) was more concerned as a philosopher with logic and foundations than with internal mathematical theorems. He investigated the concept "continuous" once in the sense of divisibles, and at another time as a self-ordering entity. We would interpret the first notion presumably as "closed" and the second perhaps as "complete". Aristotle was enthusiastically studied in the Middle Ages, and some observations can be found in Felscher [17] (Vol.II, pp.164÷169) concerning the corresponding difficulties of the scholastic philosophers.

Even Leibniz, who had to consider the problem in the context of introducting the differential and integral calculus, said "The Continuum is a Labyrinth." Today one tries to express the infinitely small quantities that Leibniz included in his concept of number, by means of the recent creation of Non-Standard Analysis. With the more exact formulation of Analysis around 1830 by Gauss, Cauchy and others, the burning question that moved many mathematicians, physicists and philosophers in the nineteenth century was "How does one describe the 'continuity' of the Number-line". "What is the continuum?" To the satisfaction of mathematicians at least, Dedekind and Cantor answered the questions around 1870, and after that the debate gradually faded away.

Shortly afterwards, Cantor used the word in a new question, the so-called Continuum problem of the theory of sets: *is there an infinite subset of* $\mathbb{R}$ *which is neither bijective with* $\mathbb{R}$ *nor with* $\mathbb{N}$? In his famous lecture of 1900, Hilbert had called for a solution to this problem, as being one of the most important problems of mathematics in the coming century; and as its development has shown, his estimate was right. We shall give a discussion of the related theory of cardinal numbers in Chapter 6.

***Literature.*** The reader who is interested in pursuing these brief observations and has further questions on it, should first refer to the cited work of Felscher, and then to the book of Becker [4].

# 3

# Irrational numbers

## INTRODUCTION

The object of this chapter is to explain the classification of the real numbers into rational, algebraic, and transcendental numbers. For completeness, properties of the decimal expansion of real numbers (presumably already known from school) will be listed in the first section. Using the quadratic irrationals and the Continued Fraction representation, we shall indicate the connections with methods used by the Greeks in pre-Euclidean mathematics. The most important theorems of this chapter are Cantor's theorem on the countability of the algebraic numbers, and Liouville's theorem on the recognition of transcendental numbers.

## 3A DECIMAL EXPANSIONS.

Decimal fractions appeared in the mathematical literature in the sixteenth century and one writer especially was the Dutchman, Simon Stevin (1548–1620). The division algorithm that we learn in school produces decimal expansions for $m/n \in \mathbb{Q}^+$ with $m,n \in \mathbb{N}$. The process can be described as follows:

$$m = a_o n + r_1 \quad \text{with} \quad a_o \in \mathbb{N} \quad \text{and} \quad 0 \leqq r_1 < n$$

$$10r_1 = q_1 n + r_2 \quad \text{with} \quad 0 \leqq q_1 < 10 \quad \text{and} \quad 0 \leqq r_2 < n$$

$$10r_2 = q_2 n + r_3 \quad \text{with} \quad 0 \leqq q_2 < 10 \quad \text{and} \quad 0 \leqq r_3 < n$$

... and so on.

One then has $m/n = a_o + q_1/10 + q_2/10^{2}+ \ldots + = a_o + 0.q_1q_2 \ldots$ For this, the following theorem applies:

THEOREM. *For all* $m, n \in \mathbb{N}$, $m/n$ *has a finite or periodic decimal expansion. In the case* $\mathrm{hcf}(m,n) = 1$, *the expansion is finite precisely when there exist* $k, \ell \in \mathbb{N}$ *and* $n = 2^k5^\ell$.

*PROOF*. Because there are only finitely many different remainders mod n, the expansion is periodic as long as the process does not break off with some $r_i = 0$. The rest of the proof is an exercise. □

We can also prove the converse of this theorem easily:

*If* $x \in \mathbb{R}$ *has a periodic (or finite) decimal expansion, then* $x \in \mathbb{Q}$.

*PROOF*. The finite case is clear. Let:

$$x = a_o + 0.q_1 \ldots q_\ell \overline{q_{\ell+1} \ldots q_{\ell+k}}$$

with the periodic part $q_{\ell+1} \ldots q_{\ell+k}$. This results in:

$$x = a_o + \frac{q_1}{10} + \ldots + \frac{q^\ell}{10^\ell} + r$$

where
$$r = \frac{1}{10^\ell}\left[\left(\frac{q_{\ell+1}}{10} + \ldots + \frac{q_{\ell+k}}{10^k}\right) \times \sum_{i=0}^{\infty} \frac{1}{10^{ki}}\right].$$

With the geometric series

$$\sum_{i=0}^{\infty} \frac{1}{10^{ki}} = \frac{1}{1 - 10^{-k}}$$

we now see that $x \in \mathbb{Q}$. □

If we consider a finite decimal expansion as one having periodic part $\overline{0}$ then we can assemble our results:

THEOREM *If* $x \in \mathbb{R}$ *then* $x \in \mathbb{Q}^+$ *iff the decimal expansion of* $x$ *is periodic.*

With this it is very easy to produce an *irrational* number, as a decimal expansion. However, such irrational numbers are usually not very interesting.

We now analyse the so-called 'pure' and 'mixed' periodic expansions of rational numbers. Thus a 'pure' expansion occurs when the periodic part starts with the first digit after the decimal point.

Let $m, n \in \mathbb{N}$ and $\text{hcf}(m, n) = 1$. We can then represent $n$ in the form $n = 2^k 5^\ell w$ with $\text{hcf}(w, 10) = 1$.

From the additive representation of the hcf (see Exercise 3(b), p.52) we can obtain $x, y \in \mathbb{Z}$ with $xw + (2^k 5^\ell)y = 1$ and then we have

$$\frac{m}{n} = m \cdot \frac{1}{n} = m \frac{xw + (2^k 5^\ell)y}{n}$$

$$= m \left(\frac{x}{2^k 5^\ell} + \frac{y}{w}\right)$$

$$= \frac{mx}{2^k 5^\ell} + \frac{my}{w} .$$

Here, the first summand has a finite expansion; and we will show that the second has a pure periodic expansion (in the sense explained above). This means that whenever there is a mixed periodic expansion, a finite expansion is added to a periodic expansion. Now to the second summand:

*Let* $a, w \in \mathbb{N}$ *and* $\text{hcf}(a, w) = 1 = \text{hcf}(10, w)$. *Then the decimal expansion of* $a/w$ *is purely periodic.*

To prove this let $a/n = a_o + 0.q_1 q_2 \ldots$ be the expansion.

We now use a lemma from Number Theory. Since $\text{hcf}(10, w) = 1$, $10$ is an element of the multiplicative group $\mathbb{Z}_w^*$ of the residues prime to, and less than, $w$. Since this group is finite, $10$ is of finite order in $\mathbb{Z}_w^*$. There exists therefore some $s \in \mathbb{N}$ with $10^s \equiv 1 \pmod{w}$. Now, the division algorithm for $a/w$ looks like:

$$a = a_o w + r_1 \quad \text{with} \quad 0 \leq r_1 < n$$

$$10r_1 = q_1 w + r_2 \ldots, \quad \text{so} \quad 10r_1 \equiv r_2 \pmod{w}$$

$$10r_2 = q_2 w + r_3 \ldots, \quad \text{so} \quad 10r_2 \equiv r_3 \pmod{w}$$

$$\ldots \text{ and so on.}$$

For each $t \in \mathbb{N}$ we get:

$$f_{t+1} \equiv 10r_t \equiv 10^2 r_{t-1} \equiv \ldots \equiv 10^t r_1 \pmod{w}.$$

The Number-theoretic Lemma quoted above, yields some $s \in \mathbb{N}$ with $10^s \equiv 1 \pmod{w}$. With this $s$ we obtain:

$$r_{s+1} \equiv 10^s r_1 \equiv 1r_1 \equiv r_1 \quad \text{since} \quad 10^s \equiv 1 \pmod{w} .$$

Therefore the first remainder $r_1$ appears again (and with it $q_1 \ldots$) after $s$ steps; so the periodic part begins with $q_1$. If we had chosen $s$ to be the least $s'$ having the named properties, then $s'$ would also be the length of the periodic part. □

<u>Exercise 1</u>.

(a) Prove that every real number has a ternary (base 3) expansion. Which numbers have a ternary expansion which is (a) terminating (b) periodic (c) unique? Give brief justifications for your answers. If C denotes the set of all those numbers in $\langle 0,1 \rangle$ which have a ternary expansion containing only 0's and 2's, prove that every point of C is a limit point of C. Explain how C can be considered to be both large and small (in appropriate senses).

(b) Prove that the real number

$$x = \sum_{n=0}^{\infty} \frac{1}{(2^n n!)^2}$$

is irrational. [Hint: for a sufficiently large integer N, Nx would lie in $\mathbb{N}$ if x were rational.]

(c) Prove that $(a_n)$ is a Cauchy sequence of real numbers when

$$a_n = \sum_{k=0}^{n} \frac{(-1)^k}{k!} .$$

Let

$$b_n = \sum_{k=n+1}^{\infty} \frac{(-1)^k}{k!} .$$

Prove that

$$0 < \underset{k=0}{(-1)^{n+1}} b_n < \frac{1}{(n+1)!}$$

and deduce that $e^{-1}$ is irrational. (Southampton University)

## 3B ALGEBRAIC NUMBERS.

A number $\alpha$ is **algebraic** iff $\alpha$ is a zero of some polynomial $p(t)$ with rational coefficients. Thus $p(\alpha) = 0$, and $\alpha$ is also called a **root** of the polynomial. It is therefore necessary to include complex roots as well. Non-algebraic numbers are called **transcendental** because they cannot be included within the methods of classical algebra of $\mathbb{Q}$; they therefore lie outside this area of thought.

We need not distinguish between polynomials and polynomial functions because they occur here in infinite fields, in particular in $\mathbb{Q}$. (See Griffiths-Hilton [23], Ch.22, and also Lang, [42] Chapter V, Section 1.).

If $\alpha$ is a zero of $f \in \mathbb{Q}[x]$ with

$$f(x) = a_n x^n + \ldots + a_o \quad (a_i \in \mathbb{Q}) ,$$

then without loss of generality we can take f as irreducible; for if f could be factorised as $f = gh$, $\alpha$ would be a zero of one of the factors. For further standardization, we often take $a_n = 1$ as well and then call $f$ the **minimal polynomial** of $\alpha$ and $n$ the **degree** of $\alpha$. The set **A** of the (complex) algebraic numbers forms a field (see Lang [42], Chapter V, Section 1, Example 4), that contains the real algebraic numbers $\mathbb{A}_{\mathbb{R}}$ as a sub-field.

*EXAMPLE.* Each rational number $r \in \mathbb{Q}$ is algebraic of degree 1, because $r$ is a zero of the linear polynomial $x - r$. Conversely, every algebraic number of degree 1, (i.e. every zero of a linear polynomial $a_1x + a_2$ with $a_1, a_2 \in \mathbb{Q}$) is itself rational.

*EXAMPLE.* The polynomial $x^n - a$ has either none, one or two real zeros ($\sqrt[n]{a}$), and these are algebraic numbers. (They are not necessarily of degree $n$, as $x^3 - 27$ shows.) In $\mathbb{C}$ we have $n$ different solutions $\sqrt[n]{|a|} \exp(k/n \, 2\pi i)$ with $0 \leqq k < n$.

*EXAMPLE.* The algebraic number $\sqrt[k]{a}$ with $a \in \mathbb{N}$ is either in $\mathbb{N}$ or an irrational. The reason is simple: for a proper fraction $r/s \in \mathbb{Q}$ with $\text{hcf}(r, s) = 1$, we have $\text{hcf}(r^k, s^k) = 1$, and therefore $(r/s)^k$ is not in $\mathbb{N}$. (The argument uses the uniqueness of prime factorization, but can also be proved another way on the basis of Euclid's algorithm, as Euclid VII, 27 shows). In particular, the numbers $\sqrt{2}$, $\sqrt{3}$, $\sqrt{5}$, ..., $\sqrt[3]{2}$, $\sqrt[3]{3}$, $\sqrt[3]{4}$, ... are all irrational.

Cantor determined the "number" of algebraic numbers, using his concept of countability:

THEOREM. *The set* **A** *of algebraic numbers is countable.*

*PROOF.* Each algebraic number $\alpha$ is a zero of a polynomial $f \in \mathbb{Q}[x]$. If we multiply $f$ by a suitable number $n \in \mathbb{N}$ (perhaps the common "denominator" of the coefficients), then the zeros will not change, but $f$ now lies in $\mathbb{Z}[x]$. Then we can also say: **A** consists of the common zeros of all polynomials of $\mathbb{Z}[x]$. Let

$$f(x) = a_n x^n + \ldots + a_o \in \mathbb{Z}[x] .$$

We define the 'height' $h(f)$ of $f$ by

$$h(f) = n + |a_n| + |a_{n-1}| + \ldots + |a_o| \in \mathbb{N} .$$

Because one can represent a natural number $k$ in only a finite number of ways as the sum of numbers from $\mathbb{N}_0$ there are only finitely many, say $w$, polynomials $f$ with $h(f) = k$ for a given height $k$. Their degree is $\leqq k$, so each of the $w$ polynomials has at most $k$ zeros. For a given height $k$ we have, therefore, at most $wk$ algebraic numbers. By going through the successive heights $k \in \mathbb{N}$ in turn one obtains an enumeration for **A**. □

COROLLARY. (Cantor). **Transcendental numbers exist.** For, $\mathbb{R}$ is unncountable while **A** and $\mathbf{A} \cap \mathbb{R} \subset \mathbf{A}$ are countable, so there must be other elements in $\mathbb{R}$ apart from algebraic numbers.

However, this argument doesn't provide us with a transcendental number, definitely in our hands. To find at least one such is the goal of the later sections. Meanwhile, our classification of the numbers now looks like this:

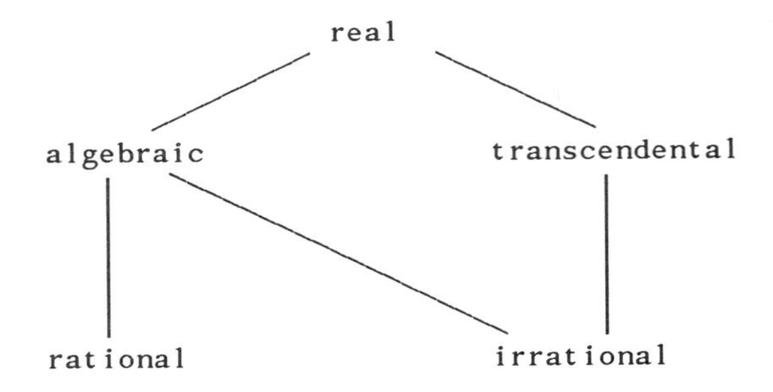


## 3C QUADRATIC IRRATIONAL NUMBERS.

These are roots of those quadratic polynomials $x^2 + bx + c$, which can not be factorized into factors $(x - r)$, $(x - s)$ with $r, s \in \mathbb{Q}$. They are therefore algebraic numbers, and can be recognized from their continued fraction expansions, by the following theorem on the Continued Fraction sequence of quadratic irrational numbers.

<u>THEOREM</u>. *Let* $[a_0; a_1, a_2, a_3, \ldots]$ *be the Continued Fraction sequence of* $\alpha \in \mathbb{R}$. *Then* $\alpha$ *is a quadratic irrational number only when* $[a_0; a_1, a_2, \ldots]$ *is periodic (and infinite).*

*<u>PROOF</u>*. See Hardy-Wright [27], Chapter 10, or other books on number theory. □

As a related example, we consider the segment ratio $t : s$, where $t = AD$ is a diagonal and $AB = s$ is a side of a regular pentagon. By the Exchange algorithm, we note (cf. Fig.20):

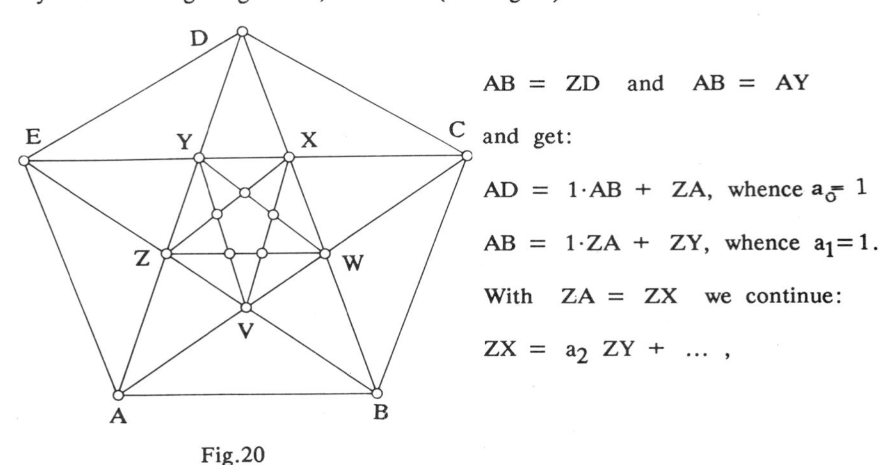


$AB = ZD$ and $AB = AY$

and get:

$AD = 1 \cdot AB + ZA$, whence $a_0 = 1$

$AB = 1 \cdot ZA + ZY$, whence $a_1 = 1$.

With $ZA = ZX$ we continue:

$ZX = a_2\ ZY + \ldots$ ,

Fig.20

but here we are precisely in the same situation as before : ZX and ZY are diagonals and sides in the smaller regular pentagon VWXYZ. This means that the process is periodic, and we have:

$$\kappa(t : s) = [1; 1, 1, 1, 1, \ldots] = [1; \overline{1}]$$

This is the simplest imaginable infinite Continued Fraction; its successive approximating numerators and denominators are the famous Fibonacci numbers.

By the quoted theorem on periodic Continued Fractions the ratio $x = t : s$ must satisfy a quadratic equation. This equation is easily found with the help of the Continued Fraction expansion of $x$. We have:

$$x = 1 + \cfrac{1}{1 + \cfrac{1}{1 + \cfrac{1}{\ddots}}}$$

$$= x$$

and the part below the bar is just the expansion of $x$ once more, i.e. $x = 1 + 1/x$. Hence $x$ is quadratic irrational since it satisfies the equation $x^2 - x - 1 = 0$; and the formula for the roots contains $\surd 5$, which is not in $\mathbb{Q}$.

Exercise 2.

(a) Derive this equation for $x$ from the geometrical figure.

(b) Use the above argument for $x$ to prove that if $\alpha \in \mathbb{R}$ has a periodic Continued Fraction expansion then $\alpha$ is a quadratic irrational.

The second, semi-geometric example concerns the Continued Fraction expansion for $w = \surd 3$ for which our experiences with the calculator have already suggested the expansion

$$\kappa(w) = [1; 1, 2, 1, 2, 1, 2, \ldots]$$

with periodic part $\overline{1, 2}$.

Let then $w^2 = 3$ and $\ell = 1$, so

$$w = a_0 e + r_1 \quad \text{with} \quad 0 \leq r_1 < \ell$$

$$\ell = a_1 r_1 + r_2 \quad \text{with} \quad 0 \leq r_2 < r_1$$

$$r_1 = a_2 r_2 + r_3 \quad \text{with} \quad 0 \leq r_3 < r_2$$

$$r_2 = a_3 r_3 + r_4 \quad \ldots$$

$$\ldots$$

The values $a_0 = 1$ and $a_1 = 1$; $a_2 = 2$; $a_3 = 1$ can easily be read off from a reasonably exact drawing (see Fig. 21), whenever one does not want to carry out the following estimates. (With this one can also calculate $w$ directly with the help of Pythagoras' theorem.)

We first determine $a_0$, $a_1$ and $a_2$. Since $w^2 = 3$ and $\ell^2 = 1$, we must have $a_0 = 1$ and $r_1 > \frac{1}{2}\ell$. From $\ell > r_1 > \frac{1}{2}\ell$ it follows that $a_1 = [\ell/r_1] = 1$. To determine $a_2 = [r_1/r_2]$ we consider

$$w = 1 + \frac{1}{1 + r_2/r_1}$$

If $r_1/r_2 \leqq 2$, then we would have $w^2 \leqq (5/3)^2 < 3$.

If $r_1/r_2 \geqq 3$, then we would have $w^2 \geqq (7/4)^2 > 3$.

Because $w^2 = 3$ neither of these are possible, and so we have

$$2 < r_1/r_2 < 3 \quad \text{and} \quad a_2 = [r_1/r_2] = 2 .$$

Now we consider the Exchange process geometrically on a rectangle with sides $w$ and $\ell$ : see Fig.21.

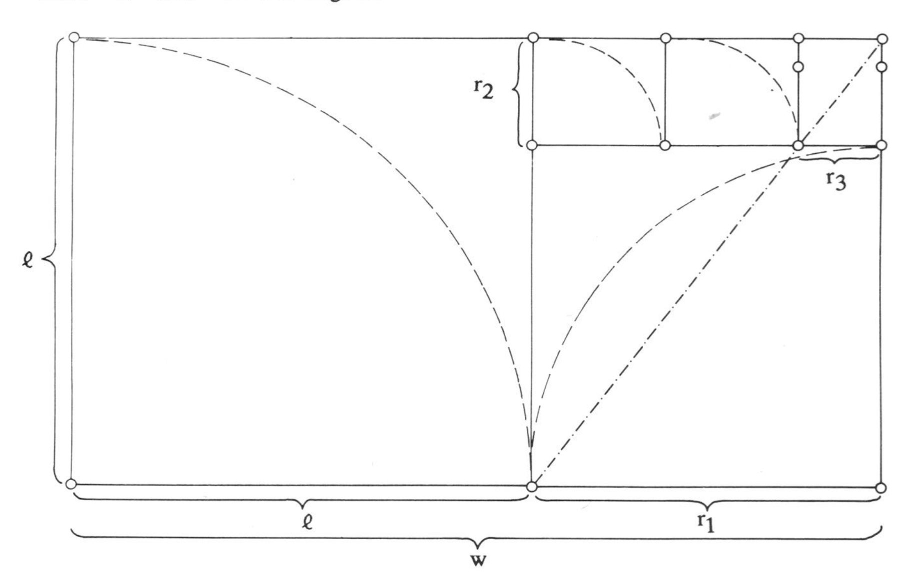


Fig. 21

If we look at the small rectangle, with sides $r_2$ and $r_3$, it seems to be similar to the bigger rectangle with sides $\ell$ and $r_1$. If we can prove this, then the following double step of the exchange process will repeat itself in the smaller rectangle, and the expansion will be periodic. The double step is:

$$\ell = r_1 + r_2$$

$$r_1 = 2r_1 + r_3$$

and we want to prove

$$r_2 : r_3 = \ell : r_1 ,$$

which is equivalent to

$$\ell r_3 = r_1 r_2 .$$

From the successive steps of the exchange process we derive

$$r_1 = w - \ell \quad \text{and} \quad r_2 = 2\ell - w \quad \text{and} \quad r_3 = 3w - 5\ell$$

and can then quickly verify the assertion $\ell r_3 = r_1 r_2$ by observing that $w^2 = 3\ell^2$. Result: the Continued Fraction sequence of $\surd 3$

is $[1; 1, 2, \overline{1, 2}]$.

Apart from the method of writing, we have derived our conclusions in these two examples completely within the frame of Euclid's mathematics. If the proof, of the periodicity of the Exchange process for $s : t$, is carried out in this or similar ways, then $s : t$ must be irrational, — or better in the Greek terminology, the associated segments $s$ and $t$ must be incommensurable by Euclid's criterion [X, 2].

For these and further examples, one can find more on the methodology of Euclid's proof in, for example, van der Waerden, *Science Awakening*, [71] pp.238/39 and very explicitly in D. Fowler: "Book II of Euclid's Elements and a pre-Eudoxan Theory of Ratio" [19], pp.5–36.

Observe that with this process, we have a wholly new proof for the irrationality of $\surd 3$, as compared with our previous one using splittings into prime factors.

## 3D TRANSCENDENTAL NUMBERS.

By Cantor's theorem there exist in $\mathbb{R}$ more transcendental than algebraic numbers. The question remains however, how we can decide of a concretely given real number — say the number $\pi$ or the Euler constant C — to decide whether it is transcendental or not. For this purpose, we look for criteria which are equivalent to the definition — the so-called *characterizations*, together with other types of properties. In 1851, Liouville gave such a criterion for the transcendental numbers.

As preparation, we consider for a fixed $q \in \mathbb{N}$ and $x \in \mathbb{R}$ the position of $x$ with respect to the 'lattice' of numbers $k/q$ with $k \in \mathbb{Z}$.

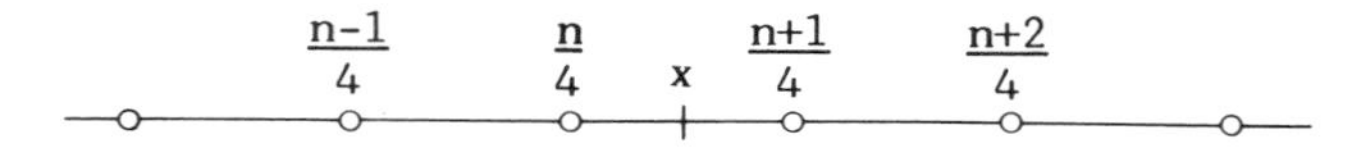


Fig. 22

We can at once say: given $x \in \mathbb{R}$ and $q \in \mathbb{N}$ there exists a $k \in \mathbb{Z}$ with

$$\left| x - \frac{k}{q} \right| \leqq \frac{1}{2q} .$$

If we make the lattice finer, by increasing $q$, then the distance from $x$ to its nearest lattice point will become small. That is rather crude; interesting assertions can only be expected if in the approximation to $x$, we fix the denominator $q$ to be *as small as possible.*

From the theory of continued fractions (see again Hardy– Wright, Chapter 10) one finds that the approximating fractions of the Continued fraction of $x$ are the best possible rational approximations to $x$ in the following sense: if $p_n/q_n$ is the $n^{th}$ approximating fraction for $x$, we have

a) $$\left| x - \frac{p_n}{q_n} \right| < \frac{1}{q_n q_{n+1}} < \frac{1}{q_n^2}$$

(which can be improved somewhat), and

b) if $$0 < q \leqq q_n \quad \text{and} \quad \frac{p}{q} \neq \frac{p_n}{q_n} ,$$

then $$\left| x - \frac{p_n}{q_n} \right| < \left| x - \frac{p}{q} \right| .$$

Part b) says that we cannot obtain as good an approximation with a denominator $< q_n$ as we have with the continued fraction denominator. Moreover, if $q = q_n$ then the numerator $p_n$ gives the optimal approximation.

<u>THEOREM</u> (Liouville). *Let* $\alpha \in \mathbb{R}$ *be an algebraic irrational number with associated polynomial* $f \in \mathbb{Z}[x]$, *so that*

$$f(x) = a_n x^n + \ldots + a_o \quad \textit{with} \quad a_i \in \mathbb{Z} \quad \textit{and} \quad f(\alpha) = 0 .$$

*Then there is a constant* $K > 0$ *such that for all* $p \in \mathbb{Z}$, $q \in \mathbb{N}$ *we have*

$$\left| \alpha - \frac{p}{q} \right| > \frac{K}{q^n}$$

*The constants* $K$ *and* $n$ *depend on* $f$ *(but not on* $p$ *or* $q$*).*

Liouville's Theorem says that algebraic numbers cannot be especially well approximated by rational numbers: see Fig.23.

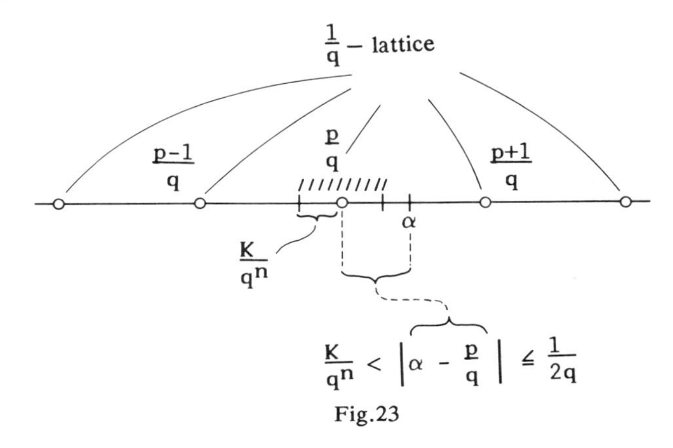


Fig.23

The algebraic number $\alpha$ can therefore not lie in the region of breadth $2K/q^n$ round $p/q$. Hence it is clear that by truncating the eventually infinite Continued Fraction for $p/q$ we derive a still better approximation: with smaller $q$ the forbidden domain for $\alpha$ becomes bigger. So we have: the smaller the denominator, the further distant $\alpha$ is.

After the proof of the theorem, we shall illustrate how it works with two examples.

The idea of the proof can easily be seen in a sketch (see Fig.24). Let $f$ be the polynomial with the zero $\alpha$ and with $p$, $q$ given. We want to estimate the distance $|\alpha - p/q|$. For brevity, write $r = p/q$. We are interested in the distance $d = |\alpha - r|$ .

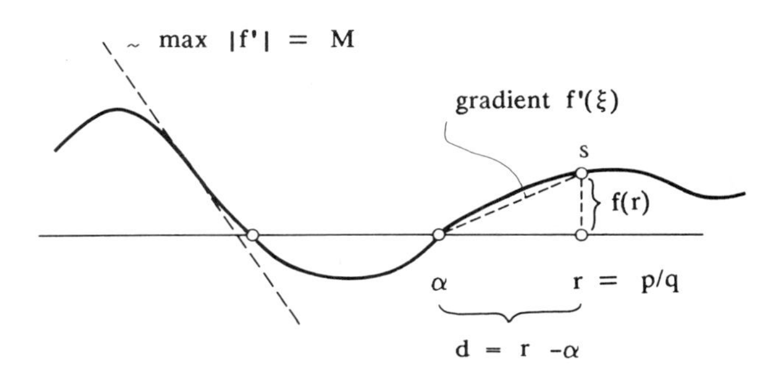


Fig.24

We orient ourselves first relative to the drawing which is so arranged that one can forget about the modulus sign. In the triangle with corners $\alpha$, $r$ and $s = (r,f(r))$, we find the gradient $m$ of the line joining $\alpha$ and $s$ to be $m = f(r)/d$ so that

$$d = f(r)/m \,, \qquad r = p/q \,.$$

The given situation allows us to estimate $m$ and $f(r)$ and hence to prove the assertion of the theorem.

Beginning the detailed work, we can assume that $r$ lies in the interval $\langle \alpha - 1, \alpha + 1 \rangle$ and that no other zero $\beta$ of $f$ is nearer to $r$ than $\alpha$. Thus, in particular, $f(r) \neq 0$. For the gradient $m$ we know from the mean value theorem that there exists $\xi \in \langle \alpha, r \rangle$ with $m = f'(\xi)$. As $f'$ is continuous there is in the interval $\langle \alpha - 1, \alpha + 1 \rangle$ a maximal value $M$ of $f'$ and $m \leqq M$. We put $K = 1/(M+1)$ and have

$$|\alpha - r| = |f(r)/m| < K|f(r)| \,.$$

For the function value $f(r)$ we obtain

$$|f(r)| = |a_n(r))^n + \ldots + a_o|$$

which is, since $r = p/q$

$$|a_n p^n + a_{n-1} p^{n-1} q + \ldots + a_o q^n| / q^n$$

$$\geqq 1/q^n$$

because $f(r) \neq 0$ and $a_n p^n + \ldots + a_o q^n$ is a whole number.

Collecting things together, we obtain:

$$|\alpha - r| > K|f(r)| \geqq K/q^n$$

as asserted. (For complex $\alpha$ one finds the corresponding theorem in H. Stark : *An Introduction to Number Theory* [68]. □

We clarify the state of affairs for ourselves, first with the example of the square root $\sqrt{n}$ of a natural number. Suppose then, $n < 10{,}000$ and $\alpha = \sqrt{n}$ is the zero of the polynomial $f(x) = x^2 - n$. The maximal gradient of $f$ we estimate crudely by: $f'(10{,}000) = 20{,}000 = 2 \cdot 10^4$. Now we choose $q = 100$. Then Liouville's theorem tells us

$$\left|\alpha - \frac{p}{100}\right| > \frac{1}{100^2 . 2 . 10^4} = 0.000\ 000\ 005$$

Therefore there can be no $\alpha$ with $\alpha^2 = n < 10^4$ so near to a decimal number $a_o.d_1 d_2$ (with $a_o \in \mathbb{N}$ and $0 \leqq d_1, d_2 \leqq 9$) that the distance is smaller than $5 \times 10^{-9}$. In other words, $\alpha$ cannot look like this:

$$a_o . d_1\ d_2\ 0\ 000\ 000\ d_{10}\ d_{11} \ldots$$

so there can be no such long sequence of zeros at these positions in the decimal expansion.

Conversely, one uses such especially long sequences of zeros in the decimal representation of a certain specific number, in order to prove its transcendence.

THEOREM. *The number*

$$\alpha = \sum_{k=1}^{\infty} \frac{1}{10^{k!}} = \frac{1}{10} + \frac{1}{10^2} + \frac{1}{10^6} + \frac{1}{10^{24}} + \ldots + \frac{1}{10^{k!}} \ldots$$

$$= 0.1\ 1\ 0\ 0\ 0\ 1\ 0\ 0\ \ldots\ 0\ 1\ 0\ \ldots$$

*is transcendental.*

*PROOF*. If $\alpha$ were algebraic, then there would be a polynomial $f \in \mathbb{Z}[x]$ of degree $n$ with $f(\alpha) = 0$; consequently, there would be numbers $0 < K \in \mathbb{R}$ and $n \in \mathbb{N}$ with

$$\left|\alpha - \frac{p}{q}\right| > K/q^n \quad \text{for all} \quad p \in \mathbb{Z},\ q \in \mathbb{N}.$$

The long sequences of zeros in the decimal expansion of $\alpha$ will now allow us to construct an appropriate $p/q$ that will lead to a contradiction. Looking for an approximating fraction for $\alpha$ we consider the partial sums

$$\alpha_N = \sum_{k=1}^{N} \frac{1}{10^{k!}} .$$

Then
$$\alpha_N = p/q \quad \text{with} \quad q = 10^{N!}.$$

We now choose $N$ so big that we have

$$N > n \quad \text{and} \quad \frac{2}{10^{N!}} < K ,$$

and then fix $N$. Therefore

$$a - \alpha_N = \alpha - \frac{p}{q}$$

$$= \sum_{k=N+1}^{\infty} \frac{1}{10^{k!}}$$

$$= \frac{1}{10^{(N+1)!}} \times 1.0\ \ldots\ 0\ 1\ 0\ \ldots$$

$$< \frac{1}{10^{(N+1)!}} \times 2 .$$

Therefore we have

$$\alpha - \frac{p}{q} < \frac{2}{(10^{N!})^{(N+1)}} = \frac{2}{(10^{N!})^{N}.10^{N!}}$$

$$= \frac{2}{q^{N}.10^{N!}} < \frac{K}{q^{N}}$$

which contradicts Liouville's theorem. Hence $\alpha$ can not be algebraic. □

This $\alpha$ is certainly a very wretched transcendental number — it has no other use than to be just a transcendental number that is easy to display — just like those irrational numbers which are given by certain non-periodic decimal expansions. Well-known transcendental numbers are $e$ and $\pi$, and the proof of their transcendence can be found for example in Hardy and Wright, for $\pi$ also in the book *Foundations of Algebra* of Schafmeister and Wiebe [66]).

## 3E MULTIPLES OF IRRATIONAL NUMBERS MODULO 1.

Given an irrational number $\alpha \in \mathbb{R}$ we form the multiples $\alpha$, $2\alpha$, $3\alpha$, $4\alpha,\ldots$ and thence the sequence

$$\alpha_1 = \alpha - [\alpha]$$

$$\alpha_2 = 2\alpha - [2\alpha]$$

$$\alpha_3 = 3\alpha - [3\alpha]$$

$$\ldots\ldots$$

$$\alpha_i = i\alpha - [i\alpha]$$

$$\ldots\ldots$$

where, as usual $[x]$ means the greatest whole number $\leqq x$.

For each $i$ we have $i\alpha \neq [i\alpha]$, otherwise $\alpha$ would be rational. Therefore for all $i$ we have $0 < \alpha_i < 1$. For all $i, j \in \mathbb{N}$ with $i \neq j$ we have $\alpha_i \neq \alpha_j$; for otherwise, we could easily show again that $\alpha$ was rational; therefore, the members $\alpha_i$ of the sequence are all different from each other. The question is: how are the $\alpha_i$ distributed within the interval $\langle 0, 1\rangle$? It is answered in the following theorem.

<u>THEOREM</u>. *If* $\alpha \in \mathbb{R}$ *is irrational, then the set of all the numbers* $\alpha_i = i\alpha - [i\alpha]$, *with* $i \in \mathbb{N}$, *is dense in the interval* $\langle 0, 1\rangle$.

<u>*PROOF*</u>. We have to show that between any two elements $r, s$ with $0 \leqq r < s \leqq 1$ there is at least one $\alpha_i$. For this purpose, we choose $N \in \mathbb{N}$ so large that $1/N < \frac{1}{2}(s-r)$. Then there exists $k \in \mathbb{N}$ with $r \leqq k/N < (k+1)/N < s$.

If we can show that in every sub-interval of the form $\langle k/N\ ,\ (k+1)/N \rangle$ we can always find an $\alpha_i$, then we are through. In one of these sub-intervals, there must be at least two different $\alpha_p$, $\alpha_q$, — for there are only N such sub-intervals, but infinitely many $\alpha_i$. Assuming without loss of generality, that $\alpha_p > \alpha_q$, we therefore have

$$0 < \alpha_p - \alpha_q < \frac{1}{N}$$

so from the definition of $\alpha_i$ this gives

$$\alpha_p - \alpha_q = p\alpha - [p\alpha] - q\alpha + [q\alpha]$$

$$= (p-q)\alpha - ([p\alpha] - [q\alpha]) < \frac{1}{N} \ .$$

Consider the case $p > q$. The whole number $[p\alpha] - [q\alpha]$ lies at a distance less than $1/N$ away from $(p-q)\alpha$, so we have

$$[p\alpha] - [q\alpha] = [(p-q)\alpha] \ .$$

Substituting $j = p-q$, we obtain

$$\alpha_j < \frac{1}{N} \ .$$

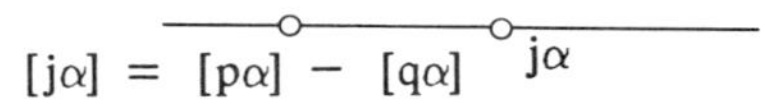

$[j\alpha] = [p\alpha] - [q\alpha]$   $j\alpha$

Fig.25

If however $p < q$, then $p - q = -j < 0$, so $[p\alpha] \leq [q\alpha]$ and the situation looks like this:

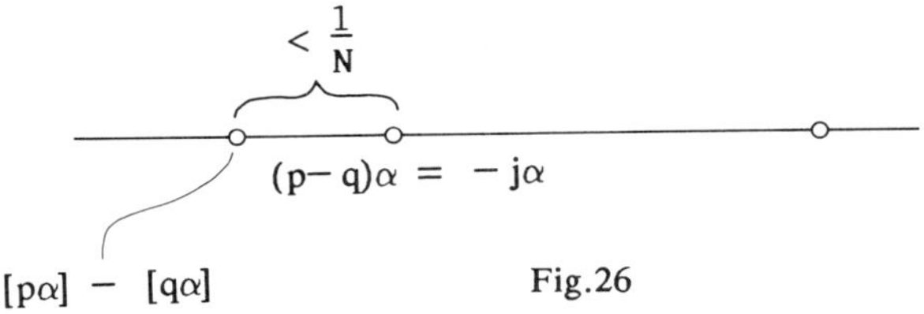

$< \frac{1}{N}$

$(p-q)\alpha = -j\alpha$

$[p\alpha] - [q\alpha]$

Fig.26

In either case, $j\alpha$ satisfies $1 - 1/N < \alpha_j < 1$.

In the first case, we derive from

$$j\alpha = [j\alpha] + \alpha_j \quad \text{and} \quad 2j\alpha = 2[j\alpha] + 2\alpha$$

that $\alpha_{2j} = 2\alpha_j$ because $2\alpha_j < 1$. Repeating this further, we have

$$\alpha_{rj} = r\alpha_j \quad \text{as long as} \quad r\alpha_j < 1 \ .$$

However, these elements lie at a distance $< 1/N$ from each other, so at least one $\alpha_{rj}$ must lie in each of the open intervals $\langle k/N , (k+1)/N \rangle$ .

In the second case, the elements $\alpha_{rj}$ wander downwards from 1 with constantly decreasing distances. The theorem is thereby proved. □

This theorem has an interesting consequence, which we include as an exercise.

Exercise 3.

(a) If $G$ is a subgroup of $(\mathbb{R}, +)$ which is generated by elements 1 and $c$, then: either $G$ is cyclic or the elements of $G$ lie densely in $\mathbb{R}$. Prove this.

(b) (A further curious consequence.) Suppose $x$ is a natural number with the decimal representation $x = a_n 10^n + \ldots + a_o$ with $a_i \in \{0,\ldots,9\}$. Show that there is aways a power of 2 of which the expansion begins with the same sequence of digits. (In other words, there exist $k, r \in \mathbb{N}_0$ with $x10^r \leqslant 2^k < (x+1)10^r$.) For the proof, observe that with base 10, $\log 2$ is irrational and then apply our theorem.

(c) Show that $\log_m(n)$ is rational iff $m, n$ possess the same prime factors and all ratios of powers of identical primes are equal. (This gives an easy way of manufacturing irrationals, e.g. $\log_2(2k+1)$, $k \in \mathbb{N}$. See: T. Crilly, *Mathematical Gazette*, Vol.70, p.219, 1986).

(d) Let $\alpha$ be any irrational real number. Show that the set $\{m + n\alpha \mid m, n \in \mathbb{Z}\}$ is dense in the unit interval.

# 4

# The complex numbers

INTRODUCTION.

The main theorem to be discussed in this chapter is the so-called *Fundamental Theorem of Algebra* with its different variants which will be enumerated in Sections 4C/D. Before discussing it, we follow a programme analogous to that of Chapter 2, and first present several possibilities in Section 4A for defining the system $\mathbb{C}$ of Complex numbers. In Section 4B, we make remarks on some properties of the additive multiplicative group of Complex numbers and the role of the exponential function in this connection.

## 4A CONSTRUCTIONS OF $\mathbb{C}$.

Historically, the positive Real numbers have traditionally been connected with intuitive models, at least in the form of ratios of intervals. Moreover, the use of the Greek equivalent of the word 'irrational' leads us to conclude that incommensurable ratios were held to be alien from the time of their discovery, because they are not directly expressible by means of whole numbers. This distaste was first overcome during the Renaissance. The same doubts hung over the negative numbers: while it was self-evident to every eastern trader, that one had to calculate with debts and credits, true numbers ought to be things one could multiply together, and that seemed to make no sense for debts. (Similarly, one has today a similar feeling when, say, one may need to construct from a given topological space a so-called 'negative' of it.) Eventually, there developed corresponding demonstrations on the number line. But then by the 16th century, a new problem entered the scene, because the solution formulae for polynomials of the third and fourth degree had been discovered, and the intermediate calculations with these depend in certain cases on the roots of negative numbers, which cancel out at the end. What then should one understand by these things? The square of any 'proper' number is $\geq 0$, therefore these roots are artificial, imaginary quantities; so these calculations can not hold in reality. Although for example, Euler had already put forward far-reaching calculations with complex numbers, these doubts were only overcome around 1800 through the intuitive model of the complex numbers in the plane, developed by Wessel, Argand and Gauss. True, these mathematicians had not put forward a formal definition of the complex numbers; they had only suggested an intuitive interpretation. However, the philosophical problems concerning existence and meaning of these numbers were essentially now solved. One could "understand" what a Complex number is, and no longer needed to be in a paradoxical situation of having to calculate with fictitious variables. An adequate demonstration of a mathematical object had been given, which made it possible to undertake fruitful researches with feelings of security rather than doubt, even though there still remained the problem of giving correct formal definitions that could be developed into a strict logical system.

## 1. Hamilton's Definition: Number pairs.

Concerning the historical context of Hamilton's definition of the Complex numbers in 1833, it is best to consult Gericke [22]. For ourselves, we consider the set

$$\mathbb{R}^2 = \{(a, b) \mid a, b \in \mathbb{R}\}$$

and define in $\mathbb{R}^2$ an addition, a multiplication and a metric through the following formulae

(i) $$(x, y) + (u, v) = (x + u, y + v)$$

(ii) $$(x, y) . (u, v) = (xu - yv, xv + yu)$$

(iii) $$d[(x, y), (u, v)] = \surd[(x - u)^2 + (y - v)^2] .$$

The formulae (i) and (iii) are the definitions appropriate to a product space; for $(\mathbb{R}^2, +) \cong (\mathbb{R}, +) \times (\mathbb{R}, +)$ as a direct product of groups, and since we are using the Pythagorean formula for the distance between $(x, y)$ and $(u, v)$, we satisfy the definition for the direct product of two metric spaces. However, the definition of multiplication in (ii) does not arise from a general, standard scheme. Nevertheless, this gives us firstly $(\mathbb{R}^2\backslash\{0\}, .)$ as a commutative group with zero $\mathbf{0} = (0, 0)$. Secondly $(\mathbb{R}^2, +, .)$ is now a field which contains as a subfield the Real number field, in the form of the set of all the elements $(a, 0)$. Especially important is the equation

$$(0, 1)\cdot(0, 1) = (-1, 0).$$

We derive the usual way of writing Complex numbers through the abbreviations

$$i = (0, 1) \quad \text{and} \quad (a, b) = a + bi$$

The metric then gives us the usual norm in $\mathbb{R}^2$ by:

$$d[(0, 0), (x, y)] = |(x, y)| ;$$

and from this we can define the topology through the usual definitions of convergence, continuity, etc. We usually write $\mathbb{C}$ for the field of Complex numbers, and we then mean it always to have these rules of calculation, and this metric. One normally uses $\mathbb{R}^2$ to denote the point-set of the plane, (for example when we are thinking in terms of co-ordinate geometry).

2. <u>Definitions using Matrices</u>.
Let I, J denote the matrices

$$I = \begin{bmatrix} 1 & 0 \\ 0 & 1 \end{bmatrix}, \quad J = \begin{bmatrix} 0 & -1 \\ 1 & 0 \end{bmatrix};$$

we consider a set of matrices defined by

$$K = \{(aI + bJ \mid a, b \in \mathbb{R}\} .$$

Then K, with the usual addition and multiplication of matrices forms a commutative field, which contains $\mathbb{R}$ as a subfield in the form of all matrices $aI$. If $M = aI + bJ$ in K, then the **norm** $N(M)$ can be defined by $N(M) = \surd(a^2+b^2)$; since $\det M = a^2 + b^2$ it is an easy exercise to verify with the help of determinants that the usual axioms for a norm are satisfied. Since $J^2 = -I$, we can then identify $a + bi$ with the matrix $M = aI + bJ$, and the usual modulus $|a+bi|$ is $N(M)$.

The advantage of this method of introduction lies in the fact that one obtains the operations and the metric in a natural way from matrix algebra. A mild drawback arises in the fact that the matrices are not displayed directly as points of the plane. However, the following connection is interesting: granted an intuitive knowledge of $\mathbb{C}$, one identifies the complex number $a + bi$ with a matrix via the mapping

$$z \to (a + bi)\cdot z$$

which is a rotation of the plane induced by $a + bi$. In particular, the matrices $M = aI + bJ$ satisfying $N(M) = 1$ are identified with those rotations that move 1 to $a + bi$ (and z now lies on the unit circle).

3. <u>Cauchy's definition using Residue Classes</u>.
One would like the number i to be so defined that i is a zero of the polynomial $x^2 + 1$. Thus $x^2 + 1 = 0$ has to be satisfied. We can arrange for this by calculating in the ring $\mathbb{R}[x]$ of real polynomials "modulo $x^2 + 1$", as we now explain. The well known Remainder Theorem allows us to find, for each polynomial $f \in \mathbb{R}[x]$, polynomials $g, r \in \mathbb{R}[x]$ with

$$f(x) = g(x)\cdot(x^2 + 1) + r(x) \quad \text{and degree } r < 2 .$$

The residues mod $x^2+1$ are therefore linear polynomials of the form $r(x) = a + bx$ where $a, b \in \mathbb{R}$. If one had also $s(x) = u + vx$ then we have the sum

$$r(x) + s(x) \equiv a + u + (b + v)x \qquad [\text{modulo } (x^2 + 1)]$$

and the product

$$r(x)\cdot s(x) = au + (av + bu)x + bvx^2$$

$$\equiv au - bv + (av + bu)x \qquad [\text{modulo } (x^2 + 1)]$$

since $\qquad bvx^2 = bv(x^2 + 1) - bv$

These are again the usual formulae for the operations with Complex numbers. In particular,

$$x^2 = (x^2 + 1) - 1 \equiv -1 \qquad [\text{modulo } (x^2 + 1)]$$

and therefore $x$ plays "modulo $(x^2 + 1)$" the role of $i$.

Exercise 1.
Show that, modulo $(x^2 + 1)$, $f(x) = 5x^4 + 7x^3 - 9x^2 + 4x - 3$ $\equiv 4x - 14x^2 - 10 \equiv 4(x+1) \doteq g(x)$. Calculate $f(x)^3$, $g(x)^3$, and show that these are equal modulo $(x^2 + 1)$.

Using a certain amount of algebraic machinery, one can describe this process in the following way: in the ring $\mathbb{R}[x]$, $x^2 + 1$ is irreducible, and generates the principal ideal:

$$P = \{f(x)\cdot(x^2 + 1) \mid f \in \mathbb{R}[x]\}$$

of all polynomial multiples of $x^2 + 1$; and then $P$ is maximal. On that account the residue class ring $\mathbb{R}[x]/P$ is a field. Viewed historically, this direct process of Cauchy in 1847 is the precursor of the usual algebraic concepts that we now possess. We pause here to consider the following problem, and give a solution.

PROBLEM. What happens if instead of $x^2 + 1$ we use another irreducible quadratic $p(x) = x^2 + mx + n$ and carry through the same procedure?

*SOLUTION*. The algebraic machinery functions for this case just as well as previously, and we again obtain a field, the elements of which are residue classes of linear polynomials $a + bx$. The addition formulae are the same as before, but for multiplication we have

$$(a + bx) * (c + dx) = ac + (bc + ad)\cdot x + bdx^2$$

$$\equiv ac - bdn + (bc + ad - bdm)\cdot x \quad (\text{modulo } p),$$

since

$$bdx^2 = bd\cdot(x^2 + mx + n) - bdmx - bdn .$$

To make things clearer we now write the elements of $\mathbb{R}^2$ as column vectors. Then we have constructed a field $(\mathbb{R}^2, +, *)$ with the multiplication formula

$$\begin{bmatrix} a \\ b \end{bmatrix} * \begin{bmatrix} c \\ d \end{bmatrix} = \begin{bmatrix} ac - bdn \\ ad + bc - bdm \end{bmatrix}$$

in which $m$ and $n \in \mathbb{R}$ are subject only to the constraint that $n - (m/n)^2 > 0$, in order to guarantee the irreducibility of $p$. This gives very many possibilities for a field $(\mathbb{R}^2, +, *)$. How do these fields relate to $(\mathbb{C}, +, \cdot)$? For clarification we write $p(x)$ as

$$p(x) = (x - r^2) + s^2$$

where $$2r = -m \quad \text{and} \quad s^2 = n - r^2 > 0 .$$

(Since $p$ is irreducible then $n - r^2 = n - (m/2)^2 > 0$.)

Here we see clearly that the situation we started from is a special case with $r = 0$ and $s = 1$. In the general case, the multiplication formula for $*$ now reads:

$$\begin{bmatrix} a \\ b \end{bmatrix} * \begin{bmatrix} c \\ d \end{bmatrix} = \begin{bmatrix} ac - bd(r^2 + s^2) \\ ad + bc + 2rbd \end{bmatrix}$$

One may perhaps feel that this multiplication can scarcely be contemplated because the formula is too complicated. A better reason is however, that the fields $(\mathbb{C}, +, \cdot)$ and $(\mathbb{R}^2, +, *)$ are not essentially different: they are isomorphic. On that account it does not matter whether we use a version with complicated multiplication or not. For the proof of isomorphism we need to find a mapping

$$\psi : (\mathbb{R}^2, +, *) \to (\mathbb{C}, +, \cdot)$$

which is bijective and preserves addition and multiplication. A hint as to the definition of $\psi$ comes from the observation that $p(x)$ has $\begin{bmatrix} 0 \\ 1 \end{bmatrix}$ and $r+si$ as zeros in $(\mathbb{R}^2, +, *)$ and $(\mathbb{C}, +, \cdot)$ respectively; and these zeros must somehow be related to each other. This can be arranged by using a matrix $S$ to construct the linear transformation $\psi: \mathbb{R}^2 \to \mathbb{C}$ with $\psi(u) = S\cdot u$ for every $u$, where

$$u = \begin{bmatrix} a \\ b \end{bmatrix} \in \mathbb{R}^2 , \qquad \text{and} \qquad S = \begin{bmatrix} 1 & r \\ 0 & s \end{bmatrix} .$$

The bijectivity of $\psi$ results from the fact that $s \neq 0$, and it preserves addition because the mapping defined by the matrix is linear. To show that $\psi$ preserves multiplication one has to show that if in $\mathbb{R}^2$

$$u = \begin{bmatrix} a \\ b \end{bmatrix} , \quad v = \begin{bmatrix} c \\ d \end{bmatrix}$$

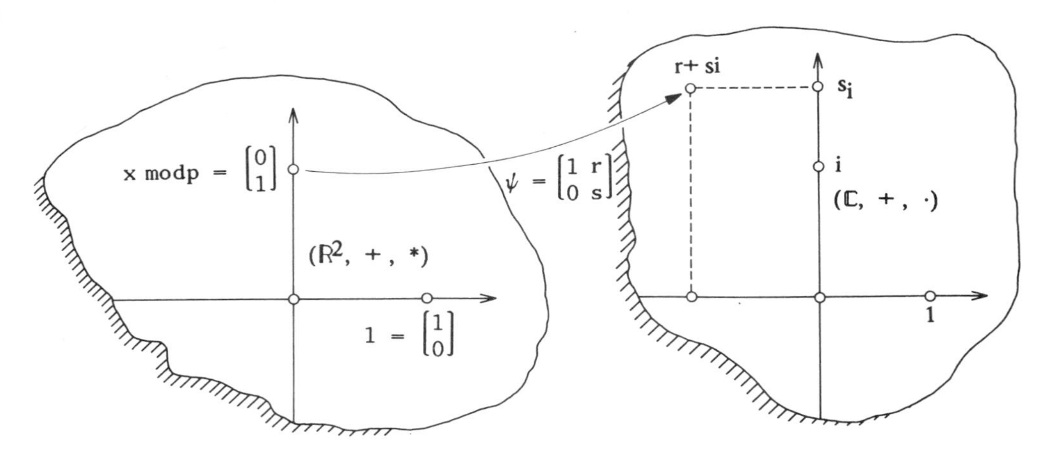


Figure 27

then in $(\mathbb{C}, +, \cdot)$

$$\psi(u*v) = \psi(u)\cdot\psi(v)$$

i.e. that the matrix equation

$$S\cdot(u*v) = (S\cdot u)\cdot(S\cdot v)$$

is valid, and one can verify this easily by direct calculation.

Exercise 2.

(a) Work out the multiplication formula, and the matrix of $\psi$, when the polynomial $p$ is

(i) $x^2+x+1$, (ii) $x^2+3x+5$.

Work out $p(x)$ when $\begin{bmatrix} a \\ b \end{bmatrix}$ (N.B. $p(x)$ is now $x*x+ mx+ n$.)

(b) Show that if $z = x + iy$ in $\mathbb{C}$, then $z$ satisfies the quadratic equation $z^2 = 2iyz + \rho$ where $\rho = |z|^2$. Hence express $z^3$, $z^4$ etc in the form $f(y,\rho)z + g(\rho)$.

From now on we shall use the usual notation for Complex numbers and therefore for example, write $z = x + iy$ and $\bar{z} = x - iy$ instead of $\left[\begin{smallmatrix} x \\ y \end{smallmatrix}\right]$ etc. We shall also take it as well-known that the mapping $\kappa : z \to \bar{z}$ is an automorphism and – regarding it as a reflection in the real axis, – that $\kappa$ is an isometry (i.e. preserves distances).

## 4B SOME STRUCTURAL PROPERTIES OF ℂ.

### 1. The Additive Group $(\mathbb{C}, +)$.

We have introduced addition in $\mathbb{C}$ in such a way that $(\mathbb{C}, +) = (\mathbb{R}^2, +)$ is the vector group of the plane $\mathbb{R}^2$. On that account $(\mathbb{R}, +)$ is the vector group of the line. We now have an intuitively obvious difference. Can we express this difference as a purely algebraic property of the groups? What group-theoretic properties are there in which the line and plane differ? The blunt answer is that no such differences exist as long as one confines oneself to the additive structure. In order to clarify this situation somewhat, we consider the multiplicative group $(\mathbb{Q}^+, \cdot)$ of the positive rational numbers. We claim:

THEOREM. $(\mathbb{Q}^+, \cdot)$ *is isomorphic to its "direct square"*, i.e.

$$(\mathbb{Q}^+, \cdot) \cong (\mathbb{Q}^+, \cdot) \times (\mathbb{Q}^+, \cdot).$$

*PROOF*. We use the prime factor splitting of rational numbers with positive and negative exponents in the form:

$$\frac{m}{n} = p_1^{\alpha_1} \ \dots \ p_r^{\alpha_r} \quad \text{with } \alpha_1 \in \mathbb{Z} .$$

For simplicity we list the prime numbers in each splitting in the order $p_1 = 2$, $p_2 = 3$, $p_3 = 5, \dots$, and then we have almost all exponents $\alpha_i = 0$. Therefore, for example

$$\frac{5^2 \cdot 13}{7^3 \cdot 17} = 2^0 \cdot 3^0 \cdot 5^2 \cdot 7^{-3} \cdot 11^0 \cdot 13^1 \cdot 17^{-1} \cdot 19^0 \cdot 23^0 \ \dots$$

Having thus fixed the sequence of prime numbers, it now suffices to note only the sequence of exponents themselves, as in the formula

$$\varphi: \frac{5^2 \cdot 13}{7^3 \cdot 17} \to (0,\ 0,\ 2,\ -3,\ 0,\ 1,\ -1,\ 0,\ 0,\ \dots) .$$

The multiplication of rational numbers corresponds, under this mapping, to the addition of the corresponding exponent sequences. These sequences form a group which is an infinite direct sum, the group $\sum(\mathbb{Z}, +)$ of exponent-sequences. Hence we have shown that this group is isomorphic to $G = (\mathbb{Q}^+, \cdot)$. With this mode of writing, the isomorphism asserted in the claim above is especially simple to express. We merely indicate its effect by means of the diagram:

$$\begin{array}{ccc} [(\alpha_1, \alpha_2, \alpha_3, \dots) \ , \ (\beta_1, \beta_2, \beta_3, \dots)] & \in & G \times G \\ \downarrow \varphi & & \downarrow \varphi \\ (\alpha_1, \beta_1, \alpha_2, \beta_2, \alpha_3, \beta_3, \dots) & \in & G \end{array}$$

and leave to the reader the details of showing that $\varphi$ is an isomorphism of groups. □

One recognises best the essential kernel of the previous argument if we consider the group $\Sigma(\mathbb{Z}, +)$ as a $\mathbb{Z}$-module (i.e. a 'vector space' but with scalars from $\mathbb{Z}$) and which has the 'basis vectors', (1, 0, 0, 0,...), (0, 1, 0, 0,...), (0, 0, 1, 0, 0,...) and so on. Each element of $\sum(\mathbb{Z}, +)$ is a finite linear combination of these basis elements. In a natural way, the basis itself can be identified with $\mathbb{N}$; and for $(\mathbb{Q}^+, \cdot)$ itself, the $i^{th}$ prime number corresponds to the $i^{th}$ basis element. The isomorphism we have constructed owes its property simply to the fact that there is a bijection between $\mathbb{N}$ and the even, or the odd, numbers.

We can now understand the analogous relationships for $(\mathbb{R}, +)$ and $(\mathbb{R}^2, +)$. Thus we regard $(\mathbb{R}, +)$ and $(\mathbb{R}^2, +)$ as vector spaces over $\mathbb{Q}$, so the only scalars allowed are to be rational numbers. Using the techniques of Set–theory, it will be shown in Section 6F that these vector spaces have equipotent bases, and hence the spaces are isomorphic and of infinite dimension relative to $\mathbb{Q}$; therefore also their additive groups are isomorphic. Recall that the *real* vector spaces $(\mathbb{R}, +)$ and $(\mathbb{R}^2, +)$ are *not* isomorphic, on account of their different *finite* dimensions.

Exercise 3.

(a) Calculate the exponent–sequences $\varphi(x)$ when $x$ is (i) 450/121, (ii) 630/1001, and use these numbers to check the formula $\varphi(x+y) = \varphi(x)+\varphi(y)$.

(b) Now show that $\varphi$ is a group–isomorphism.

From the isomorphism $(\mathbb{R}^2, +) \cong (\mathbb{R}, +)$, we obtain directly $(\mathbb{R}^n, +) \cong (\mathbb{R}, +)$; hence the additive groups of the finite–dimensional real spaces are therefore all isomorphic to each other.

With addition alone there is therefore too little structure to obtain the intuitive understanding of the differences that we are looking for. (More on this in Section 6F.)

**2. The Multiplicative Group** $(\mathbb{C}^\times, \cdot)$.

In contrast to the state of affairs with addition, the group $(\mathbb{C}^\times, \cdot)$ can not be isomorphic to $(\mathbb{R}^\times, \cdot)$ or $(\mathbb{R}^+, \cdot)$. It contains in particular elements of finite order (as for example $w = (1+i)/\sqrt{2}$ with order 8 (i.e. $w^8 = 1$), and there are no such elements in $(\mathbb{R}^\times, \cdot)$. In order to split the multiplicative group of complex numbers into its constituent parts, we recall the modulus and argument of a complex number $\neq 0$.

(i) **The Modulus.** The *modulus function* $\beta : \mathbb{C} \to \mathbb{R}$ is defined using the distance $d$ (see (iii) on p.76) by

$$\beta(z) = |z| = \sqrt{z\bar{z}} = d(0,z)$$

Now $\beta(z) = 0 \Leftrightarrow z = 0$, so we can restrict the modulus to $\beta : \mathbb{C}^\times \to \mathbb{R}^+$ and derive from the property

$$\beta(zw) = \beta(z) \cdot \beta(w)$$

a homomorphism $\beta : (\mathbb{C}^\times, \cdot) \to (\mathbb{R}^+, \cdot)$ .

(ii) **The Argument.** In the complex plane, the unit circle is the set $\mathbb{S} = \{z \in \mathbb{C} \mid |z| = 1\}$. Using multiplication of complex numbers, this forms a subgroup $(\mathbb{S}, \cdot)$ of $(\mathbb{C}^\times, \cdot)$. The *argument function* $\alpha : \mathbb{C}^\times \to \mathbb{S}$ is then defined by

$$\alpha(z) = z/|z|, \quad (z \neq 0) .$$

Hence we derive from the property

$$\alpha(zw) = \alpha(z) \cdot \alpha(w)$$

a homomorphism $\qquad \alpha : (\mathbb{C}^{\times}, .) \to (\$, .)$

(iii) **Polar Coordinates.** These two homomorphisms yield for us a direct splitting

$$(\mathbb{C}^{\times}, \cdot) \cong (\mathbb{R}^{+}, \cdot) \times (\$, \cdot)$$

which is given by the mapping

$$\varphi : z \to (\beta(z), \alpha(z)) .$$

Its bijectivity is obtained from the inverse mapping

$$(\beta(z), \alpha(z)) \to \beta(z) \cdot \alpha(z) = z ,$$

and the isomorphism conditions can be checked immediately. The mappings $\alpha$ and $\beta$ are nothing other than the projections on the "axes", of this splitting of $z$ into the components $\beta(z)$, $\alpha(z)$ – see Fig.28.

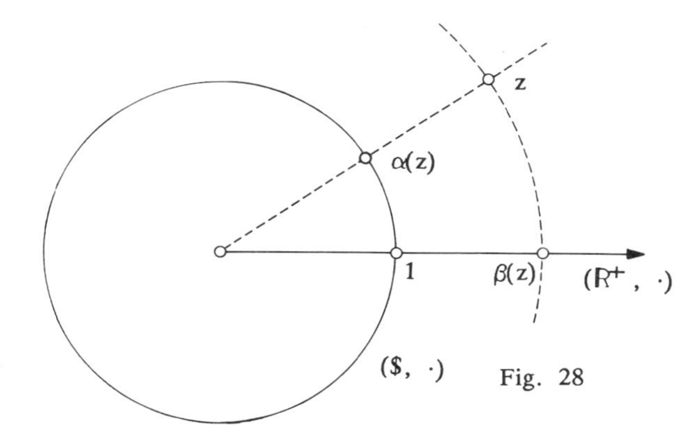


Fig. 28

3. The Multiplicative Group ($,·).

The group $(\$,\cdot)$ has the remarkable property that in it we have the formula: $z^{-1} = \bar{z}$. It is a homomorphic image of $(\mathbb{R}, +)$ under the mapping

$$\text{cis} : y \to \cos y + i \sin y .$$

This so-called "Winding function" wraps the real axis onto the unit circle, like a thread on a bobbin. The proofs of the surjectivity of $\text{cis} : \mathbb{R} \to \$$ and the bijectivity of $\text{cis} : \langle 0, 2\pi \rangle \to \$$ are problems of Analysis, and they depend on the definitions of sine and cosine. We can see that cis is a homomorphism, in the following way:

$$\text{cis}\,(x + y) = \cos\,(x + y) + i \sin\,(x + y)$$

$$= \cos x \cos y - \sin x \sin y$$

$$+ i\,(\sin x \cos y + \cos x \sin y)$$

$$= (\cos x + i \sin x)\cdot(\cos y + i \sin y)$$

$$= \operatorname{cis} x \cdot \operatorname{cis} y\ .$$

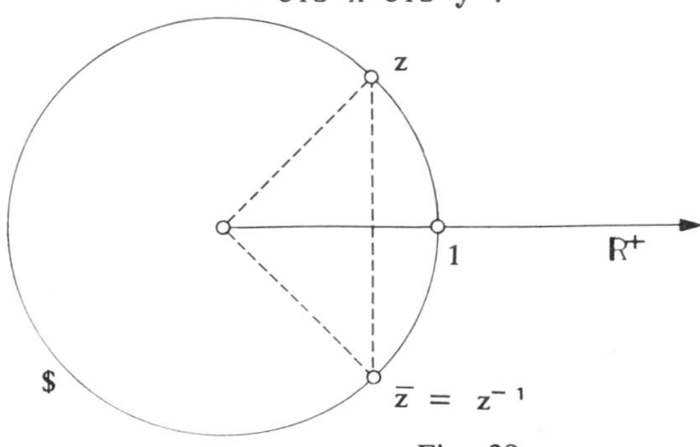


Fig. 29

<u>Exercise 4</u>.

(a) Prove that $\operatorname{cis}(-x) = 1/\operatorname{cis} x$, $\operatorname{cis} 2x = (\operatorname{cis} x)^2$ and $\operatorname{cis}(nx) = (\operatorname{cis} x)^n$ when $n \in \mathbb{N}$. (This is 'De Moivre's Theorem'.) Hence prove that one solution of the equation $y^n = \operatorname{cis} x$ is $y = \operatorname{cis}(x/n)$.

(b) Show that $\operatorname{cis} x = 1$ iff $x = 2k\pi$, $k \in \mathbb{Z}$. Hence show that if $y^n = \operatorname{cis} a$ then $y = \operatorname{cis}(a/n + 2k\pi/n)(k = 0,1,2,\ldots,n-1)$, and these solutions form the vertices of a regular polygon in $\mathbb{C}$. [Thus, $n^{th}$ roots always exist in $\mathbb{C}$. The numbers $\operatorname{cis}(2k\pi/n)(k = 0,\ldots,n-1)$ are usually called the $n^{th}$ **roots of unity**.]

The homomorphism property depends therefore essentially on the addition theorems for sine and cosine; so these formulae are really nothing other than the expression of the fact that if we wind the real line on the unit circle, the addition of real numbers goes over into the addition of angles. The latter expresses itself in the multiplication formulae for the elements of $. Moreover, one can easily convince oneself of the addition theorems if one understands them as formulae for the multiplication of complex numbers of modulus 1.

As a homomorphism, cis possesses a kernel, and this consists of all integral multiples of $2\pi$; for, $\operatorname{cis} 2\pi k = 1$ for every $k \in \mathbb{Z}$. Often it would be more convenient if the kernel were $\mathbb{Z}$ itself, and by a change of variable we can arrange this, through the following observations. One first changes the definition of cis to:

$$\operatorname{cis} x = \cos 2\pi x + i \sin 2\pi x$$

which signifies nothing more than that we first stretch the real line by the factor $2\pi$, and then do the winding. We use this variant of cis for the study of the group EW of all roots of unity in $. (E and W stand for the German *Einheit* = unit, *Wurzel* = root). Denote by

$$EW_n = \{z \mid z^n = 1\}$$

the set of zeros of the polynomial $z^n - 1$. Using the new cis function, we can write these solutions straight down: for $n\cdot(k/n) = k \in \mathbb{Z}$ we have

$$\begin{aligned} 1 &= \cos 2\pi k + i \sin 2\pi k \\ &= \operatorname{cis} k \\ &= \operatorname{cis} (n\cdot k/n) \\ &= (\operatorname{cis} k/n)^n \quad \text{(by Exercise 4)}, \end{aligned}$$

and therefore $\operatorname{cis} k/n$ is a zero of $z^n - 1$. If $0 \leqq k < n$ then $\operatorname{cis} k/n$ has altogether $n$ different roots, and these are just the zeros of $z^n - 1$; so we have

$$EW_n = \{\cos 2\pi \frac{k}{n} + i \sin 2\pi \frac{k}{n} \mid 0 \leqq k < n\} .$$

Another way of saying this is

$$EW_n = \text{image of } \{0, \frac{1}{n}, \frac{2}{n}, \ldots, \frac{n-1}{n}\} \text{ under cis} .$$

In this way the circle $ is divided into n equal parts of length $2\pi/n$, and we therefore* call $z^n - 1$ the $n^{th}$ *cyclotomic polynomial*. Hence, we call the subfield of $\mathbb{C}$ that is generated by adjunction of $\operatorname{cis} 1/n$ to $\mathbb{Q}$, the $n^{th}$ *cyclotomic field*. Viewed geometrically in the plane, the $n^{th}$ roots of unity form a regular polygon of n sides. If now we put together all the roots of unity of the different degrees, we obtain the set

$$EW = \bigcup \{EW_n \mid n \in \mathbb{N}\}$$

of all roots of unity. Therefore

$$EW = \{\operatorname{cis} \frac{k}{n} \mid k \in \mathbb{Z} \quad \text{and } n \in \mathbb{N}\}$$

so we have

$$EW = \operatorname{cis} \mathbb{Q} ;$$

and for the multiplicative group, we have the isomorphism

$$(EW, \cdot) \approx (\mathbb{Q},+)/(\mathbb{Z},+)$$

---

* We have here some etymology: cf. such medical words as *appendectomy* referring to cutting.

since the image of a homomorphism is always the quotient of the domain by the kernel. This group plays an important role in the theory of Abelian groups. One can find more about ($, ·) etc in Salzmann's notes [65]. There it is shown clearly that also in this case, the group structure alone cannot characterise the full nature of this group; it is only by combining group theory with topology, that we can adequately explain the phenomena.

## 4. The Complex Exponential Function.

Suppose we define the exponential function by means of power series. Then as for real numbers, we know that the following functional equation is satisfied:

$$e^{z+w} = e^z \cdot e^w$$

Since $e^z \neq 0$ for all $z \in \mathbb{C}$, this then says that

$$\exp : (\mathbb{C}, +) \to (\mathbb{C}^{\times}, \cdot)$$

is a homomorphism (moreover, exp is an analytic and holomorphic function, etc). Writing $z = x + iy$ with $x, y \in \mathbb{R}$, we see that the above functional equation shows

$$\begin{aligned} \exp(x + iy) &= \exp x \cdot \exp iy \\ &= \exp x \cdot (\cos y + i \sin y) \\ &= \exp x \cdot \mathrm{cis}\, y \end{aligned}$$

this time using the initial definition of cis. This relation had already been found by Euler, who developed it further. If we consider it group-theoretically, it says that in $(\mathbb{C}, +)$, the real and imaginary axes (each isomorphic to $(\mathbb{R}, +)$) are mapped onto the two axes $(\mathbb{R}^+, \cdot)$ and ($, ·) of $(\mathbb{C}^{\times}, \cdot)$: see Fig.30).

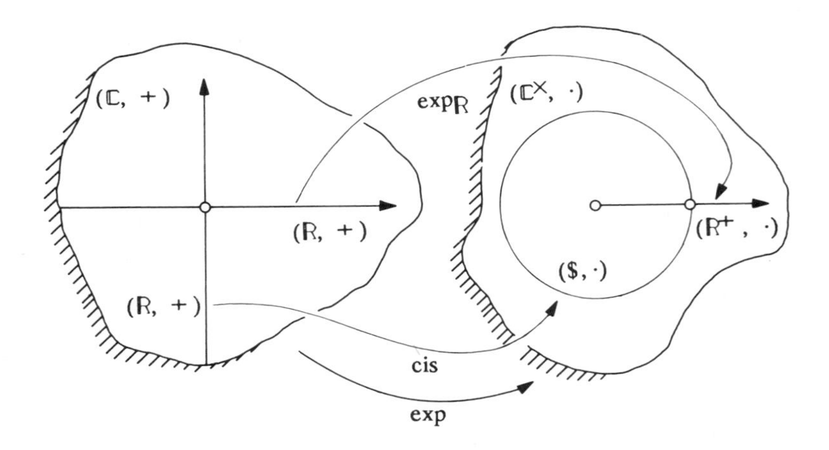


Fig.30

<u>Exercise 5</u>.

(a) Use exp to describe the mapping of the strip

$$\mathrm{Str} = \{z \in \mathbb{C} \mid 0 \leqslant \operatorname{Im} z < 2\pi\}$$

and hence explain how, if $\exp z = w$, the $n$ different $n^{th}$ roots of $w$ can be formed in terms of $z/n$.

(b) Define $\cos z$, $\sin z$ by the usual formulae

$$\cos z = (e^{iz}+e^{-iz})/2 \ , \ \sin z = (e^{iz}-e^{-iz})/2i$$

for all $z \in \mathbb{C}$. Find all the zeros of $\cos z$ in $\mathbb{C}$, and similarly for $\sin z$. Prove that the addition formulae still hold, and investigate the behaviour of $\cos z + i \sin z$. [Assume the usual properties of $e^x$, $e^{iy}$ when $x,y$ are real and $z = x+iy$.]

5. <u>Remarks on Square Roots in $\mathbb{C}$</u>.

Everybody knows what is meant by $i = \surd{-1}$, but the square root function cannot be defined easily in the Complex numbers as long as we avoid the use of Riemann surfaces. Indeed, if we were to expect the same behaviour as the real square root with its known homomorphic property $\surd ab = \surd a \cdot \surd b$, we would easily obtain a paradox. For, on the one hand we would have

$$\surd{-1} \cdot \surd{-1} = i \cdot i = -1$$

and on the other hand the alternative

$$\surd{-1} \cdot \surd{-1} = \surd((-1)\cdot(-1)) = \surd 1 = 1 \ .$$

A "good" definition of square roots of complex numbers will therefore require care. This is not merely the result of an arbitrariness about the assignment of square roots, as is shown by the following exercise:

<u>Exercise 6</u>.

Prove that there is no homomorphism $\varphi : (\mathbb{C}^\times, \cdot) \to (\mathbb{C}^\times, \cdot)$ with the property

$$(\varphi(z))^2 = z \quad \text{for all } z \in \mathbb{C}^\times \ .$$

For a penetrating further discussion connected with the question of the complex root function, see for example, G. Pickert: "Scientific Foundations of the Concept of Function". *MU* <u>*15*</u> *(1969)*, [55] pp.40–98.

The difficulties of the root function can be seen intuitively if one thinks of the unit circle as made of two halves, each looped as shown in Fig.31.

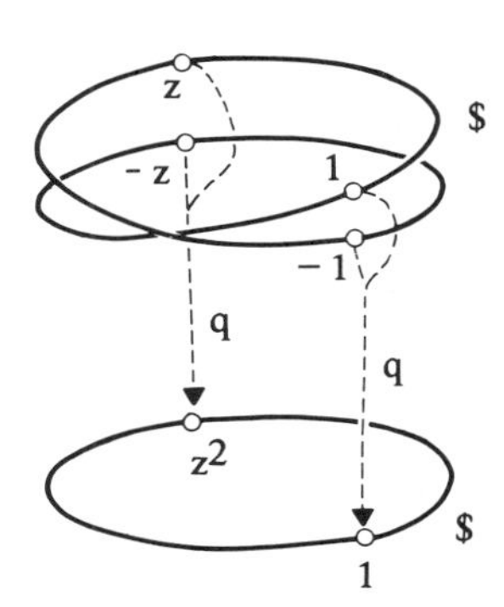


Fig. 31

With the function $q : z \to z^2$, if $z$ runs once round the half circle, then $z^2$ runs once round the whole circle. The upper loop is thus wound round, as when one puts an elastic band onto a cardboard cylinder. Conversely, a root function would need to map from below to above, and yet there is no obvious possibility of choosing one of the two inverse images of $z^2$ in a consistent way. By contrast, with the *real* square function, there is a consistent and intuitively obvious choice of square root (see Fig.32), if we insist that $\surd 1 = 1$ and choose this as the starting point, without passing beyond zero. (Thus we restrict the domain of the square root to the positive reals, in order to keep it both single-valued and continuous, and to have the algebraic advantage that $\surd 1 = 1$.)

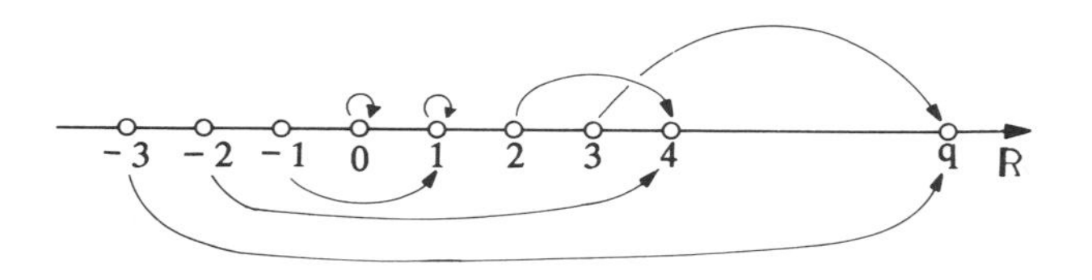


Fig.32

## 4C THE FUNDAMENTAL PROPERTIES OF $\mathbb{C}$.

### 1. The Topological Completeness of $\mathbb{C}$.

The topological completeness of $\mathbb{C}$ (more precisely, with respect to a metric) can be seen through the following assertion.

THEOREM. *Every Cauchy sequence* $(z_n)$ *has a limit point in* $\mathbb{C}$.

This means that $\mathbb{C}$ is a complete metric space and one need not add new limit points to it as in the passage from $\mathbb{Q}$ to $\mathbb{R}$. The underlying reason is simply that the metric for the topology in $\mathbb{C}$ has been chosen as the product metric for the product topology of $\mathbb{R}^2$. Correspondingly the theorem is a simple consequence of the completeness of $\mathbb{R}$.

*PROOF*. Let the real and imaginary parts of $z_n$ be $a_n$, $b_n$; so $z_n = a_n + ib_n$. Thus we obtain real sequences $(a_n)$, $(b_n)$, each of which is a Cauchy sequence, because (for example) $|a_n - a_j| \leqq |z_n - z_j|$ and $(z_n)$ is a Cauchy sequence. Hence, these sequences have limit points in $\mathbb{R}$: say $a_n \to u$ and $b_n \to v$. Then given $\varepsilon > 0$ there is $N \in \mathbb{N}$ with $|a_n - u| < \varepsilon/2$ and $|b_n - v| < \varepsilon/2$ for all $n > N$. With $w = u + iv$ we then have

$$|z_n - w| = |a_n + ib_n - u - iv|$$

$$\leqq |a_n - u| + |ib_n - iv|$$

$$< \frac{\varepsilon}{2} + \frac{\varepsilon}{2} = \varepsilon, \text{ for all } n > N$$

Therefore $w$ is a limit point of $(z_n)$, as required. □

Because of this property the usual theorems about continuous functions and so on hold also in $\mathbb{C}$. So, for example, if $M$ is a compact (= closed bounded) subset of $\mathbb{C}$, then every continuous function $f : M \to \mathbb{R}$ takes a minimum and a maximum value in $M$ (for a proof, see analysis books, for example W. Maak: *Differential and Integral Calculus*, [47] Section 30). This is the essential foundation for the first part of the proof of the fundamental theorem of algebra that we give in the next Section.

If we ignore the metric and consider only the topology, then it follows that fundamental topological properties of $\mathbb{R}$ carry over to $\mathbb{R}^2$ in the product topology. So, for example, $\mathbb{R}^2$ is locally compact, connected and has a countable base for the open sets (compare for example, L. Führer: *General Topology with Applications* [21]). These properties display $\mathbb{R}^2$ as a 'backdrop' for $\mathbb{C}$. For complex analysis (= Function Theory), the concept of differentiability for complex functions is both decisive and fundamental, but it is *not* simply a product concept. Such product concepts do occur in real n-dimensional analysis — for example, with partial derivatives. Complex differentiability with its far-reaching consequences, for example the Cauchy Integral Theorem, rests upon a theory of a wholly different character.

The topological completeness of $\mathbb{C}$ says that $\mathbb{C}$ is closed under the operation of taking limits of Cauchy sequences. Next we show that it is also closed under the operation of 'taking zeros' of polynomials.

## 2. The Algebraic Closure of $\mathbb{C}$.

This closure is expressed via the so-called Fundamental Theorem of Algebra (FA):

(FA) *Each polynomial* $p$ *with complex coefficients has a zero in* $\mathbb{C}$.

It is usual to describe $p$ by writing: $p \in \mathbb{C}[t]$. From this, one can formulate variants:

(FA') *Each polynomial* $p \in \mathbb{C}[t]$ *splits into linear factors, which means: there exist* $\alpha_1,\ldots,\alpha_n \in \mathbb{C}$ *(not necessarily distinct) with*

$$\begin{aligned} p(t) &= a_n t^n + \ldots + a_o \\ &= a_n(t - \alpha_1)(t - \alpha_2) \ \ldots \ (t - \alpha_n) \ . \end{aligned}$$

(FA) implies (FA'), because we can make zeros of polynomials correspond to linear factors. Also (FA') implies (FA), since any of $\alpha_1,\ldots,\alpha_n$ will do for one of the required zeros.

Another variant of (FA) is:

(FA") *The only irreducible polynomials in* $\mathbb{C}[t]$ *are the linear polynomials*

(For irreducible polynomials and their splitting into prime factors over a field K, see for example, Lang, [42], Chapter IV, Section 3.

Before the proof of the theorem, we observe a consequence for real polynomials which Gauss chose as the title for his dissertation.

1. **COROLLARY**. *Every real polynomial* $f \in \mathbb{R}[t]$ *can be factorised completely into quadratic and linear factors.*

This means that the only irreducible polynomials in $\mathbb{R}[t]$ with leading coefficient 1 are the

linear: $t - a$ with $a \in \mathbb{R}$, and the

quadratic: $t^2 + at + b$ with $a, b \in \mathbb{R}$ and $b-(a/2)^2 > 0$.

(Also from this corollary, one can obtain (FA) again and so we have a further variant.) We shall use (FA') to prove the corollary.

*PROOF of the corollary.* Suppose $f(t) = a_n t^n + \ldots + a_o$ with $a_i \in \mathbb{R}$. We can treat $t$ as a complex number, because $f(t) \in \mathbb{C}[t]$ and has zeros in $\mathbb{C}$. If $\alpha$ is a zero, then so also is $\bar{\alpha}$ is because

$$\begin{aligned} f(\bar{\alpha}) &= a_n \bar{\alpha}^n + \ldots + a_o, \quad (\text{since } \bar{a}_i = a_i) \\ &= \bar{a}_n \bar{\alpha}^n + \ldots + \bar{a}_o \\ &= \overline{f(\alpha)} = \bar{0} = 0 . \end{aligned}$$

It is possible that $\bar{\alpha} = \alpha$, and in that case $\alpha$ is real. Using (FA') we derive for the splitting of f(t) into linear factors in $\mathbb{C}[t]$:

$$\begin{aligned} f(t) = a_n &(t - \alpha_1)(t - \bar{\alpha}_1) \ldots \\ &(t - \alpha_r)(t - \bar{\alpha}_r)\cdot(t - \beta_1) \ldots (t - \beta_s) \end{aligned}$$

with $\alpha_i \notin \mathbb{R}$ and $\beta_i \in \mathbb{R}$. But

$$\begin{aligned} (t - \alpha_1)\cdot(t - \bar{\alpha}_1 &= t^2 - (2 \text{ real part } \alpha_1)\cdot t + \alpha_1 \bar{\alpha}_1 \\ &= t^2 + a_1 t + b_1 \text{ with } a_1, b_1 \in \mathbb{R} . \end{aligned}$$

Hence,

$$\begin{aligned} f(t) = a_n &(t^2 + a_1 t + b_1) \ldots \\ &(t^2 + a_r t + b_r)\cdot(t - \beta_1) \ldots (t - \beta_s) \end{aligned}$$

which is the required splitting in $\mathbb{R}[t]$. □

For the fundamental theorem of algebra, there are many proofs, and new variants keep appearing. Also, Gauss himself repeatedly returned to the them. In the course of time, the arguments have been ever further simplified. For literature-citations, see for example H. Leinfelder: "The Fundamental Theorem of Algebra" [44], pp.187÷194, or J.L. Brenner and R.C. Lyndon: "Proof of the Fundamental Theorem of Algebra" [9], pp.253–256.

The proof we give splits into two parts. First it will be shown that for a polynomial $f$, the real valued function $g: z \to |f(z)|$ takes a minimum; and then at this minimum, $g$ can only be zero.

<u>Part 1</u>. We first need to prove a lemma concerning the behaviour of polynomials for large $|z|$.

<u>LEMMA</u>. *For all non-constant polynomials* $f$ *and all* $0 < K \in \mathbb{R}$ *there exists* $R \in \mathbb{R}$ *such that if* $|z| > R$, *then* $|f(z)| > K$.

(In other words: $\lim_{|z| \to \infty} |f(z)| = \infty$ ) .

<u>*PROOF*</u>. Let $f(z) = a_n z^n + \ldots + a_o$ with $n > 0$ and $a_n \neq 0$. We have for $z \neq 0$:

$$f(z) = z^n \cdot (a_n + \ldots + \frac{1}{z^n} a_o)$$

$$= z^n(a_n + \frac{1}{z} \cdot (a_{n-1} + \ldots + \frac{1}{z^{n-1}} a_o))$$

$$= z^n(a_n + w), \quad \text{say.}$$

Since $n > 0$, we can choose the "radius" $R_1$ so large that $|w| < \frac{1}{2}|a_n|$ for $|z| > R_1$. Then because $|v + w| \geqslant |v| - |w|$, we have the inequalities

$$|f(z)| = |z^n| \cdot |a_n + w|$$

$$\geqq |z^n| \cdot \big|\, |a_n| - |w| \,\big|$$

$$\geqq |z^n| \cdot \big|\, |a_n| - \tfrac{1}{2}|a_n| \,\big| \quad \text{because} \quad |w| \leqq \tfrac{1}{2}|a_n|$$

$$\geqq \tfrac{1}{2}|z^n| \cdot |a_n| .$$

For any given constant $K$, we now choose $R \geqslant R_1$ so that $R > K \cdot 2/|a_n|$. Then we have

$$|f(z)| \geqq \tfrac{1}{2}|z^n| \cdot |a_n| > K \cdot R^{n-1} \quad \text{for all } z \text{ with } |z| > R .$$

If we also take $R > 1$, then since $n > 1$, we have shown: given $K > 0$, there is an $R$ with $|f(z)| > K$ for all $z$ with $|z| \geqslant R$, as required. □

Exercise 7.
(a) If $a, b \in \mathbb{C}$, show that, if $|z| \geq 1$ then

$$\left|\frac{a}{z} + \frac{b}{z^2}\right| \leq (|a| + |b|)/|z| .$$

(b) Find a number $R_1$ as in the proof, when

$$f(z) = iz^5 + (2+3i)z^4 - 5z^3 + (1-i)z^2 + 7z - 9i$$

Returning to a non-constant polynomial $f$ as in the Lemma, let us obtain $R$ by choosing $K = 2|a_0|$. Then if $M$ is the closed disc consisting of all $z$ with $|z| \leq R$, we have a continuous function $g: z \to |f(z)|$ that maps $M$ into $\mathbb{R}$. As remarked earlier, there exists $\alpha \in M$ such that $g(\alpha)$ is the minimum value of $g$ on $M$. In particular, then, $g(\alpha) \leq g(z)$ for every $z$ in $M$, and

$$g(\alpha) \leq g(0) = |f(0)| = |a_0| .$$

Outside $M$, $g(z) > 2|a_0|$ by the Lemma. Hence $g(\alpha) \leq g(z)$ for all $z \in \mathbb{C}$. Therefore $g(\alpha)$ is the absolute minimum of $g$ in $\mathbb{C}$, and we next show that $g(\alpha)$ is, in fact, zero.

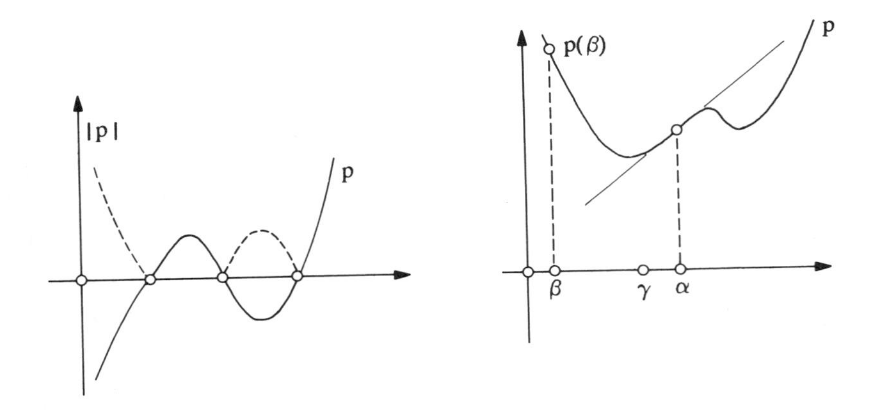


Fig.33

Part 2. We first give the idea of the proof by considering Fig.33, which applies to a real polynomial $p \in \mathbb{R}[t]$. First we see that $|p|$ becomes minimal at the zeros of $p$. The second picture shows how we eventually conclude the converse of this assertion. For, at the point $\alpha$, $p(\alpha)$ is not minimal, so if the tangent is not horizontal, we can move in a suitable direction along the tangent and find a $\gamma$ with $|f(\gamma)| < |f(\alpha)|$.

Unfortunately, our sketch also shows that even with a horizontal tangent, one cannot guarantee a zero among the real numbers. In the complex numbers, however, this same process does in fact lead us to a zero, and this is the essential idea of the proof. Only two technical things remain to be observed:

(i) we should not move too far (say as far as $\beta$) from $\alpha$ along the tangent, so we shall need to specify a suitable step–length, which will be guaranteed by the following Lemma below:

(ii) we can easily find the complex version of the tangent by writing $t^j = (\alpha + (t-\alpha))^j$ and expanding $p$ in powers of $(t - \alpha)$. This gives

$$p(t) = c_n(t - \alpha)^n + \ldots + c_1(t - \alpha) + c_o$$

with $c_o = p(\alpha)$ and linear part $c_1(t - \alpha) + c_o$ as the "tangent". However, it could happen that $c_1 = 0$, in which case we have to find the appropriate direction for moving away from $\alpha$ by using the $j^{th}$ roots of a complex number.

By contrast with Part 1, we begin with a lemma that describes the growth of polynomials for small $|z|$.

<u>LEMMA</u>. *Suppose* $g(z)$ *is a polynomial of the form* $g(z) = c_n z^n + \ldots + c_j z^j + c_o$, ($c_j \neq 0$ *and* $n \geq j \geq 1$). *Thus* $g(z) = h(z) + c_j z^j + c_o$. *Then there exists* $R > 0$ *such that, for all* $r$ *with* $0 \leq r \leq R$ *we have*

$$|h(z)| = |c_n z^n + \ldots + c_{j+1} z^{j+1}| < \tfrac{1}{2}|c_j z^j| .$$

<u>*PROOF*</u>. We choose $R < 1$ so small, that if $|z| < R$ then

$$|c_n z^{n-j} + \ldots + c_{j+1} z| < \tfrac{1}{2}|c_j|$$

It then follows that for every $0 < |z| < R$ we have the required inequality:

$$|h(z)| < \tfrac{1}{2}|c_j| \cdot |z^j| \qquad \square$$

<u>Exercise 8</u>.
(a) Show that, if $a, b \in \mathbb{C}$ and $|z| < 1$ then

$$|az + bz^2| < |z| \cdot (|a| + |b|) .$$

(b) Calculate the number $R$ in the proof, when $n = 5$, $j = 2$, and $c_5 = 1 + i$, $c_4 = -7$, $c_3 = 2 + 5i$.

After these preparations, we now come to the central part of the proof of the Fundamental Theorem. For this, suppose that f is not constant.

<u>Claim</u>. *If* $|f(\alpha)|$ *is minimal, then* $f(\alpha) = 0$.

We prove the contrapositive:

*Suppose* $f(\alpha) \neq 0$, *then* $|f(\alpha)|$ *is not minimal.*

We first expand $f(z)$ in powers of $(z - \alpha)$, to give

$$f(z) = c_n(z - \alpha)^n + \ldots + c_1(z-\alpha) + c_o .$$

Since $f(\alpha) = c_o$ then $c_o \neq 0$. In order to see how the basic idea depends on complex numbers we assume first that $c_1 \neq 0$. Now choose $z$ so that $z - \alpha = -\frac{1}{2}c_o/c_1$. Then $c_1(z - \alpha) + c_o = \frac{1}{2}c_o$. But then we could lessen the modulus of $f(z)$ if we could also be sure that $h(z) = c_n(z-\alpha)^n + \ldots + c_2(z - \alpha)^2$ had modulus $< \frac{1}{2}c_o$. We deal with that by using the above lemma concerning small $|z|$ (here naturally for small $|z - \alpha|$).

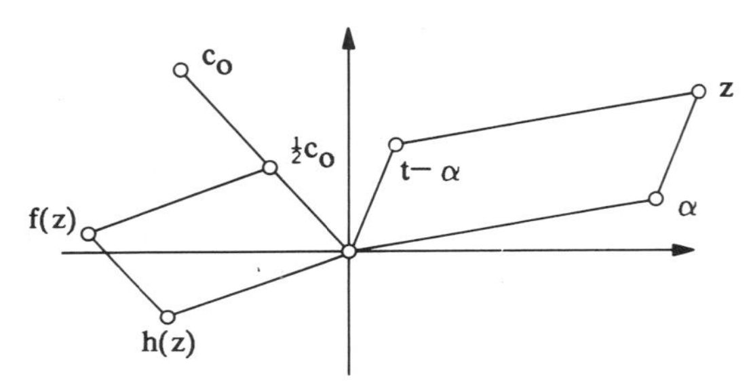


Fig.34

In the general case, we are not sure that $c_1 \neq 0$, but we do have

$$f(z) = c_n (z - \alpha)^n + \ldots + c_j(z - \alpha)^j + c_o$$

with $c_j \neq 0$, because $f$ is not constant. In order to get the same effect as previously, we choose $z$ so that

$$(z - \alpha)^j = w^j = -\frac{c_o}{c_j}$$

and it is here that we use the already known existence of $j^{th}$ roots in $\mathbb{C}$. Further, we have $w \neq 0$, because $c_o \neq 0$. (Observe: that $w = 0$ iff $z = \alpha$.) For the intuitive meaning of $w$, $tw$, etc, see Fig.35.

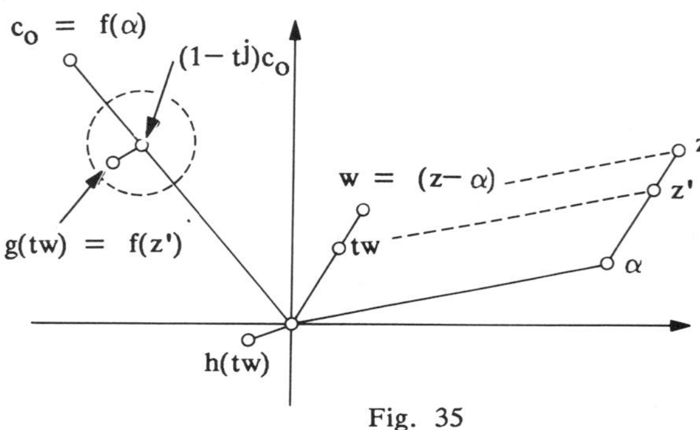


Fig. 35

In order to be able to apply the Lemma above, we consider $tw$ with $t$ real, $0 \leqq t < 1$; then

$$\begin{aligned} g(tw) &= c_n\cdot(tw)^n + \ldots + c_{j+1}\cdot(tw)^{j+1} + c_j\cdot(tw)^j + c_o \\ &= h(tw) + t^j\cdot c_j w^j + c_o \\ &= h(tw) - t^j\cdot c_o + c_o \qquad \text{by choice of } w \\ &= h(tw) + (1 - t^j)\cdot c_o \; . \end{aligned}$$

Writing $z' = tw + \alpha$ we have

$$\begin{aligned} f(z') &= c_n\cdot(z' - \alpha)^n + \ldots + c_o \\ &= c_n\cdot(tw)^n + \ldots + c_o \end{aligned}$$

whence

$$\begin{aligned} |f(z')| = |g(tw)| &\leqq |h(tw)| + |1 - t^j|\cdot|c_o| \\ &\leqq |h(tw)| - t^j\cdot|c_o| + |c_o| \; . \end{aligned}$$

By the lemma, we can now choose $0 < t < 1$ so small that $|h(tw)| < \frac{1}{2}|c_j\cdot(tw)^j|$. Therefore

$$\begin{aligned} f(z') &< \tfrac{1}{2}|c_j\cdot(tw)^j| - t^j\cdot|c_o| + |c_o| \\ &< \tfrac{1}{2}t^j\cdot|c_o| - t^j\cdot|c_o| + |c_o| \\ &< |c_o| - \tfrac{1}{2}t^j\cdot|c_o| \\ &< |c_o| = |f(\alpha)| \; . \end{aligned}$$

This proves the fundamental theorem. □

Let us look again at the basic ideas of the proof. First we showed the existence of a minimum for $|f(z)|$ using topological considerations. In the second part, we used the existence of $j^{th}$ roots of complex numbers in order that, given $|f(\alpha)| \neq 0$, we could find a $z'$ (near $\alpha$) with $|f(z')| < |f(\alpha)|$. From this we deduced that the minimum modulus — which we know exists — can only be zero.

As remarked previously, there are many proofs of the fundamental theorem, that use other basic ideas. (See for example, Armitage and Griffiths [2] p.171).

<u>Exercise 9</u>. (For the reader who has taken a course on Functions of a Complex Variable). Relate Part 2 of the proof to such ideas as the maximum modulus principle, and the maximum principle for harmonic functions.

## 4D THE FUNDAMENTAL THEOREM AS AN ASSERTION ABOUT EXTENSION FIELDS OF $\mathbb{R}$.

We have seen that through multiplication of complex numbers, $(\mathbb{R}^2, +, \cdot)$ is turned into the field $(\mathbb{C}, +, \cdot)$ which is a two-dimensional vector space over $\mathbb{R}$ and contains $\mathbb{R}$ as a sub-field. This suggests the question whether it is possible also to define in $\mathbb{R}^n$ multiplications of the same kind. Following his definition of $\mathbb{C}$, Hamilton worried about this problem for a whole year along these lines, but was not able to find corresponding formulae.

In the next chapter, we shall go explicitly into the description of the quaternions that he discovered in this connection, but we shall first give the grounds for showing why his search was truly in vain. For a better understanding of the nature of the problem, we look first at an example of its relevance to $\mathbb{Q}$ in place of $\mathbb{R}$. One would therefore like to define multiplication $*$ in such a way that $(\mathbb{Q}^n, +, *)$ becomes a field.

Now, using vector addition, and multiplication with rational scalars, we naturally have an n-dimensional vector space over $\mathbb{Q}$. So, the required structure can easily be arranged, with the help of an irreducible polynomial $f \in \mathbb{Q}$ of degree $n$; for example if $f(x) = x^n - p$ where $p$ is a prime number, then by Eisenstein's criterion, $f(x)$ is irreducible over $\mathbb{Q}$. Then $\mathbb{Q}[x]/(f)$, the quotient of $\mathbb{Q}[x]$ by the ideal $(f)$, is an n-dimensional extension field of $\mathbb{Q}$ as required. This is a residue class ring of the type we met in Section 4A; thus we calculate here "modulo f" by putting $x^n = p$; so we can write $q = p^{1/n}$ for (the residue class of) $x$ and we obtain the elements of $\mathbb{Q}[x]/(f)$ in the form

$$a_{n-1} \cdot q^{n-1} + \ldots + a_1 \cdot q^1 + a_o$$

with $a_i \in \mathbb{Q}$. Thus we consider $q^k$ as the $k^{th}$ basis vector. The multiplication $*$ of these vectors $(a_{n-1}, \ldots, a_o)$ follows by the usual rules for multiplying real numbers and polynomials (indeed one can then regard the field $\mathbb{Q}^n, +, *)$ thus constructed as a sub-field of $\mathbb{R}$). These and the following considerations can be worked out wholly along the lines of Lang, [42], Chapter IV, Section 1. Admittedly, we could not directly cite the theorems of Lang, because he confines his theory to sub-fields of $\mathbb{C}$.

Now to our original question concerning the spaces $\mathbb{R}^n$. Let us assume that we have found a multiplication $*$ such that $(\mathbb{R}^n, +, *) =$ K is an extension field of $\mathbb{R}$. Naturally the old addition should be retained so that K becomes a vector space over $\mathbb{R}$ just as we had previously for $\mathbb{Q}$. We choose a vector $\alpha \in K$ with $\alpha \notin \mathbb{R}$, where we assume $n \geqslant 2$; and we consider the powers $\alpha^o = 1$, $\alpha$, $\alpha^2 = \alpha * \alpha, \ldots, \alpha^n = \alpha * \alpha^{n-1}$. In the n-dimensional space $\mathbb{R}^n$ these n+1 vectors are linearly dependent. There exists therefore real numbers $a_i$, not all zero with

$$a_n\alpha^n + \ldots + a_1\alpha + a_o = 0 .$$

Thus $\alpha$ is a zero of the polynomial $f \in \mathbb{R}[x]$, where

$$f(x) = a_n x^n + \ldots + a_1 x + a_o$$

which is not the zero polynomial. By the theorem of Gauss mentioned above, we can split $f$ into its quadratic or linear irreducible factors and then $\alpha$ must be a zero of one of these factors. The zeros of the linear factors belong to $\mathbb{R}$, which we had expressly excluded for $\alpha$. Thus there is a quadratic polynomial $p$ with $p(\alpha) = 0$.

We can therefore adjoin $\alpha$ to $\mathbb{R}$ to form an intermediate field $\mathbb{R}(\alpha)$ such that

$$\mathbb{R}(\alpha) \cong \mathbb{R}[x]/(p) \cong \mathbb{C}$$

as we did in Section 4A. This gives us $\mathbb{C}$ as a sub÷field of K. Hence we may use complex numbers as scalar multipliers for the vectors of K, and then K becomes a vector space over $\mathbb{C}$. The dimension of K over $\mathbb{C}$ is half as large as the dimension of K over $\mathbb{R}$. Therefore $\dim(K : \mathbb{C}) = r = n/2$.

If $n$ were larger than 2, we could choose some $\beta \in K$ with $\beta \notin \mathbb{C}$ and operate with $\beta$ as we did previously with $\alpha$. Out of this we get a polynomial $g \in \mathbb{C}(z)$ with zero $\beta$,

$$g(z) = c_r z^r + \ldots + c_o$$

The zeros of $g$ belong to $\mathbb{C}$, by the fundamental theorem of algebra; therefore $\beta \in \mathbb{C}$, yet $\beta \notin \mathbb{C}$. This contradiction shows that $n \not> 2$, so $\mathbb{R}^n = \mathbb{C} = \mathbb{R}^2$. Thus we have established the following:

*Consequence of the Fundamental Theorem* (Frobenius 1878):
**If $n \geqslant 3$ it is impossible to define a multiplication so that $(\mathbb{R}^n, +, *)$ becomes an n– dimensional field extension of $\mathbb{R}$.**

Our considerations in Section 4A also show that for $n = 2$ there is only one single possibility for such a multiplication – at least to within isomorphism.

*REMARK*: Instead of the double step from $\alpha$ to $\beta$ given in the last proof, we could use the so–called "Theorem of the primitive element", which guarantees a $\gamma \in K$, with $K = \mathbb{R}(\gamma)$ – and then consider the minimal polynomial of $\gamma$. For a direct elementary proof, see "When is $\mathbb{R}^n$ a field?" by R.M. Young, in Math. Gazette (72), 1988, pp.128÷9.

From the above Theorem of Frobenius, we shall now obtain the fundamental theorem again. Let us assume $f \in \mathbb{C}[z]$ is a polynomial of the $n^{th}$ degree, which without loss of generality we may assume to be irreducible. By the algebraic process used before, we derive an extension field $K = \mathbb{C}[z]/(f)$ of dimension $n$ over $\mathbb{C}$. Then K would be an extension field of $\mathbb{R}$ of dimension $2n$. But now, the Theorem of Frobenius tells us that $2n = 2$, so $n = 1$ and $f$ is linear. This means that the only irreducible polynomials in $\mathbb{C}[z]$ are the linear ones. Therefore we have obtained the further variant of the fundamental theorem:

(FA"') *To within isomorphism, the only proper finite dimensional field extension of* $\mathbb{R}$ *is* $\mathbb{C}$.

At the same time this gives us a characterization of $\mathbb{C}$.

After this result, if we are looking for number-domains which extend $\mathbb{C}$, there remain to us two possibilities. Either we deny ourselves the finite dimension over $\mathbb{R}$ – which happens with the non-standard numbers in Chapter 7 – or we must relax one or other of the field axioms, and that leads us to Hamilton's quaternions in the next Chapter.

# 5

# Quaternions

## INTRODUCTION.

Quaternions lie in none of the usual number domains and are therefore introduced here from the beginning. In the form of certain $4 \times 4$ matrices, they allow an uncomplicated definition and development of their properties, and then they can be considered as elements of $\mathbb{R}^4$ with a multiplication carried over from matrices. In Section 5E we show how to use quaternions for representing the rotations of $\mathbb{R}^3$, and we consider the group–theoretic questions which arise in that connection. A theorem of Frobenius is given in Section 5F, to characterizes the quaternions as the only proper, finite dimensional, skew field over $\mathbb{R}$. It is natural then to mention the Cayley numbers, but these are discussed only briefly, whereas we give more space to the interesting cross–connections between the themes of this chapter and geometry.

## 5A PRELIMINARY REMARKS.

As we saw in the previous chapter, Hamilton's efforts to find a multiplication in $\mathbb{R}^3$ could never have been successful. In spite of that, he made a great discovery; it was not in $\mathbb{R}^3$, but in $\mathbb{R}^4$ that one could introduce a meaningful multiplication which moreover was connected to the rotations in $\mathbb{R}^3$. In short, if we consider "four–variable expressions" of the form:

$$a_0 + a_1 i + a_2 j + a_3 k$$

and multiply them using the distributive and associative laws and Hamilton's multiplication table for i, j, k given by

$$i^2 = j^2 = k^2 = -1$$

$$ij = -ji = k$$

$$jk = -kj = i$$

$$ki = -ik = j ,$$

then we obtain a multiplication of the sort we are looking for. Of the historical circumstances of Hamilton's discovery, we shall say a little more at the end of Section 5C.

Exercise 1.

(a) Using the above rules, show that $(1+i)\cdot(2j+3k) = -j+5k$.

(b) Calculate $(2j+3k)\cdot(1+i)$, $(1+i+2j+3k)\cdot(-5+i-3j+k)$, and the squares of these products.

(c) Show that if $A = a+bi+cj+dk$, then $A^2 = (a^2-b^2-c^2-d^2) + 2a(b+c+d)$ and hence that $A^2 - 2aA + N(A)\cdot 1 = 0$, where $N(A) = a^2 + b^2 + c^2 + d^2$. Investigate the solutions (if any) of the equations $A^2 = 1 + i + 2j + 3k$, $A^2 = -5 + i - 3j + k$

(d) Show that given $B = b_o + b_1 i + b_2 j + b_3 k$, the equation $A^2 = B$ has no solution for $A$ if $1 + b_o < 0$.

Some calculations can be saved if we introduce the quaternions as a special type of $4 \times 4$ matrices. This occurs in a close analogy to the matrix definition in Section 4A of the complex numbers by which

$$i = \begin{bmatrix} 0 & -1 \\ 1 & 0 \end{bmatrix} \quad \text{and} \quad a + bi = \begin{bmatrix} a & -b \\ b & a \end{bmatrix}.$$

In the present case we put

$$I = \begin{bmatrix} 0 & -1 & 0 & 0 \\ 1 & 0 & 0 & 0 \\ 0 & 0 & 0 & -1 \\ 0 & 0 & 1 & 0 \end{bmatrix}, \quad J = \begin{bmatrix} 0 & 0 & -1 & 0 \\ 0 & 0 & 0 & 1 \\ 1 & 0 & 0 & 0 \\ 0 & -1 & 0 & 0 \end{bmatrix}, \quad K = \begin{bmatrix} 0 & 0 & 0 & -1 \\ 0 & 0 & -1 & 0 \\ 0 & 1 & 0 & 0 \\ 1 & 0 & 0 & 0 \end{bmatrix}$$

Further, let

$$E = \begin{bmatrix} 1 & 0 & 0 & 0 \\ 0 & 1 & 0 & 0 \\ 0 & 0 & 1 & 0 \\ 0 & 0 & 0 & 1 \end{bmatrix}$$

be the $4 \times 4$ unit matrix. A simple calculation now shows that these matrices I, J, K satisfy Hamilton's rules, given above, for i, j, k provided we replace 1 and $-1$ by E and $-E$.

*REMARK.* The matrices $\pm E$, $\pm I$, $\pm J$, $\pm K$ form an interesting group of 8 elements, the so-called **quaternion group**.

By a quaternion, we now mean a matrix A of the form $a_0E + a_1I + a_2J + a_3K$, for which $a_0, a_1, a_2, a_3 \in \mathbb{R}$.

We start the indexing with zero because (as we show later) the quaternions of the special form $a_1I + a_2J + a_3K$ can be identified with vectors of $\mathbb{R}^3$, and we wish to have a compatible notation.

If we write out the quaternion $A = a_0E + a_1I + a_2J + a_3K$ in full, we have

$$A = \begin{bmatrix} a_0 & -a_1 & -a_2 & -a_3 \\ a_1 & a_0 & -a_3 & a_2 \\ a_2 & a_3 & a_0 & -a_1 \\ a_3 & -a_2 & a_1 & a_0 \end{bmatrix}$$

Let H be the set of these matrices. (We use H because of Hamilton, $\mathbb{Q}$ being already reserved for the rational numbers).

PROPOSITION. *With the matrix operations of addition and multiplication,* $(H, +, \cdot)$ *forms a ring.*

*PROOF.* Matrix addition is carried out "component-wise". As a quaternion-matrix is completely determined by its *first* column, we have at once $(H, +) \approx (\mathbb{R}^4, +)$. The addition corresponds simply to vector addition in $\mathbb{R}^4$.

For multiplication, we shall first show that the product of two quaternion-matrices is a third. Let A be as before, and $B = b_0E + b_1I + b_2J + b_3K$. The product

$$(a_0E + a_1I + a_2J + a_3K) \cdot (b_0E + b_1I + b_2J + b_3K)$$

can be calculated in the ring of all $4 \times 4$ matrices, using the distributive law. If we recall also the rules $I^2 = -E, \ldots, IJ = K, \ldots$ etc, then we obtain a product of the form

$$AB = c_0E + c_1I + c_2J + c_3K$$

which is again a qauternion-matrix, with first column

$$\begin{bmatrix} c_o \\ c_1 \\ c_2 \\ c_3 \end{bmatrix} = \begin{bmatrix} a_ob_o - a_1b_1 - a_2b_2 - a_3b_3 \\ a_ob_1 + a_1b_o - a_2b_3 - a_3b_2 \\ a_ob_2 - a_1b_3 + a_2b_o + a_3b_1 \\ a_ob_2 - a_1b_3 + a_2b_o + a_3b_1 \end{bmatrix} \qquad \text{(i)}$$

This shows that the set **H** of quaternions is multiplicatively closed, so the distributive and associative laws carry over to **H** because they hold in the matrix ring. The neutral element is the unit matrix
$E = 1E + OI + OJ + OK \in \mathbf{H}$. This proves the Proposition. □

It should not be forgotten that **quaternion multiplication is not commutative.**

*Simplifying the notation.*
As we have already remarked, each quaternion is already fixed as soon as we know its first column. Therefore we can write, more briefly:

$$A = \begin{bmatrix} a_o \\ a_1 \\ a_2 \\ a_3 \end{bmatrix}, \quad E = \begin{bmatrix} 1 \\ 0 \\ 0 \\ 0 \end{bmatrix}, \quad I = \begin{bmatrix} 0 \\ 1 \\ 0 \\ 0 \end{bmatrix}, \quad \ldots$$

We can now completely forget the matrices, and define multiplication by

$$\begin{bmatrix} a_o \\ a_1 \\ a_2 \\ a_3 \end{bmatrix} \begin{bmatrix} b_o \\ b_1 \\ b_2 \\ b_3 \end{bmatrix} = \begin{bmatrix} c_o \\ c_1 \\ c_2 \\ c_3 \end{bmatrix}$$

with the $c_i$ given in (i) above. The quaternions are then just the points of $\mathbb{R}^4$ (written as column vectors). This is understood in the sense analogous to the relation between $\mathbb{C}$ and $\mathbb{R}^2$. We thus retain the original form of writing elements of **H** as $a_o + a_1I + a_2J\ a_3K$, with the corresponding rules of calculation.

Only in cases of possible ambiguity shall we revert to writing quaternions as matrices, and otherwise we work solely with this column representation. However, for convenience in printing, it is preferable to use rows rather than columns, so we use the operation † of matrix transposition to write, for example, the left hand side of equation (i) as $(c_o, c_1, c_2, c_3)^\dagger$.

In this mode we assign to each quaternion A its **conjugate** $A^*$; thus *if* $A = (a_0, a_1, a_2, a_3)^\dagger$ *then* $A^* = (a_0, -a_1, -a_2, -a_3)^\dagger$

Inspection of the corresponding square matrices shows that $A^*$ is the transposed matrix $A^\dagger$, just as with $z$ and $\bar{z}$. The analogue of the equation $z\bar{z} = a^2 + b^2$ in $\mathbb{C}$ is, in $\mathbf{H}$

$$AA^* = (a_0{}^2 + a_1{}^2 + a_2{}^2 + a_3{}^2, 0, 0, 0)^\dagger = A^*A.$$

Hence A and $A^*$ *commute*: their order of multiplication can be interchanged. Further, we have:

$$AA^* = 0 \quad \text{iff} \quad A = 0 .$$

It is usual to call the real number

$$N(A) = a_0{}^2 + a_1{}^2 + a_2{}^2 + a_3{}^2 \in \mathbb{R}$$

the **Norm** of A. The modulus $\beta(A) = \surd N(A)$ is then nothing other than the length of the vector $(a_0, a_1, a_2, a_3)^\dagger$ in $\mathbb{R}^4$.

Next, we can calculate the inverse of $A \neq 0$. We have

$$\left(\frac{1}{N(A)} A^*\right)\cdot A = E , \quad \text{so} \quad A^{-1} = \frac{1}{N(A)}\cdot A^* ,$$

and hence $A^{-1}$ exists in $\mathbf{H}$, and is therefore a quaternion.

We summarise the results obtained so far in a theorem. First, we define $\mathbf{H}$ to be a **skew-field** (i.e. in $\mathbf{H}$, all the axioms of a field hold, except for the commutative law of multiplication). (Many authors use the term 'Division algebra' for a skew-field: see also Section 5F below.) Then we have proved:

THEOREM. $(\mathbf{H}, +, \cdot)$ *is a skew-field.* □

Just as in $\mathbb{C}$, norm and modulus in $\mathbf{H}$ are multiplicative. We have:

PROPOSITION. $N(AB) = N(A)\cdot N(B)$, *and* $\beta(AB = \beta(A)\cdot\beta(B)$.

*PROOF*. Going back to the matrices, we have

$$\begin{aligned}
(AB)\cdot(AB)^* &= (AB)\cdot(AB)^\dagger && \text{(transposed matrix)}\\
&= (AB)\cdot(B^\dagger A^\dagger) && \text{(rule for transposition)}\\
&= A(BB^\dagger)\cdot A^\dagger\\
&= A(BB^*)\cdot A^*\\
&= A(bE)\cdot A^* && \text{where } bE = BB^* = N(B)E\\
&= AA^*\cdot(bE) && \text{(E and scalars commute with matrices)}\\
&= AA^*\cdot BB^*
\end{aligned}$$

Hence $N(AB) = N(A)N(B)$, and by taking square roots we have established the Proposition. □

Now suppose that $A, B, C$ are such that

$$A = (a_o. a_1, a_2, a_3)^\dagger, \; B = (b_o, b_1, b_2, b_3)^\dagger,$$

and
$$AB = C = (c_o, c_1, c_2, c_3)^\dagger .$$

Then we obtain at once from the proposition the pretty formula for multiplying sums of squares:

$$(a_o{}^2 + a_1{}^2 + a_2{}^2 + a_3{}^2)\cdot(b_o{}^2 + b_1{}^2 + b_2{}^2 + b_3{}^2)$$

$$= c_o{}^2 + c_1{}^2 + c_2{}^2 + c_3{}^2$$

and the $c_i$ are formed from the a's and b's according to (i) above. As a numerical example, take $A = (1, 2, 3, 4)^\dagger$, $B = (5, 6, 7, 8)^\dagger$; then by (i), $C = (-60, 12, 30, 24)^\dagger$, so the equation $N(AB) = N(A)\cdot N(B)$ gives

$$(1^2 + 2^2 + 3^2 + 4^2)\ (5^2 + 6^2 + 7^2 + 8^2)$$

$$= 60^2 + 12^2 + 30^2 + 24^2 = 5220 .$$

Since $N(AB) = N(BA)$, then $(5, 6, 7, 8)^\dagger(1, 2, 3, 4)^\dagger$ gives us a second splitting of 5220 into a sum of four squares.

Exercise 2.

(a) Calculate this second splitting.

(b) Calculate $(1, 2, 3, 4)^\dagger\ (4, 3, 2, 1)^\dagger$ to express 900 as a sum of four squares

(c) If $A = (1, 2, 3, 4)^\dagger$ calculate $A^{-1}$.

(d) Prove that $(AB)^* = B^*A^*$ in two ways.

For such "integral" quaternions, we can formulate these deductions by saying that in $\mathbb{N}_0$, the set $M$ of all sums of four squares is *multiplicatively closed.* This is an important corner-stone for the proof of Lagrange's theorem: every natural number is the sum of four squares (so $M = \mathbb{N}_0$). For a proof of this theorem, see Hardy and Wright [27] Chap. 2D, where the reader will also find further material on the divisibility properties of integral quaternions; that theory does not work too well, owing to the failure of commutativity in $\mathbf{H}$.

## 5B EMBEDDING $\mathbb{R}$ AND $\mathbb{C}$ IN $\mathbf{H}$.

The real numbers can be isomorphically embedded in $\mathbf{H}$, through the mapping

$$a_o \to (a_o, 0, 0, 0)^\dagger = a_o E .$$

We can therefore — just as we do with $\mathbb{C}$ — speak of the real axis in $\mathbf{H}$, but it is more usual to call it the 'axis of scalars'. Instead of $a_oE$ we shall often write simply $a_o$, just as with the complex numbers. Without reference to the coordinate representation, we can pick out $\mathbb{R}$ from $\mathbf{H}$ in a 'structural' way:

THEOREM. *The set of all* $A \in \mathbf{H}$, *such that*

$$\textit{for all } X \in \mathbf{H}, \quad AX = XA$$

*is the set of quaternions* $a_oE$. In other words:

*The centre of* $\mathbf{H}$ *is* $\mathbb{R}$.

(Here, the **centre** is defined as for a group; it consists of these elements that commute with every element.)

*PROOF*. (a) That a quaternion $a_oE$ commutes with every quaternion, can be verified either by use of the multiplication formula (i), or more simply by going back to the matrices: the diagonal matrix $a_oE$ commutes with every $4 \times 4$ matrix.
(b) Suppose $A$ commutes with all quaternions. We test $A$ with I, J, K, so

$$AI = \begin{bmatrix} a_o \\ a_1 \\ a_2 \\ a_3 \end{bmatrix} \begin{bmatrix} 0 \\ 1 \\ 0 \\ 0 \end{bmatrix} = \begin{bmatrix} -a_1 \\ a_o \\ a_3 \\ -a_2 \end{bmatrix} \qquad IA = \begin{bmatrix} 0 \\ 1 \\ 0 \\ 0 \end{bmatrix} \begin{bmatrix} a_o \\ a_1 \\ a_2 \\ a_3 \end{bmatrix} = \begin{bmatrix} -a_1 \\ a_o \\ -a_3 \\ a_2 \end{bmatrix}$$

Hence, since $A$ commutes with I, we must have $a_3 = -a_3$ and $a_2 = -a_2$, so $a_3 = a_2 = 0$. Testing with J, K similarly, we obtain $a_1 = 0$. Therefore $A = a_oE$. □

Exercise 3.
(a) Use the last argument to show that quaternions of the form $B = (0, 0, b_2, b_3)^\dagger$ are exactly those satisfying $BI = -IB$.

(b) Same exercise with I replaced by J.

We can work with the complex numbers in a manner similar to what we have just done with the reals. Recalling once more the representation

$$z = a + bi = \begin{bmatrix} a & -b \\ b & a \end{bmatrix}$$

we derive an embedding of $\mathbb{C}$ in $\mathbf{H}$ through the mapping

$$z = a + bi \to \begin{bmatrix} a & -b & 0 & 0 \\ b & a & 0 & 0 \\ 0 & 0 & a & -b \\ 0 & 0 & b & a \end{bmatrix} = aE + bI$$

Instead of using $I$ with $I^2 = -1$ we could operate similarly with $J$ or $K$ and derive two further embeddings through

$$a + bi \to aE + bJ$$

$$a + bi \to aE + bK$$

Any two vectors (= quaternions) $X$, $Y$ in $\mathbb{R}^4$ will generate a sub-space which we denote by $\langle X, Y\rangle$; in detail

$$\langle X, Y\rangle = \{rX + sY \mid r, s \in \mathbb{R}\}$$

In particular, we obtain the "coordinate planes"

$$\langle E, I\rangle, \ \langle E, J\rangle \text{ and } \langle E, K\rangle$$

and each is isomorphic to $\mathbb{C}$ if we use within it the quaternion multiplication. We have even more:

THEOREM. *Every (two-dimensional) plane in* $H = \mathbb{R}^4$ *which contains the real axis is (with quaternion multiplication) isomorphic to* $\mathbb{C}$.

*PROOF*. A plane of the given type can be written in the form

$$\langle E, X\rangle = \{aE + bX \mid a, b \in \mathbb{R}\}$$

where $X \neq aE$ for all $a \in \mathbb{R}$. As a two-dimensional sub-space, $\langle E, X\rangle$ is additively isomorphic to $(\mathbb{R}^2, +)$. Therefore we need only worry about the multiplication. To this end, we shall show first that $X$ satisfies a quadratic equation with *real* coefficients. We have, in fact,

$$\begin{aligned} 0 &= (X + X^*)\cdot(X - X) \\ &= (X + X^*)\cdot X - (X+X^*)\cdot X \\ &= X^2 - (X + X^*)\cdot X + XX^* \\ &= X^2 - (2x_o)\, X + (x_o{}^2 + x_1{}^2\ x_2{}^2 + x_3{}^2) \end{aligned}$$

when $X = (x_o, x_1, x_2, x_3)^\dagger$. But $2x_o$ and $N(X) = XX^*$ are real, as we require. Hence we have shown:

(a) $\langle E, X\rangle$ is multiplicatively closed, since
$X^2 = 2x_oX - (x_o{}^2 + x_1{}^2\ x_2{}^2 + x_3{}^2) \in \langle E, X\rangle$.

(b) The equation in (a) shows that elements of $\langle E, X\rangle$ are multiplied modulo the quadratic polynomial that belongs to X. Therefore we can use our arguments from the end of Section 4.A and immediately obtain $\langle E, X\rangle \approx \mathbb{C}$, as claimed above. This proves the theorem. □

(In particular, therefore, $\langle E, X\rangle$ with quaternion multiplication is a *commutative* ring, as one can verify directly and easily.)

Some subspaces of **H** which are isomorphic to $\mathbb{C}$ are shown in Figure 36:

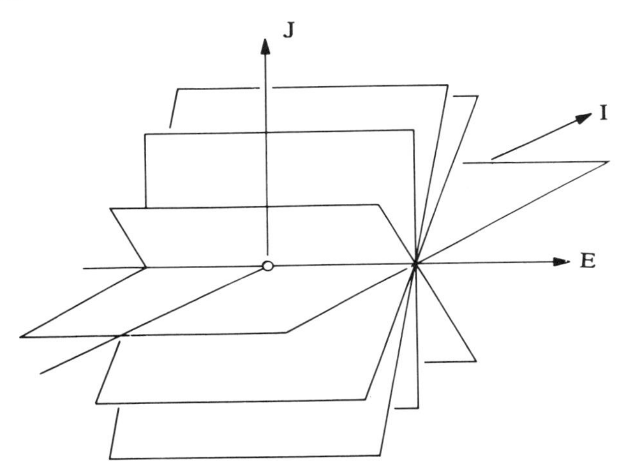


Fig. 36

## 5C QUATERNIONS AND VECTOR CALCULATIONS.

Hamilton had found himself forced to operate with "four variable expressions" although he still wanted to multiply only in the three dimensional space. In 1840, higher-dimensional spaces were not yet considered in Mathematics. Indeed quaternions were, generally speaking, one of the impulses that led mathematicians to go beyond three dimensions. His concern with three dimensions led Hamilton to split quaternions into scalar and vector parts:

$$A = (a_o, a_1, a_2, a_3)^{\dagger}$$

$$= (a_o, 0, 0, 0,)^{\dagger} + (0, a_1, a_2, a_3)^{\dagger} .$$

Here the **scalar part** of $A = a_oE + a_1I + a_2J + a_3K$ is $a_oE = a_o$, and the **vector part** is $a_1I + a_2J + a_3K$. This shows the appropriateness of the indexing that we have used from the beginning. Quaternions with scalar parts 0 he called **pure** quaternions and of these he considered I, J, K as basis vectors in $\mathbb{R}^3$. From this even today, physicists call $a_1\mathbf{i} + a_2\mathbf{j} + a_3\mathbf{k}$ a vector in $\mathbb{R}^3$. In this connection also, Hamilton introduced many concepts of vector analysis such as gradient, curl and divergence etc.

We follow Hamilton and identify the set of pure quaternions $(0, a_1, a_2\ a_3)^\dagger$ with $\mathbb{R}^3$ itself.

*EXAMPLE.* For a quaternion of this form we have $A^2 = (-a_1{}^2 - a_2{}^2 - a_3{}^2)E$. Therefore in the space of pure quaternions, the solution set of the quadratic equation $X^2 = -1$ is simply the set of all vectors of length 1, and this is the 2-sphere $S^2$ in $\mathbb{R}^3$. The fact that we have here infinitely many solutions for a quadratic equation naturally arises from the fact that $\mathbf{H}$ is not commutative.

<u>Exercise 4</u>.

(a) Prove that the only quaternions $A$ which are square roots of $-I$ are unit quaternions (i.e. $N(A) = 1$).

(b) For any quaternion $A$, show that its scalar part $Sc(A)$ is $\frac{1}{2}(A+A^*)$ and its vector part is $\frac{1}{2}(A-A^*)$.

(c) Using the fact that $(AB)^* = B^*A^*$ show that if $B$ is a unit quaternion then $Sc(A^*BA) = Sc(A)$.

From the multiplication of pure quaternions, we derive the usual products of the vector calculus. If we use the conventional notation and write **a**, **b** and so on for pure quaternions (= vectors of $\mathbb{R}^3$) and $(\mathbf{a}\,|\,\mathbf{b})$, $\mathbf{a}\times\mathbf{b}$ for the usual scalar and vector products to distinguish them from the quaternion product $\mathbf{a}\cdot\mathbf{b}$, then we have:

$$\mathbf{a}\cdot\mathbf{b} = \begin{bmatrix} 0 \\ a_1 \\ a_2 \\ a_3 \end{bmatrix} \begin{bmatrix} 0 \\ b_1 \\ b_2 \\ b_3 \end{bmatrix} = \begin{bmatrix} -a_1b_1 - a_2b_2 - a_3b_3 \\ a_2b_3 - a_3b_2 \\ -a_1b_3 + a_3b_1 \\ a_1b_2 - a_2b_1 \end{bmatrix}$$

$$= -(\mathbf{a}\,|\,\mathbf{b}) + \mathbf{a}\times\mathbf{b}$$

The *scalar product* is the scalar part of the quaternion product, and the *vector product* is the vector part of the quaternion product of *pure* quaternions. All these notations, still in use today, stem from Hamilton. The parallel related concepts of inner and outer product go back to Grassmann who at the same period wrote his book *Theory of extension* which however, remained unnoticed until about 1870. (His notation $(\mathbf{a}, \mathbf{b})$ is often used for $(\mathbf{a}\,|\,\mathbf{b})$.)

With the help of quaternion multiplication, we can separate the two products of vectors. As we have seen, for *pure* quaternions **a**, **b** we have

$$\mathbf{b}\cdot\mathbf{a} = -(\mathbf{b}\,|\,\mathbf{a}) + \mathbf{b}\times\mathbf{a}$$

and since $\qquad (\mathbf{b}\,|\,\mathbf{a}) = (\mathbf{a}\,|\,\mathbf{b})$ and $\mathbf{b}\times\mathbf{a} = -\mathbf{a}\times\mathbf{b}$

then $\qquad (\mathbf{a}\,|\,\mathbf{b}) = -\frac{1}{2}(\mathbf{a}\cdot\mathbf{b}+\mathbf{b}\cdot\mathbf{a})$

$$a \times b = -\tfrac{1}{2}(a \cdot b - b \cdot a)$$

Exercise 5.

(a) Calculate $(\mathbf{a} | \mathbf{b})$ and $\mathbf{a} \times \mathbf{b}$ when $\mathbf{a} = (1, 2, 3)$, $\mathbf{b} = (-7, 5, 4)$ in $\mathbb{R}^3$.

(b) Show that a quaternion $q$ is a pure quaternion if and only if $\bar{q} = -q$ and is a unit pure quaternion if and only if $q^2 = -1$. Hence show that a quaternion $p$ is pure if and only if $p^2$ is a negative real number.

Describe the connection between the product of two pure quaternions and vector and scalar products. Hence or otherwise show that if $p_1$ and $p_2$ are non-zero pure quaternions then $p_1 p_2 = p_2 p_1$ if and only if $p_2 = \lambda p_1$ for some $\lambda \in \mathbb{R}$.

Find the centralizer cent(q) of an arbitary non-zero quaternion $q = s + p$, where $s \in \mathbb{R}$ and $p \neq 0$ is pure, in the multiplicative group of non-zero quaternions. Show that cent(q) is isomorphic to the multiplicative group of non-zero complex numbers. [Southampton University, 1983]

**Historical Notice: Hamilton on the Brougham Bridge in Dublin**
Concerning Hamilton's discovery, there is a very beautiful work by B.L. Van der Waerden (*Hamilton's Discovery of Quaternions*, [72]). Van der Waerden describes in detail Hamilton's agonised efforts, in particular for establishing what he called the "Law of Moduli", N(AB) = N(A) N(B). We cite only some characteristic points, the sources of which can be found in Hamilton's collected works. (In the citations, the texts of Van der Waerden are in normal type, while those of Hamilton are in italics.)

"Through all these documents we can follow Hamilton's single-minded steps of thought precisely. A rare exception in which we can observe what went on in the mind of a mathematician when he poses a problem for himself, brings it step by step to solution and then through a blinding flash, so modifies the problem that it becomes soluble." (p.4)

"Hamilton knew and used the geometrical representation of complex numbers. In his published works, however, he showed how to define the complex numbers as number pairs (a, b) which had to be added and multiplied according to definite rules.

Binding himself to that mode of representation, Hamilton could pose for himself the problem: To find how we can multiply number triples (a, b, c) analogously to the number pairs (a, b).

The wish, to discover the multiplication rule for triples, had been with Hamilton, as he himself said, for a long time. But, in October 1843, this wish became much stronger and more earnest. *The desire,* as he put it, *to discover the law of the multiplication of triplets regained with me a certain strength and earnestness*". (p.5)

"Hamilton wrote a letter to his son about his first researches:

*Every morning in the early part of the above-cited month [October 1843], on my coming down to breakfast, your brother William Edwin and yourself used to ask me, 'Well , Papa, can you multiply triplets?' Whereto I was always obliged to reply, with a sad shake of the head, 'No, I can only add and subtract them'."* (p.6)

"And now comes the idea which gave a new twist to the whole problem. In the letter to Graves, Hamilton expresses the idea very clearly:

*And here dawned on me the notion that we must admit, in some sense, a fourth dimension of space for the purpose of calculating with triplets."* (p.8)

"From Hamilton's letter to his son, we infer [...] the express circumstances under which the idea came to him like a flash of lightning. Directly after the already quoted words "No, I can only add and subtract them" Hamilton goes further:

*But on the 16th day of the month [October 1843] – which happened to be a Monday and a Council day of the Royal Irish Academy – I was walking in to attend and preside, and your mother was walking with me, along the Royal Canal ...; and although she talked with me now and then, yet an undercurrent of thought was going on in my mind, which gave at last a result, whereof ... I felt at once the importance. An electric circuit seemed to close, and a spark flashed forth, the herald (as I foresaw immediately) of many long years to come of definitely directed thought and work...*

*I pulled out on the spot a pocket-book, which still exists, and made an entry there and then. Nor could I resist the impulse – unphilosophical as it may have been – to cut with a knife on a stone of Brougham Bridge, as we passed it, the fundamental formula with the symbols i, j, k:*

$$i^2 = j^2 = k^2 = -ijk = -1$$

*which contains the solution of the Problem, but of course, as an inscription, has long since mouldered away."* (p.10/11)

"What then passed directly before and on that remarkable stroll by the Royal Canal, he wrote down on the very same day in his notebook thus:

*I believe I now remember the order of my thought. The equation $ij = 0$ was recommended by the circumstance that*

$$(ax - y^2 - z^2)^2 + (a + x)^2 (y^2 + z^2)$$
$$= (a^2 + y^2 + z^2)(x^2 + y^2 + z^2)$$

*I therefore tried whether it might not be true that*

$$(a^2 + b^2 + c^2)\ (x^2 + y^2 + z^2)$$

$$= (ax - by - cz)^2 + (ay + bx)^2 + (az + cx)^2$$

*but found that this equation required, in order to make it true, the addition of* $(bz - cy)^2$ *to the second member. This* _forced_ *on me the non–neglect of* $ij$, *and* _suggested_ *that it might be equal to* $k$, *a new imaginary.*

Because of the underlining of the verbs "forced" and "suggested", Hamilton emphasizes that there were in his mind two quite different procedures going on. The first procedure was a necessary logical step which arose directly from the calculations. One cannot put $ij = 0$ because otherwise, the law of the moduli would not hold. The second step was the idea which sprang to him suddenly along the canal like a spark ("an electrical circuit seemed to close, and a spark flashed forth"), namely we had to be able to accept $ij$ as a new imaginary unit." (p.11)

## 5D THE MULTIPLICATIVE GROUP OF QUATERNIONS.

We can ask ourselves several questions as to how aspects of the structure of $\mathbb{C}$ will be found in H. Thus*:

*What course, parallel to the handling of complex numbers, can one follow if one wants to think about the multiplicative group of Quaternions?*

Just as each complex number has a representation by means of polar co-ordinates $(r, \theta)$, so there is an analogue for quaternions, but with $\theta$ in a sphere of dimension 3 (instead of 1). For the 'r', the modulus $\beta(A)$ of a quaternion A has been introduced as the length of the vector $(a_0, a_1, a_2, a_3)^\dagger$, with modulus $\beta(A) = \surd(a_0{}^2 + a_1{}^2 + a_2{}^2 + a_3{}^2)$. As already remarked, we then have

$$\beta(AB) = \beta(A) \cdot \beta(B)$$

and

$$\beta(A) = 0 \Leftrightarrow A = 0$$

As to the '$\theta$' we introduce by analogy with $\mathbb{C}$ the quaternion

$$\alpha(A) = \frac{1}{\beta(A)} \cdot A \quad \text{for} \quad A \neq 0 .$$

---

* _Translator's note_: With these questions, the author is following the lines of a passage from James Joyce *Ullyses* [36], pp.586÷621 – a book apparently familiar to many Germans. Artmann has ingeniously (and humorously) noted that he is here confronted with the same type of didactical problem as Joyce's hero, Bloom. A further overtone is its setting in Dublin, and hence the Irish connections with Hamilton.

One observes that $\beta(\alpha A) = 1$, so $\alpha(A)$ is a quaternion of length 1. In $\mathbb{R}^4$ the vectors of length 1 form a 3-sphere $\mathbb{S}^3$ (= set of points with distance 1 from the origin). Thus, we have assigned to A the pair $(\beta(A), \alpha(A))$ of 'co-ordinates', in $\mathbb{R}^+ \times \mathbb{S}^3$, and we now study this assignment.

The sphere $\mathbb{S}^3$ plays in H a role that is analogous to the unit sphere (1-sphere) $\mathbb{S} = \mathbb{S}^1$ in $\mathbb{C}$. Here

$$\mathbb{S}^3 = \{A \in H \mid \beta(A) = 1\} .$$

Also, we denote by $(H^\times, \cdot)$ the multiplicative group of non-zero quaternions.) Just as $\mathbb{S}$ sits in $(\mathbb{C}^\times, \cdot)$, we have:

<u>THEOREM</u>. $(\mathbb{S}^3, \cdot)$ *is a sub-group of* $(H^\times, \cdot)$.

<u>*PROOF*</u>. We verify the usual sub-group criteria. The neutral element E lies in $\mathbb{S}^3$; if $X, Y \in \mathbb{S}^3$, then also $X, Y \in \mathbb{S}^3$, because $\beta(XY) = \beta(X)\beta(Y)$. If X belongs to $\mathbb{S}^3$, then so does $X^{-1} = (1/\beta(X)) \cdot X^*$ because $\beta(X) = \beta(X^*) = 1$. □

Moreover, we have here as with $\mathbb{S}^1$ in $\mathbb{C}$: if $X \in \mathbb{S}^3$ then $X^{-1}$ is the conjugate quaternion $X^*$.

*What then is the difference (as groups) between* $\mathbb{S}^3$ *and* $\mathbb{S}^1$*?*

As we shall see below, the group $\mathbb{S}^3$ is non-commutative. (Of course, as topological spaces, $\mathbb{S}^3$ and $\mathbb{S}^1$ have dimensions 3 and 1 respectively.)

Pursuing further the 'polar coordinates' $(\beta(A), \alpha(A))$ we note first that by the multiplicative property of $\beta$, noted above, we have

$$\beta : (H^\times, \cdot) \rightarrow (\mathbb{R}^+, \cdot) \quad \textit{is a homomorphism.}$$

Similarly, $\alpha$ is multiplicative; for if $A, B \in H^\times$, then

$$\alpha(AB) = \frac{1}{\beta(AB)} AB$$

$$= \frac{1}{\beta(A)} A \frac{1}{\beta(B)} B$$

$$= \alpha A \ \alpha B$$

i.e. $\alpha : (H^\times, \cdot) \rightarrow (\mathbb{S}^3, \cdot)$ *is a homomorphism.*

*What results from this?*

THEOREM. *As a direct product of groups.*

$$(H^{\times}, \cdot) \cong (R^{+}, \cdot) \times (S^{3}, \cdot) .$$

The proof is simple, using the mapping

$$\varphi : A \to [\beta(A), \alpha(A)] .$$

In this splitting into a direct product, we can assume the factor $(R^{+}, \cdot)$ is known and we need only concentrate our further considerations on the structure of $(S^{3}, \cdot)$.

Exercise 6.

(a) Complete the line of thought begun in Exercise 1(c) to show that A has a square root iff $\alpha(A)$ has; and the square roots of A are of the form $(\sqrt{\beta(A)})\cdot X$, where X is any square root of $\alpha(A)$. Investigate the solution of quadratic equations in H.

(b) Use the earlier result on the centre of $(H^{\times}, \cdot)$ to show that the centre of $(S^{3}, \cdot)$ consists of $\{I, -I\}$.

(c) Show that every unit pure quaternion is a square root of $-1$. Prove that every unit quaternion can be written in the form $\cos\alpha + y\sin\alpha$, where y is a unit pure quaternion. Use induction to derive the formula

$$(\cos\alpha + y\sin\alpha)^{n} = \cos n\alpha + y\sin n\alpha, \qquad (n \geq 1).$$

Let $A = \cos\pi/n + I\cdot\sin\pi/n$ and $B = J$. Let $G_n$ be the subgroup of the multiplicative group of unit quaternions generated by A and B. Show that

$$A^{n} = B^{2} = (AB)^{2} = -I$$

and deduce that $G_n$ has order 4n. Find the centre Z of $G_n$ and identify the quotient group $G_n/Z$. [Southampton University, 1984].

## 5E QUATERNIONS AND ORTHOGONAL MAPPINGS IN $R^3$.

We can also ask ourselves several questions about the structure of quaternion–theory itself, and its relations with Geometry and Physics. Thus:

*What should readers of this section note for themselves, where do the ideas arise, what should they remember? In short, what is the content of the paragraphs?*

In this section, $R^3$ is always represented as the space of pure quaternions of the form $(0, x_1, x_2, x_3)^{\dagger}$. The orthogonal mappings of $R^3$ into itself are important in Geometry and Physics, and they can be expressed, within the algebra of quaternions, in a rather simple way.

The following exposition derives from Appendix IV, "Quaternions and Rotations" of the book of J. Dieudonné [11]. (For the necessary linear algebra, compare for example H.J. Kowalsky [41], § 24, Rotations). A more complete answer to the above question will be an Exercise for the reader, at the end of the section.

The complex numbers $z \in \$$ induce rotations of $\mathbb{R}^2$ through the mapping $w \to zw$ (for $w \in \mathbb{C}$). In a more complicated way, one derives from $\$^3$ the rotations in $\mathbb{R}^3$ and in another form also those in $\mathbb{R}^4$. We begin with an observation concerning $\mathbb{R}^4$.

*1) How can one, by means of quaternions, produce even such linear mappings as the rigid motions in* $\mathbb{R}^4$*?*

Since we require *linear* mappings, we exclude the translations

$$x \to x+a, \ (a \neq 0).$$

Let $A, B \in \$^3$. With these two quaternions, we associate the mapping

$$\gamma : \mathbb{R}^4 \to \mathbb{R}^4, \qquad \text{where} \qquad \gamma : X \to AXB$$

in which we regard the quaternion $X$ as a vector in $\mathbb{R}^4$. Then we leave the reader to verify that $\gamma$ has the properties

$$\gamma(X + Y) = \gamma X + \gamma Y, \quad \gamma(rX) = r\gamma X \ (r \in \mathbb{R}) \ ,$$

so $\gamma : \mathbb{R}^4 \to \mathbb{R}^4$ is linear.

The modulus $\beta$ satisfies

$$\beta(AXB) = \beta A \cdot \beta X \cdot \beta B$$

$$= \beta X \qquad \text{since} \qquad \beta A = 1 = \beta B \ ;$$

therefore $\gamma$ preserves length, and hence is an orthogonal mapping of $\mathbb{R}^4$. (In Dieudonné pp.171÷82 one can read how to derive in this way *simultaneous* rotations of $\mathbb{R}^4$ and how this is an exceptional manifestation of $\mathbb{R}^4$ among all the spaces $\mathbb{R}^n$). We shall not, however, let this worry us about $\mathbb{R}^4$.

*2) If instead of the etherial* $\mathbb{R}^4$*, we consider the concrete* $\mathbb{R}^3$*, what must we do to have a rigid motion then?*

Let
$$x = (0, x_1, x_2, x_3)^{\dagger}$$

be a pure quaternion. Given $S \in \$^3$ we consider the mapping

$$\varphi_S : x \to SxS^{-1}$$

and we show that $\varphi_S$ maps $\mathbb{R}^3$ into $\mathbb{R}^3$ again, indeed as a rotation; and moreover that in this way, we can simultaneously derive all rotations of $\mathbb{R}^3$. First, we show: $\varphi_S \mathbf{x} \in \mathbb{R}^3$. For, if $x_o$ is a scalar quaternion, then $\varphi_S x_o = Sx_oS^{-1} = x_o SS^{-1} = x_o$, because the scalars lie in the centre of H. Therefore, the scalar axis is left pointwise fixed. Since $\varphi_S$ is a special mapping of the sort considered in 1) above, it is orthogonal. Therefore the orthogonal complement of the real axis, namely the space $\mathbb{R}^3$ of pure quaternions, must be mapped into itself. (One can also calculate this from scratch in the form

$$S(0, x_1, x_2, x_3)^\dagger S^{-1} = (y_0, y_1, y_2, y_3)^\dagger$$

and show that $y_0 = 0$). We have therefore seen that $\varphi_S$ is an orthogonal linear mapping

$$\varphi_S : \mathbb{R}^3 \to \mathbb{R}^3 .$$

*What theoretical advantage do we derive if we work with* $R \in \mathbf{H}^\times$ *instead of with* $S \in \$^3$*?*

None. In spite of this there does result some information which will be important later. Given $T \in \mathbf{H}^\times$ with $T = tS$ and $t \neq 0$ we have

$$TXT^{-1} = SXS^{-1}$$

because the scalars $t$ and $t^{-1}$ lie in the centre of $\mathbf{H}$. Thus, by restricting ourselves to $S \in \$^3$, we lose nothing essential. If $t = -1$, we have $\varphi_{-S} = \varphi_S$ for all $S \in \$^3$.

*3) How are the reflections distinguished among the orthogonal mappings of* $\mathbb{R}^n$*? Which is the most useful notation to use? What simple representation by means of quaternions do we derive?*

Each orthogonal mapping of $\mathbb{R}^n$ can be written as a product of reflections in hyperplanes; in particular, in $\mathbb{R}^3$, we need only one, two or three such reflections in planes. (Kowalsky, § 24 or Dieudonné, p.132.) Hence we can first try to represent these reflections with the help of quaternions. Suppose $\mathbf{a} \neq \mathbf{0}$ is a vector of $\mathbb{R}^3$. Then the reflection in the plane perpendicular to $\mathbf{a}$ through $\mathbf{0}$ can be written in the form

$$\sigma : \mathbf{x} \to \mathbf{x} - 2 \cdot \frac{(\mathbf{x}|\mathbf{a})}{(\mathbf{a}|\mathbf{a})} \mathbf{a}$$

where $(\cdot|\cdot)$ denotes the usual scalar product. (Reason: If $\mathbf{y} \perp \mathbf{a}$, then one calculates that $\sigma\mathbf{y} = \mathbf{y}$, and for $\mathbf{a}$ itself, we have $\sigma\mathbf{a} = -\mathbf{a}$.) (See Fig. 37.)

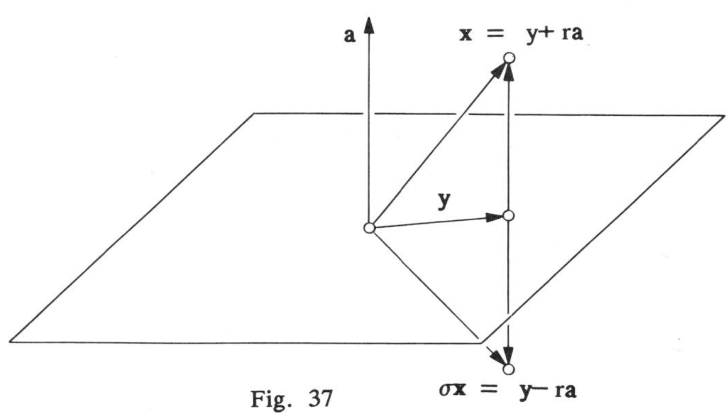


Fig. 37

We now express $\sigma\mathbf{x}$ in terms of the quaternion product itself; being vectors in $\mathbb{R}^2$, $\mathbf{a}$ and $\mathbf{z}$ are also pure quaternions, so we can multiply them in the formula for $\sigma\mathbf{x}$, to obtain

$$\sigma x = x - 2 \frac{(x|a)}{(a|a)} a$$

$$= x - 2 \frac{(x|a)}{-\frac{1}{2}(a \cdot a + a \cdot a)} a$$

$$= x + 2 (x|a) a^{-1} \quad \text{(inverse quaternion } a^{-1}\text{)}$$

$$= x - (x \cdot a + a \cdot x) a^{-1}$$

$$= x - x \cdot a \cdot a^{-1} - a \cdot x \cdot a^{-1}$$

$$= - a \cdot x \cdot a^{-1} .$$

Setting $\quad a = (0, a_1, a_2, a_3)^{\dagger}$

we can omit the dot from the product and write

$$\sigma x = - AxA^{-1} = -\varphi_A x$$

where we can assume still that $A \in \$^3$, because the length of $A$ cancels out.

<u>Exercise 7</u>.

(a) Can the calculation be simplified if we choose $A \in \$^3$, that is, we assume $|a| = 1$ and $(a|a) = 1$?

(b) Find the matrix of $\sigma$ (relative to the usual basis I, J, K of $\mathbb{R}^3$) when $\mathbf{a} = (1, 2, 3)$.

*4) The enclosures of reticence* having *fallen, what does this quaternionic representation of the rotations of* $\mathbb{R}^3$ *tell us?*

<u>THEOREM</u>. *To each rotation* $\delta$ *in the space* $\mathbb{R}^3$ *of pure quaternions there is a quaternion* $S \in \$^3$ *with* $\delta = \varphi_S$

*i.e.* $\quad \delta x = SxS^{-1}$ *for all* $x \in \mathbb{R}^3$

<u>*PROOF*</u>. Each rotation $\delta$ in $\mathbb{R}^3$ can be written as a product $\delta = \tau\sigma$ of two reflections in planes which intersect in a line forming the axis of rotation. Then as calculated above, there are pure quaternions $A$, $B \in \$^3$

$$\sigma x = - AxA^{-1} \qquad \tau y = - ByB^{-1}$$

Hence
$$\sigma x = \tau(-AxA^{-1})$$
$$= -B(-AxA^{-1})B^{-1}$$
$$= BAxA^{-1}B^{-1}$$
$$= BAx(BA)^{-1}$$

and writing $S = BA \in \$^3$ :
$$\delta x = SxS^{-1} \qquad \square$$

*5) In how many different ways can we represent a rotation?*
We have already seen that
$$\varphi_S = \varphi_{-S} ,$$
that is, we can write the rotation $\delta$ by using either of the quaternions $S$, $-S$ in $\$^3$. *What other possibilities are there?*

Let us assume that we have
$$\delta x = SxS^{-1} = TxT^{-1} \quad \text{for all} \quad x \in \mathbb{R}^3 .$$

Then
$$T^{-1}Sx = xT^{-1}S$$

or, writing $A = T^{-1}S \in \$^3$ :
$$Ax = xA \quad \text{for all} \quad x \in \mathbb{R}^3 .$$

Now substitute I, J, K successively for **x** and we see by our earlier definition of the centre of **H** that A must be a scalar quaternion $(a_o, 0, 0, 0)^{\dagger}$. Since $A \in \$^3$ we must have $a_o = \pm 1$ and we derive $T = \pm S$, therefore there is no other possibility than the ones we saw above.

The representation of $\delta$ by means of quaternions is "two valued" so to speak; one can represent $\delta$ either by means of $S$ or $-S$ but with nothing else. It now remains to fix one of the two quaternions $S$ or $-S$ and in this way to make the representation single-valued. We shall simply assert here that this is impossible for the same reason as a definition of a square root function on the unit circle $\$$ in $\mathbb{C}$. (Recall that the square function $q : z \to z^2$ on $\$$ also identifies $z$ and $-z$.)

Exercise 8.

(a) Describe geometrically the rotations $\varphi_I$, $\varphi_J$, $\varphi_K$ of $\mathbb{R}^3$ that belong to I, J, K (state the axis of rotation and the angle).

(b) Suppose $A = a_o + a_1 I$ with $a_o{}^2 + a_1{}^2 = 1$ and put $a_o = \cos\alpha$ and $a_1 = \sin\alpha$. Show that $\varphi_A$ is a rotation about the I-axis through an angle of $2\alpha$.

(c) Repeat the exercise (b) with $B = b_o + b_2J$ and $C = c_o + c_3K$, where $B, C \in \$^3$. Describe the corresponding rotations of $\mathbb{R}^3$ by using the so-called Eulerian angles to give a new proof of our theorem. (See Mirsky [51] Ch.8)

*6) How can we calculate the rotation axis and the angle of turn, from the quaternion representation?*
In the last exercise we have looked at the rotations about the axis I, J, K. More generally, there is the following result.

THEOREM. *Suppose* $S = s_o + \mathbf{s} \in \$^3$ *and* $\mathbf{s} \neq \mathbf{0}$. *Then the rotation axis of* $\varphi_S$ *is the line* $r\mathbf{s}(r \in \mathbb{R})$.

*PROOF.* Since $S \in \$^3$ we have $S^{-1} = S^* = s_o - \mathbf{s}$ and we obtain for the pure quaternion $\mathbf{s}$:

$$\begin{aligned}
\varphi_S \mathbf{s} &= S\mathbf{s}S^{-1} \\
&= (s_o + \mathbf{s})\cdot\mathbf{s}\cdot(s_o - \mathbf{s}) \\
&= (s_o + \mathbf{s})\cdot(\mathbf{s}s_o - \mathbf{s}\cdot\mathbf{s}) \qquad \text{as } s_o \text{ is scalar} \\
&= (s_o + \mathbf{s})\cdot(s_o - \mathbf{s})\cdot\mathbf{s} \\
&= SS^{-1}\,\mathbf{s} = \mathbf{s} .
\end{aligned}$$

The vector $\mathbf{s}$ and all its multiples $r\mathbf{s}$ therefore remain fixed under $\varphi_S$, and they form the rotation axis as asserted. □

*Why do we not elaborate these calculations to a more precise result, and then determine the angle of turning?*

For several reasons: because it is somewhat tedious, we don't need the result later in the text, and we already know a special case from the Exercise 8(b). In case of need, one can look up the full result in Dieudonné, p.170, Prop. 2.

*7) How can we conveniently gather together the previous arguments for more general purposes?*
The rotations in $\mathbb{R}^3$ form the **special orthogonal group** $SO_3$ (which is isomorphic to the group of orthogonal matrices with determinant 1).

THEOREM. *With each* $S \in \$^3$ *we associate the rotation* $\varphi_S$ *of* $\mathbb{R}^3$ *where* $\varphi_S\mathbf{x} = S\mathbf{x}S^{-1}$ *Then the correspondence* $\Phi : S \to \varphi_S$ *is a surjective homomorphism*

$$\Phi : \$^3 \to SO_3$$

*with kernel* $\{E, -E\}$. $\Phi$ *is called the spin-covering of* $SO_3$ .

*PROOF.* a) Homomorphism. We have $\Phi(ST) = \varphi_{ST}$ and

$$\begin{aligned}\varphi_{ST}\,x &= STx(ST)^{-1}\\ &= STxT^{-1}\,S^{-1}\\ &= S(\varphi_T x)S^{-1}\\ &= \varphi_S\,\varphi_T x,\end{aligned}$$

Since this holds for all $x \in \mathbb{R}^3$, then we have the required equality

$$\varphi_{ST} = \Phi_S\,\varphi_T\ .$$

(b) The surjectivity we have already proved because to each rotation $\delta$ there exists $S \in \$^3$ with $\delta = \varphi_S = \Phi(S)$.

(c) We already know the kernel; it is the set of those $T$ with $\Phi(T) = \text{id}$ but $\text{id} = \varphi_E$ and $\Phi(T) = \varphi_T$, so $\ker(\Phi) = \{E, -E\}$. □

*8) What proof can we adduce to show that the quaternions relate more closely towards with applied than towards pure Mathematics?*

Through the spin–covering we can, so to speak, regard each rotation as having a 'sign' $+1$ or $-1$. This explains the origin of the designation 'Spin–covering', from the electron–spin of the Physicists. The quaternions also appear (in matrix form) through the 'Dirac matrices' of Quantum mechanics (cf. for example G. Eder [13]. In Eder, one finds the Dirac matrices $\gamma_o$, $\gamma_1$, $\gamma_2$, $\gamma_3$ where $\gamma_1$, $\gamma_3$ are, respectively our K, J. Admittedly, with Dirac matrices we are allowed to use <u>complex</u> scalars. The Physicists prefer to use, instead of $S^3$, the isomorphic group $SU_2$ of (complex) hermitian $2 \times 2$ matrices with determinant 1, – in symbols

$$SU_2 = \left\{ \begin{bmatrix} z & -\bar{w} \\ w & \bar{z} \end{bmatrix} \middle| z, w \in \mathbb{C} \text{ and } z\bar{z} + w\bar{w} = 1 \right\}.$$

One can easily check that the matrices of this type form a group. (For the relevant geometry, see the essay by Pickert–Steiner in Behnke–Fladt–Süss [6], Vol.I) This group is, however, isomorphic to $\$^3$, as we can verify in the following way. We put

$$z = a_o + ia_3 \quad \text{and} \quad w = a_2 + ia_1$$

and assign

$$\psi : \begin{bmatrix} z & -\bar{w} \\ w & \bar{z} \end{bmatrix} \to (a_o, a_1, a_2, a_3)^{\dagger}\ .$$

The determinant condition

$$z\bar{z} + w\bar{w} = a_o^2 + a_3^2 + a_2^2 + a_1^2 = 1$$

shows therefore, that

$$\psi : SU_2 \to \$^3$$

is bijective. If also

$$f = b_o + ib_3 \quad \text{and} \quad g = b_2 + ib_1$$

and $\quad \psi : \begin{bmatrix} f & -\bar{g} \\ w & \bar{f} \end{bmatrix} \to (b_o, b_1, b_2, b_3)^\dagger$

then by simple calculation of the product, we have

$$\psi\left[\begin{bmatrix} z & -\bar{w} \\ w & \bar{z} \end{bmatrix}\begin{bmatrix} f & -\bar{g} \\ w & \bar{f} \end{bmatrix}\right] = (a_o, a_1, a_2, a_3)^\dagger \cdot (b_o, b_1, b_2, b_3)^\dagger$$

$$= \psi\begin{bmatrix} z & -\bar{w} \\ w & \bar{z} \end{bmatrix}\psi\begin{bmatrix} f & -\bar{g} \\ g & \bar{f} \end{bmatrix}$$

so $\psi$ respects products. Therefore $\psi$ is an isomorphism, and from the standpoint of group theory, there is no material difference between $SU_2$ and $\$^3$. One can therefore work out all relevant properties using either $\psi$ or $\psi^{-1}$ The Physicists prefer $SU_2$ because they can use it analogously to the matrix groups $SU_3$ and $SU_4$ with which trio they derive information for their researches on quarks with "charm" and "colour" — and we wish them much success in that enterprise!

*9) What can it say about relations with Geometry and Topology?*

Under the mapping $\Phi : S^3 \to SO_3$, $S$ and $-S$ have the same value: they are said to be identified by $\Phi$. Now, $S$ and $-S$ are opposite points on the sphere, so we speak of their *Antipodal Identification* by $\Phi$. For each $n$, we can apply such antipodal identification to the sphere $S^n$ and obtain the real projective space $\mathbf{P}^n$, so from the topological or geometrical point of view, $SO_3$ is nothing other than the 3-dimensional real projective space. (We have said nothing about continuity etc, but everything goes as one would expect.) For the 1-Sphere $\$^1$, we have already seen the antipodal identification through $z \to z^2$, but in this simplest case, the corresponding projective line $\mathbf{P}^1$ is again a circle. With the usual spherical surface $\$^2$, the antipodal identification leads to the so-called 'cross-cap' (see, e.g. Courant and Robbins [10] p.261) and therefore to the intuitive topological picture of the projective plane $\mathbf{P}^2$. For the higher dimensions, complicated forms are to be expected.

Are there further relations with geometry, which we could well contemplate, but have suppressed here, for the reader to raise them at future opportunities? Yes: we shall treat them explicitly in Section 5H/I.

*10) Which important Group-theoretic phenomena are displayed by the Commutator Group of* $\$^3$*?*

The Commutator Group $G'$ of a group $G$ is generated by all so-called *commutators* $xyx^{-1}y^{-1}$ with $x, y \in G$. The size of $G'$ is therefore a measure of how "non-abelian" the group $G$ is, because of the theorem: if $\alpha : G \to H$ is a homomorphism of $G$ into an abelian group, then $G'$ lies in the kernel of $\alpha$. Therefore the greater $G'$ is, the smaller is the abelian image of $G$. (See Lang [42], Chapter II, Section 7). We now prove that $S^3$ is so badly non-abelian that it has not a single abelian image except the trivial group $\{1\}$.

THEOREM. *Every quaternion* $A \in S^3$ *is a commutator, i.e. there exist* $S, T \in S^3$ *with* $A = TST^{-1}S^{-1}$.

*PROOF*. We divide the proof into two parts. First the assertion is verified for a special type of $B \in S^3$, and then we lead back the assertion for an abitrary $A$ to that for $B$.

Let then $B = b_o + b_1 I \in S^3$, so $b_o^2 + b_1^2 = 1$. In $\langle E, I \rangle \cong \mathbb{C}$ we can then choose $-v + wI$ so that $v^2 + w^2 = 1$ and $(-v + wI)^2 = b_o + b_1 I = B$. We put $W = vJ + wK$ and then have $W \in S^3$. Therefore

$$\left.\begin{aligned} JWJ^{-1}W^{-1} &= JWJ^* W^* \\ &= JW(-J)(-W) \end{aligned}\right\} \text{since } J, W \text{ are pure quaternions in } S$$

$$= JWJW = (JW)^2 = (-v+wI)^2 = B$$

For the second step, let $A = a_o + \mathbf{a} \in S^3$. For the pure quaternion $\mathbf{a} = (0, a_1, a_2, a_3)^\dagger$ there is a rotation of $\mathbb{R}^3$, which maps $\mathbf{a}$ onto $\mathbf{b} = (0, b_1, 0, 0)^\dagger$, a vector of equal length on the x-axis. This rotation can be described by means of some $D \in S^3$. Then

$$\begin{aligned} DAD^{-1} &= D(a_o + \mathbf{a})D^{-1} \\ &= Da_oD^{-1} + D\mathbf{a}D^{-1} = a_o + \mathbf{b} = a_o + b_1 I \,. \end{aligned}$$

Since $A, D \in S^3$, we have also $a_o + b_1 I \in S^3$. Therefore by the first part of the proof, there exist $V, W \in S^3$ with

$$DAD^{-1} = VWV^{-1}S^{-1} \,.$$

Putting $T = D^{-1}VD$ and $S = D^{-1}$ one easily computes:

$$A = TST^{-1}S^{-1}$$

as had to be shown. □

The group $S^3$ is therefore equal to its own commutator group and it has no non-trivial abelian homomorphic image. We will deduce from this a consequence, the content of which we first clarify by means of the real square function $q : \mathbb{R}^\times \to \mathbb{R}^+$ with $q : x \to x^2$.

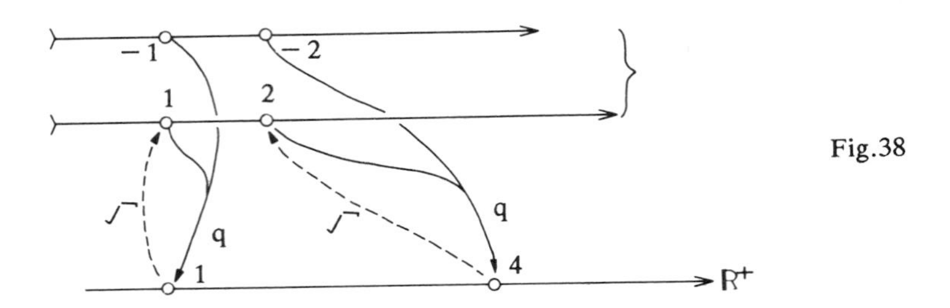


Fig.38

The real square function gives us a double covering of $\mathbb{R}^+$, for we have $(-x)^2 = x^2$. Using the root function, we then are able to "pull back" the image $\mathbb{R}^+$ to $\mathbb{R}^\times$ in such a way that we have $q(\surd y) = y$ for all $y \in \mathbb{R}^+$. This is reminiscent of a "Bloom question":

*Where had we received previous intimations of the expected results for $\$^3$, effected or projected?*

If we compare this picture with the corresponding one for $q : \$ \to \$$ in Section 4.B, then we see that obviously the root of the difference obviously lies with the fact that 0 splits $\mathbb{R}^\times$ into two separate parts. We show that the state of affairs with $\Phi : \$^3 \to SO_3$ is similar to that with $q : \$ \to \$$ in $\mathbb{C}$.

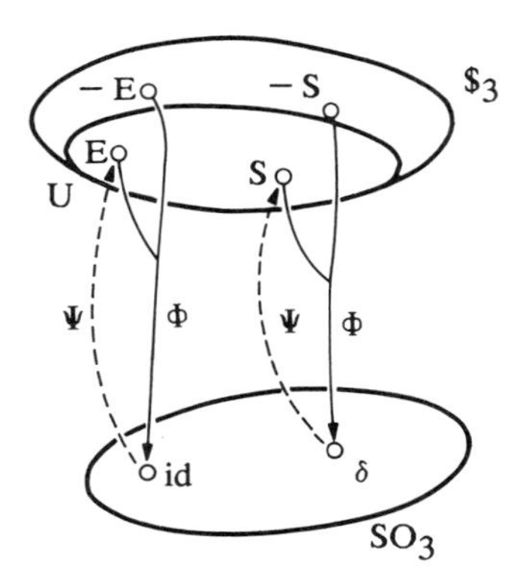


Fig. 39

Let us assume that there was a "backward" embedding, i.e. a homomorphism $\Psi : SO_3 \to \$^3$ with the property $\Phi(\Psi(\delta)) = \delta$ for each $\delta \in SO_3$. The image of $SO_3$ under $\Psi$ would then be a sub-group U of $\$^3$, with the property that for any $S \in \$^3$, than either S or $-S$ (but not both) would be contained in U. Therefore U would have in $\$^3$ precisely one coset (consisting of the elements $-U$, with $U \in U$). Hence U would be a normal subgroup of $\$^3$, and the factor group $S^3/U$ would have precisely two elements; it would therefore be abelian. We saw earlier that an abelian image of $S^3$ with *two* elements cannot exist, so there can be no backward embedding of $SO_3$ to $S^3$.

*Was this affirmation apprehended by Bloom?* — Not verbally. Substantially.
*What comforted his misapprehension?* — That as a competent keyless citizen he had proceeded energetically from the unknown to the known through the incertitude of the void. (James Joyce: *Ulysses* [36], p.618.)

Exercise 9.
This Section began with a rather involved question. Summarise the material so far given, to provide a brief but adequate answer.

## 5F THE THEOREM OF FROBENIUS.

The theorem of Frobenius characterizes the quaternions as the only finite-dimensional skew-field over $\mathbb{R}$. Therefore the absence of the commutativity law does not allow any further new multiplications on $\mathbb{R}^n$ other than the quaternion multiplication on $\mathbb{R}^4$.

For the formulation of the theorem and for later applications, we note here some concepts.

Definition. An **algebra** $\mathbf{A}$ over $\mathbb{R}$ is a vector space (with vectors a, b, c,..., — we no longer use boldface) in which a multiplication * is defined which is compatible with the vector space operations in the following sense:

(i) The distributive laws hold:

$$(a+b) * c = a * c + b * c, \qquad a * (b+c) = a * b + a * c$$

(ii) For scalars $r, s \in \mathbb{R}$ we have

$$(ra) * (sb) = r[a*(sb)] = rs[a*b]$$

Besides these fundamental properties, we usually consider adding one or more of the following properties:

(iii) there exists a neutral element $e \in \mathbf{A}$ with $0 \neq e$ and

$$e * a = a = a * e \qquad \text{for all } a \in \mathbf{A}.$$

($0 \neq e$ implies that $\mathbf{A} \neq \{0\}$)

(iv) Division is allowed, in the following sense:
Given $a, b \in \mathbf{A}$ with $a \neq 0$ there exist uniquely determined $x, y \in \mathbf{A}$ with

$$a * x = b \qquad \text{and} \qquad y * a = b$$

If (iv) holds, **A** is called a **division** algebra.

(v) The multiplication * is associative.

*REMARKS.* (1) In an algebra **A** with neutral element e, the scalar multiples re with $r \in \mathbb{R}$ form a sub-field of **A** isomorphic to $\mathbb{R}$. The basic reason is this: $\mathbb{R}e = \{re \mid r \in \mathbb{R}\}$ is a one-dimensional sub-space of **A**. Therefore, as an additive group, $\mathbb{R}e \cong (\mathbb{R}, +)$. Now, $\mathbb{R}e$ is multiplicatively closed because

$$(re) * (se) = rs(e * e) = rse$$

so the rest of the assertion follows easily.

(2) From this the elements of $\mathbb{R}e$ lie in the centre of **A**, which is to say that for every $a \in \mathbf{A}$ we have

$$(re) * a = r(e * a) = r(a * e) = a * (re) .$$

(3) An algebra **A** in which (i) – (v) hold is a **skew field.** Multiplicative inverses are obtained as solutions of the equation

$$a * x = e \text{ when } a \neq 0$$

because, if also $y * a = e$, then by application of the associative law,

$$y = y * (a*x) = (y*a) * x = x .$$

*EXAMPLES.* (1) The $n \times n$ – matrices form an associative algebra $(\mathbb{R})_n$ over $\mathbb{R}$ with neutral element E (unit matrix) in which (iv) does not hold. As a vector space, $(\mathbb{R})_n$ has dimension $n^2$ over $\mathbb{R}$.

(2) The triangular matrices of the form

$$A = \begin{bmatrix} a_1 & 0 & 0 & 0 & \cdots & 0 \\ a_2 & a_1 & 0 & & & \vdots \\ a_3 & a_2 & a_1 & & & 0 \\ \vdots & & \ddots & \ddots & & \vdots \\ & & & \ddots & \ddots & 0 \\ a_n & \cdots & a_3 & & a_2 & a_1 \end{bmatrix}$$

form an associative algebra $\mathbf{D}_n$ of dimension n over $\mathbb{R}$ in which again (iv) does not hold.

(3) The field $\mathbb{R}(x)$ of rational functions is an algebra satisfying (i) – (v) over $\mathbb{R}$, which moreover is still commutative. Simply as a vector space, $\mathbb{R}(x)$ has infinite dimension over $\mathbb{R}$.

(4) $\mathbb{C}$ is a two-dimensional algebra over $\mathbb{R}$ and **H** is a four-dimensional algebra over $\mathbb{R}$.

<u>THEOREM</u> of Frobenius. *The only algebras of finite dimension over* $\mathbb{R}$ *which are skew fields are (to within isormorphism)* $\mathbb{R}$, $\mathbb{C}$ *and* **H**.

<u>*PROOF*</u>. (We follow R.S. Palais: "The Classification of Real Division Algebras" [54], p.366÷368.) Let **D** be such an algebra. If **D** is commutative, then we have $\mathbf{D} \cong \mathbb{R}$ or $\mathbf{D} \cong \mathbb{C}$ by the third version of the fundamental theorem of algebra. We therefore assume that **D** is not commutative and must prove $\mathbf{D} \cong \mathbf{H}$.

For a better understanding of the idea of the proof, we recall some properties of quaternions. In **H**, $\mathbb{C}$ lies as a sub-field consisting of the elements of the form $a_0 + a_1I = A$, and these are just those $A \in \mathbf{H}$ with $AI = IA$. On the other hand the elements $B = b_2J + b_3K$ are precisely those with $BI = -IB$. In the first part of the proof, we will find an element like I in **D**; in the second part, we shall use a corresponding splitting of the elements in **D**.

Let **D** be our skew field with the neutral element E and $F \neq rE$ in **D**. We consider the two-dimensional sub-space $\langle E, F\rangle = \{rE + sF \mid r, s \in \mathbb{R}\}$, and need a preliminary lemma.

<u>**LEMMA**</u>. (i) $\langle E, F\rangle$ *is a maximal commutative sub-set* of **D**.

(ii) $\langle E, F\rangle$ *is also the set* $\{X \in \mathbf{D} \mid X * F = F * X\}$.

(iii) $\langle E, F\rangle$ *is a sub-field of* **D** *and isomorphic to* $\mathbb{C}$.

*PROOF of the Lemma.* Given vectors $X, Y \in \langle E, F\rangle$ we have

$$X * Y = (x_1E+x_2F) * (y_1E+y_2F)$$

$$= x_1E * y_1E + x_1E * y_2F + x_2F * y_1E + x_2F * x_2F.$$

For each $r \in \mathbb{R}$, rE belongs to the centre, so

$$X * Y = y_1E * x_1E + y_2F * x_1E + \dots$$

$$= Y * X \quad .$$

Therefore the elements of $\langle E, F\rangle$ commute. (We still do not know whether $X * Y$ actually lies in $\langle E, F\rangle$!) We choose now a sub-<u>space</u> T of **D**, which contains $\langle E, F\rangle$, is commutative and is also of maximal possible dimension. Such a sub-space exists because **D** itself has finite dimension. We will show that $T = \langle E, F\rangle$. If there were an $X \notin T$ with $X * T = T * X$ for each $T \in T$, then with a calculation as above (since scalars commute) we find that the larger sub-space $\langle T, X\rangle$ would be commutative. This gives a contradiction because T was assumed to be maximally commutative. Therefore, T is indeed a maximal sub-<u>set</u>, and an element $Z \notin T$ cannot commute with all $T \in T$. Therefore, we obtain the following two assertions:

(a) If $X, Y \in T$ then also $X * Y \in T$. Reason: as $X$ and $Y$ commute with all elements of $T$, the same holds for $X * Y$. Therefore $X * Y \in T$, for $T$ is a maximally commutative sub-set.

(b) With $0 \neq X \in T$ we also have $X^{-1} \in T$. Reason: since $X * T = T * X$ for all $T \in T$, then by multiplication with $X^{-1}$ on the right and on the left, we have $T * X^{-1} = X^{-1} * T$. Hence, $T$ is closed with respect to multiplication and the formation of inverses, so $T$ is a field. The fundamental theorem of algebra now tells us that $T$, being bigger than $\mathbb{R}$, must be isomorphic to $\mathbb{C}$. As $\mathbb{C}$ is two-dimensional over $\mathbb{R}$, the same holds for $T$ and we have $T = \langle E, F \rangle$, so the proof of the Lemma is complete. □

We now continue with the main proof of Frobenius's theorem, and we still write $T = \langle E, F \rangle$ as in the Lemma.

Since $T \cong \mathbb{C}$ there is an element $I \in T$ with $I^2 = -E$ and we can take $\mathbb{C} = \langle E, I \rangle \subseteq \mathbf{D}$. For each $Z \in \mathbb{C}$ and $D \in \mathbf{D}$ the product $Z * D$ is defined and the usual calculating rules hold for this multiplication of the "vector" $D$ with the "scalar" $Z$. Therefore $\mathbf{D}$ is a vector space over $\mathbb{C}$. (One must always multiply the scalars with the vectors from the <u>left</u>. The dimension of $\mathbf{D}$ over $\mathbb{C}$ is precisely half as big as that of $\mathbf{D}$ over $\mathbb{R}$).

We define now a linear mapping $\rho$ of $\mathbf{D}$, in which we multiply with $I \in \mathbb{C}$ from the <u>right</u>:

$$\rho : \mathbf{D} \to \mathbf{D}\,, \qquad \rho(X) = X * I\,.$$

(The linearity of $\rho$ follows from the rules of calculation in $\mathbf{D}$.)

With this linear mapping we form for ourselves a splitting of $\mathbf{D}$ into two sub-spaces analogous to that we indicated for $\mathbf{H}$ at the beginning of the proof.

If we consider $Z \in \mathbb{C}$ as a vector, then we have

$$\rho(Z) = Z * I = I * Z\,,$$

so $Z$ is an eigenvector corresponding to the eigenvalue 1 of $\rho$. However for any $X \notin \mathbb{C}$ we have $X * I \neq I * X$, otherwise $X$ would commute with all elements of $\mathbb{C}$. Here we therefore write

$$X = \tfrac{1}{2}(X - (I*X*I)) + \tfrac{1}{2}(X + (I*X*I))$$

$$= Z + Y$$

and we leave the reader to do the calculations that show

$$Z * I = I * Z \quad \text{and} \quad Y * I = -I * Y\,.$$

The vector $Y \neq 0$ is therefore a member of the set of eigenvectors corresponding to the eigenvalue $-1$; and this set forms the sub-space $\mathbf{D}^-$ of $\mathbf{D}$. From the splitting $X = Z + Y$ we have

$\mathbb{C} \cap D^- = \{0\}$ and $D = \mathbb{C} + D^-$, so the vector space $D$ is the direct sum $\mathbb{C} \oplus D^-$ of these two sub-spaces. We now show that, as a complex vector space, $D^-$ has dimension $\dim_{\mathbb{C}} D^- = 1$; therefore as a real vector space, $D$ is itself four dimensional.

We need a preliminary helpful remark:

*If* $A, B \in D^-$ *then* $A * B \in \mathbb{C}$.

Reason:
$$(A * B) * I = A * (B * I)$$
$$= A * (-I * B) = I * (A * B) .$$

Thus $I$ commutes with $A * B$, so $A * B \in \mathbb{C}$.

Now let $0 \neq A \in D^-$; define the linear mapping

$$\alpha : D \to D \quad \text{by} \quad \alpha(X) = X * A .$$

Then $\alpha$ has the inverse $\alpha^{-1}$, defined by $Y \to Y * A^{-1}$, so $\alpha$ is bijective. By our helful remark, $\alpha$ interchanges the sub-spaces $\mathbb{C}$ and $D^-$ of $D$. In particular, therefore we have:

$$\alpha : D^- \to \mathbb{C} \quad \textit{is an isomorphism,}$$

therefore $D^-$ is one-dimensional over $\mathbb{C}$.

With this, we have $\dim_{\mathbb{R}} D = 4$, as with the quaternions. We still need to find the elements $J, K$. In order to do that we take the above $A \in D^-$ and form, analogously to our previous $\langle E, F\rangle$, the field $\langle E, A\rangle \cong \mathbb{C}$. For these two-dimensional sub-spaces, we have

$$\langle E, I\rangle \cap \langle E, A\rangle = \langle E\rangle$$

since otherwise, $A$ would lie in $\langle E, I\rangle = \mathbb{C}$. Observing that $A^2 = A * A$, we have

$$A^2 \in \langle E, A\rangle \qquad \text{as } \langle E, A\rangle \text{ is a field,}$$

$$A^2 = \alpha A \in \mathbb{C} = \langle E, I\rangle \qquad \text{as } \alpha D^- = \mathbb{C} .$$

Hence
$$A^2 \in \langle E, I\rangle \cap \langle E, A\rangle = \langle E\rangle$$

so $A^2 = aE$ (say) and is real. If $a > 0$, then $A^2$ would possess in $\langle E, A\rangle \cong \mathbb{C}$ the three "square roots" $\surd aE$, $-\surd aE$ and $A \notin \langle E\rangle$, – therefore three different ones, which is impossible in $\mathbb{C}$; hence we must have $a < 0$. Taking $r = 1/\surd|a|$ and $J = rA$, we then have $J \in D^-$ and

$$J^2 = (rA)^2 = -1 .$$

From the basis $E, I$ of the real vector space $\mathbb{C}$ we derive, by multiplication with $J$, the basis $EJ = J$ and $IJ = K$ of the real vector space $D^-$. Altogether then, the set $E, I, J, K$ forms a basis of the four dimensional space $D$, and it gives the representation

$$X = x_0E + x_1I + x_2J + x_3K \quad \text{for each} \quad X \in D .$$

Since $J, K \in D^-$ we have

$$J * I = -I * J \quad \text{and} \quad K * I = -I * K ,$$

So

$$K^2 = I * J * I * J = -I * I * J * J = -1$$

and one derives analogously the whole multiplication table for $E, I, J, K$ as with the quaternions. Therefore $D$ is isomorphic to $\mathbb{H}$, and we have proved the theorem of Frobenius. □

## 5G THE CAYLEY NUMBERS (OCTAVES).

If we wish to find finite-dimensional division algebras over $\mathbb{R}$, then by the theorem of Frobenius we must at least add an additional rule of calculation. As we saw in the matrix examples at the beginning of the previous section, there are plenty of finite dimensional algebras but *with zero divisors*. However the existence of zero divisors obstructs the uniqueness of division, for then by definition:

There exists $a, b \neq 0$ with $a * b = 0 = a * 0$ .

One could also try substituting unique division by freedom from zero divisors, as one can read about in Salzmann ([65], I, p.191 onwards).

Shortly after Hamilton's publication on quaternions, Graves and Cayley gave (independently of each other) multiplication formulae for elements of $\mathbb{R}^8$. Hamilton remarked that with these formulae, the associativity law was no longer valid. Before describing the system $\mathbb{C}ay$ of the Cayley numbers we shall cite the two decisive theorems about finite-dimensional division algebras over $\mathbb{R}$.

<u>THEOREM A</u>. *If* $A$ *is a finite dimensional division algebra over* $\mathbb{R}$, *in which any two elements of* $A$ *generate an associative sub-algebra, then* $A$ *is isomorphic to* $\mathbb{R}, \mathbb{C}, \mathbb{H}$ *or* $\mathbb{C}ay$.

A proof of this theorem can be found in Salzmann [65], I, p.215.

<u>THEOREM B</u>. *If* $A$ *is a finite dimensional division algebra over* $\mathbb{R}$, *then the dimension of* $A$ *as a vector space over* $\mathbb{R}$ *is equal to 1, 2, 4 or 8.*

Essentially because of isomorphisms, only the four named algebras can appear in Theorem A. In Theorem B, only the dimensions are stipulated because one can obtain non-isomorphic variants through an easy 'distortion' of the multiplication formulae. In these variants also, the weakened associativity is no longer fulfilled. Theorem B is known and famed as the theorem of Bott and Milnor; one finds it in the work of J. Milnor [49], pp.444-449. There, Milnor remarks that Hirzebruch and Kervaire independently had come to the same conclusions. For some further remarks on this theorem see paragraph 5.H.

<u>Exercise 10</u>. List the various "Theorems of Frobenius" we have met so far, with Theorems A and B, to see in which respects one is stronger than another.

Returning to the description of the Cayley numbers, we must define a multiplication $*$ in $\mathbb{R}^8$. For this we use the quaternion multiplication in $\mathbb{R}^4$ and write the vectors of $\mathbb{R}^8$ with the help of two quaternions. Given quaternions A, B, X, Y, we put

$$\begin{bmatrix} A \\ B \end{bmatrix} * \begin{bmatrix} X \\ Y \end{bmatrix} = \begin{bmatrix} AX - YB^* \\ XB + A^*Y \end{bmatrix}$$

Here $A^*$ is the quaternion conjugate to A and — because of the non-commutativity of $\mathbb{H}$ — it is necessary to take care over the sequence of operations. Using the basis vectors $e_1, e_2, \ldots, e_8$, where

$$e_1 = (1, 0, 0, 0, 0, 0, 0, 0)^\dagger, \ \ldots,$$

$$e_8 = (0, 0, 0, 0, 0, 0, 0, 1)^\dagger,$$

Readers can, if they so wish, work out the multiplication tables of the products $e_i * e_j$ which Cayley and Graves had employed for the definition of multiplication. For $\mathbb{R}^8$ with this multiplication we write $\mathbb{C}\mathrm{ay}$. Simple calculations yield

$$\begin{bmatrix} 0 \\ E \end{bmatrix} * \left( \begin{bmatrix} J \\ 0 \end{bmatrix} * \begin{bmatrix} K \\ 0 \end{bmatrix} \right) = \begin{bmatrix} 0 \\ I \end{bmatrix}$$

$$\left( \begin{bmatrix} 0 \\ E \end{bmatrix} * \begin{bmatrix} J \\ 0 \end{bmatrix} \right) * \begin{bmatrix} K \\ 0 \end{bmatrix} = \begin{bmatrix} 0 \\ -I \end{bmatrix} .$$

so the associativity law **does not hold**. One can define a conjugation $x \to x^c$ in $\mathbb{C}\mathrm{ay}$ by

$$\begin{bmatrix} A \\ B \end{bmatrix}^c = \begin{bmatrix} A^* \\ -B \end{bmatrix}$$

and obtain, using the coordinates $a_i$ of A etc.:

$$\begin{bmatrix} A \\ B \end{bmatrix} * \begin{bmatrix} A \\ B \end{bmatrix}^c = \begin{bmatrix} AA^* + BB^* \\ 0 \end{bmatrix}$$

$$= (a_o^2 + a_1^2 + a_2^2 + a_3^2 + b_o^2 + b_1^2 + b_2^2 + b_3^2) \begin{bmatrix} E \\ 0 \end{bmatrix}$$

in analogy to the corresponding formulae in $\mathbb{C}$ and $\mathbb{H}$. (Also the multiplication formula is formed analogously to the multiplication formulae for $\mathbb{C}$ and $\mathbb{H}$).

One can now calculate that the correspondingly defined norm function $N\begin{bmatrix} A \\ B \end{bmatrix} = a_o^2 + \ldots + b_8^2$ is multiplicative, and thereby yields for example

$$(1^2 + 2^2 + 3^2 + 4^2 + 5^2 + 6^2 + 7^2 + 8^2) \times$$

$$(9^2 + 10^2 + 11^2 + 12^2 + 13^2 + 14^2 + 15^2 + 16^2) =$$

$$36^2 + 38^2 + 54^2 + 62^2 + 72^2 + 108^2 + 112^2 + 474^2 .$$

Already in 1898, Hurwitz had shown that multiplication formulae of this kind can exist only for sums of 1, 2, 4 or 8 squares. (A. Hurwitz, *Works*, Vol.II, *On the Composition of Quadratic Forms of Arbitrarily many Variables* pp. 565–571.)

More important than details of calculation (about which one can obtain information in Salzmann [65]), are the relations of these division algebras to other domains of mathematics. Two of these can be indicated briefly.

## 5H Relations with Geometry 1: Vector Fields on Spheres.

In order to describe multiplication in an algebra $A = \mathbb{R}^n$, it suffices to know only the products of vectors of length 1; for, because of the bilinearity we have, when $A = aT$ and $B = bS$:

$$A*B = (aT)*(bS) = ab(T*S) .$$

Given $A \in \mathbb{R}^n$ with $A \neq 0$, there is always such a vector $T$ with $|T| = 1$ and $A = aT$. We have already used this fact in the splitting of the multiplicative groups $\mathbb{C}^\times \cong \mathbb{R}^+ \times \mathbb{S}$ and $\mathbf{H}^\times \cong \mathbb{R}^+ \times \mathbb{S}^3$ into direct products.

The set of vectors of length 1 (= set of points at unit distance from the origin) in $\mathbb{R}^n$ is the sphere $\mathbb{S}^{n-1}$. Its dimension is one less than that of $\mathbb{R}^n$. Only the circle $\mathbb{S} = \mathbb{S}^1$ and the spherical surface $\mathbb{S}^2$ in $\mathbb{R}^3$ are available to our direct intuition. On that account, we consider the circumstances that we are about to describe, first as they apply to the unit circle $\mathbb{S} = \mathbb{S}^1$ in the complex plane. The intuition is reinforced if we agree to regard the vector $iz$ not as usual starting at the origin, but at the point $z$. As a multiplication in $\mathbb{C}$ by $i$ means nothing more than a rotation about 90°, we thus obtain through the mapping $z \to iz$ for $z \in \mathbb{S}$, a continuous field of tangential unit vectors to the sphere $\mathbb{S}^1$ (see Fig.40).

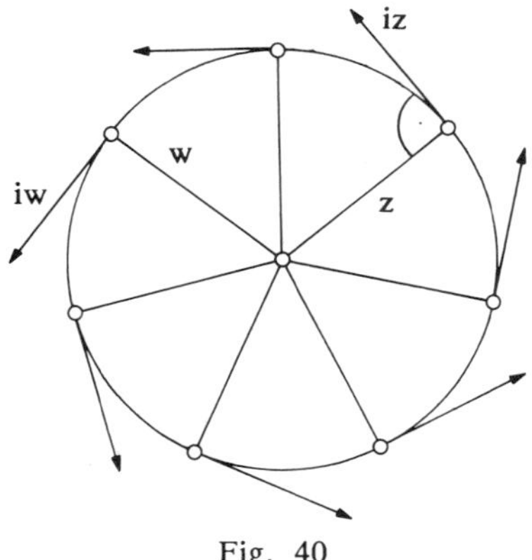


Fig. 40

A known and important theorem, the so-called 'hedgehog' theorem, says that there cannot be such a vector field on the sphere $\mathbb{S}^2$. Elementary proofs of this can be found, for example, in H. Holmann: "*Noteworthy Properties on the Spherical Surface*" [35], pp. 99–108; in J. Milnor: "*Analytical Proofs of the "Hairy Ball Theorem" and the Brouwer Fixed Point Theorem*" [50], pp. 521–524; and in M. Eisenberg and R. Guy: "*A Proof of the Hairy Ball Theorem*" [14], pp. 572–574.

By contrast with the sphere $\$^2$, one finds on the torus surface $T_2$ in fact two vector fields V and W, of the sort described above, whose elements at each point of the torus are linearly independent. A glance at the sketch suffices as a proof.

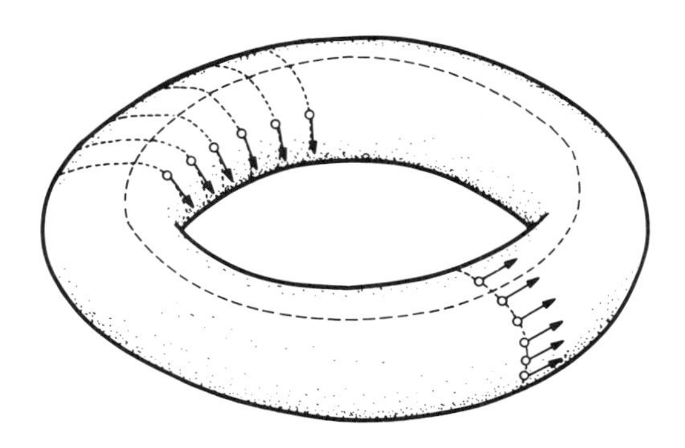

Fig. 41

On the sphere $\$^3$, the results are similar to those with $\$^1$ and the torus $T_2$. With the help of the quaternion multiplication, we can find three tangential vector fields on $\$^3$, whose elements at each point of $\$^3$ are linearly independent. (There can obviously not be more than three fields of this kind, because the tangent space at a point of $\$^3$ is three-dimensional). We proceed wholly by analogy with $\mathbb{C}$ and consider for $S = s_o + s_1I + s_2J + s_3K \in \$^3$ the quaternions

$$IS = -s_1 + s_oI - s_3J + s_2K$$

$$JS = -s_2 + s_3I + s_oJ + s_1K$$

$$KS = -s_3 - s_2I + s_1J + s_oK .$$

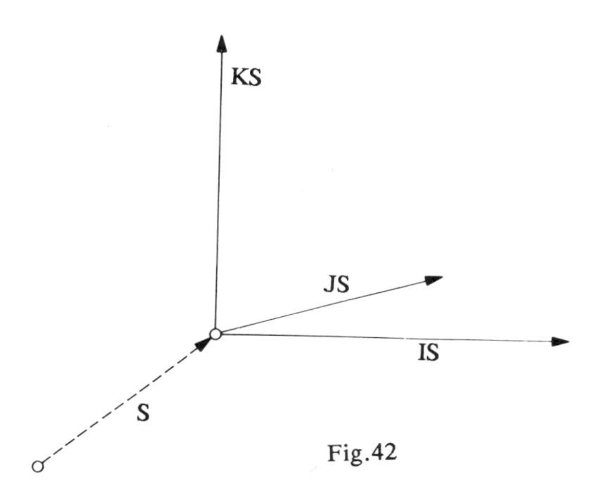


Fig.42

These three vectors are unit vectors, all three perpendicular to S (form the scalar product), therefore tangential to $\$^3$ and mutually perpendicular, therefore linearly independent. Furthermore, IS, JS and KS depend continuously on S.

By the same process, one uses the Caley numbers to derive seven tangential fields of unit vectors to the sphere $S^7$. Still more generally but with the same methods, one can derive from any n–dimensional division algebra over $\mathbb{R}$, n–1 continuous linearly independent fields to the sphere $S^{n-1}$. By a deep theorem of Adams, however, $S^1$, $S^3$ and $S^7$ are the *only* spheres $S^k$ which possess k such fields that are linearly independent at each point (Survey Article: E. Thomas: "*Vector Fields on Manifolds*" [70], pp. 643–683.) With spheres $S^k$, when k is even, then just as with $S^2$ there is no field, and one can read about this in the above cited work of Milnor in the American Mathematical Monthly.

## 5I RELATIONS WITH GEOMETRY 2: AFFINE PLANES.

The dropping of various different rules of calculation, which has led us from the field of real numbers to the Cayley numbers, lies in a very beautiful correspondence with the problem of introducing coordinates in affine and projective planes. Before we can describe this material, we must introduce a few concepts which intuitively are easily understood.

__Definition__. An Affine plane **A** consists of a set of points and a set of lines, for which there is defined a relation "lies on" that satisfies the following axioms (wherein $\ell$ "goes through" P means that P "lies on" $\ell$):

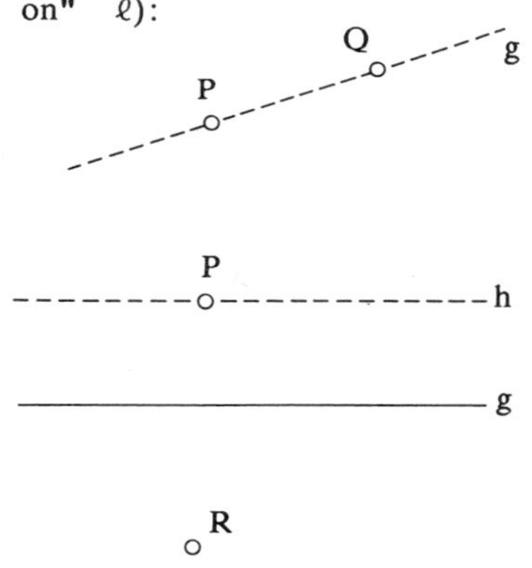


1. Through two different points goes precisely one line. We call two lines **parallel**, if there is no point which lies simultaneously on both of them. Also, we say that every line is parallel to itself.

2. (*Parallel axiom*) Given a line g and the point P, which does not lie on g, there is precisely one parallel h to g through P.

3. (*Richness axiom*) There are at least three different points which do not lie on one line.

Fig.43

(Anyone who knows coordinate geometry, knows also many affine planes. For, if $(K, +, \cdot)$ is a field, one can derive an affine plane $\mathbf{A}(K)$ belonging to K as follows. The "points" of $\mathbf{A}(K)$ are the number pairs (x,y) with $x, y \in K$. A "line" of $\mathbf{A}(K)$ is a set of "points" of one of the forms

$$[m, k] = \{(x, y) \mid y = mx + k\}$$

or

$$[k] = \{(x, y) \mid x = k\}$$

where m, k $\epsilon$ K. It is easily checked that the axioms are fulfilled by these definitions. Here the relation "lies on" is to be read as "is an element of".

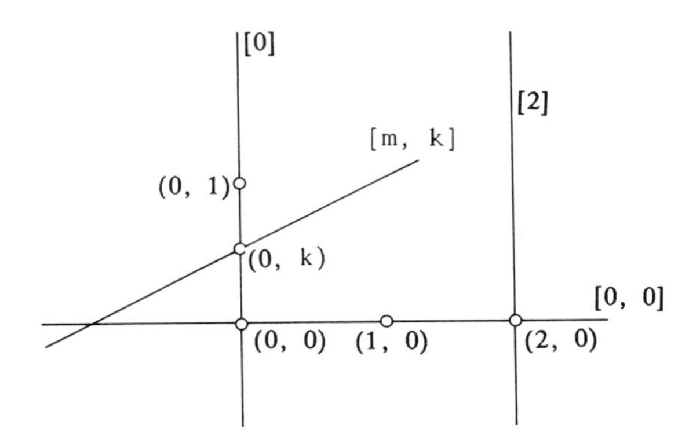


Fig. 44

An affine plane which can be represented as in the example with the help of a field, is said to allow **coordinatisation**. There are many affine planes however, which are not of this type. Nevertheless in every case, one can define a structure (K, + , ·) and then represent the given plane **A** as isomorphic to an **A**(K). But, in general, the field axioms are <u>not fulfilled</u> for (K, + , ·). As elements of K one takes the points of a fixed line, and chooses on it a pair of points, to act as zero and unit; this will play the role of the ordinary x–axis. One then defines addition and multiplication through geometrical constructions, like those in Fig.45, which are self explanatory. The details need not delay us here.

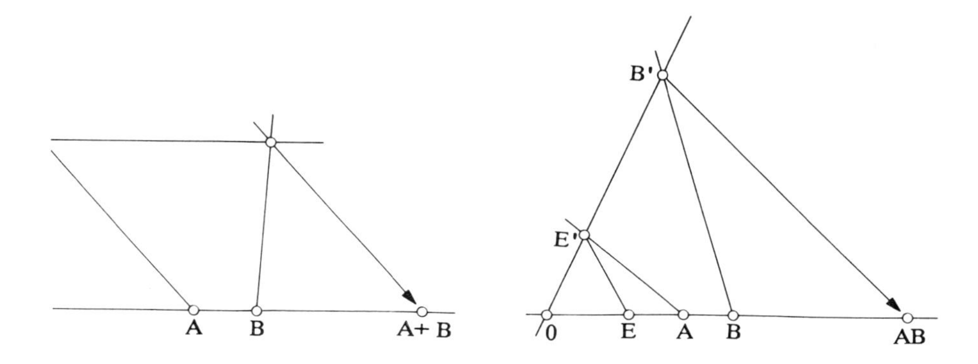


Fig. 45

One can use such constructions, to try to decide whether the multiplication (say) is commutative. Doing this, one derives the following configuration:

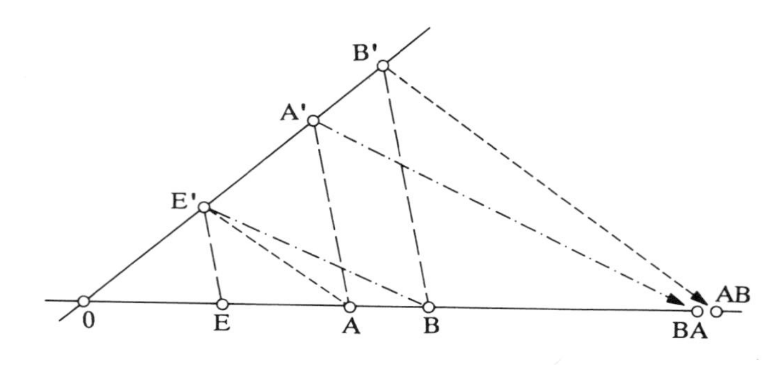


Fig. 46

If the points AB and BA must coincide, one says "the figure is closed". Correspondingly we can derive geometric consequences of the axioms of the affine plane, which say that certain configurations must "close" and we call these "closure theorems". In what follows, we shall show that fields, skew fields and generalised Cayley numbers can be characterised through such closure theorems. In the following sketches, the closure theorems are drawn in their so-called "projective" form in order to avoid having to formulate subsidiary hypotheses. (For a brief introduction to "projective" concepts, see Griffiths and Hilton [23] Ch. 16, or Courant and Robbins, [10] Ch. 4.)

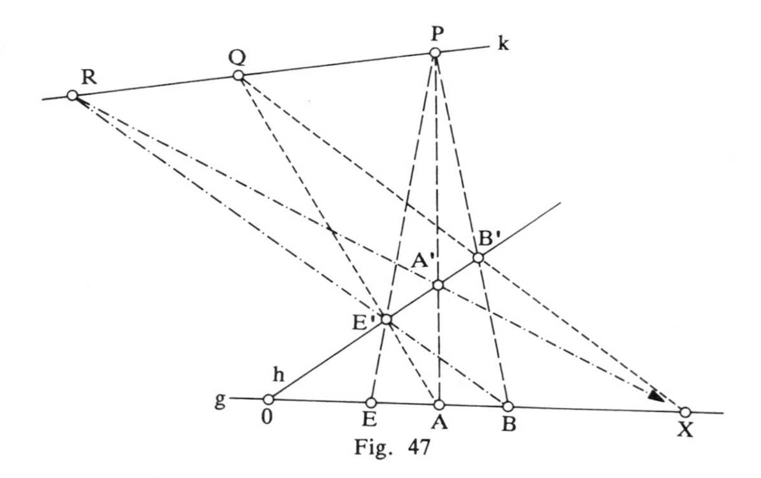


Fig. 47

(i) Closure Theorem of **Pappus**.
The lines g, h and k are given. By projection from the centre P, the points E', A' and B' correspond to the points E, A and B. The further steps of construction are given in the drawing (Fig.47). Then the figure must close at X.

Hilbert (*Foundations of Geometry*, 1st edition 1899) showed – with additions due to various successors:

$(K, +, \cdot)$ is a field (and $\mathbf{A} \cong \mathbf{A}(K)$) $\Leftrightarrow$ In $\mathbf{A}$ the theorem of Pappus holds as an additional axiom.

(ii) Closure theorem of Desargues.
The triangles ABC and A'B'C' lie in *perspective* from the centre Z. Then the figure closes at X. (The triangles are also said to lie *axially*.)

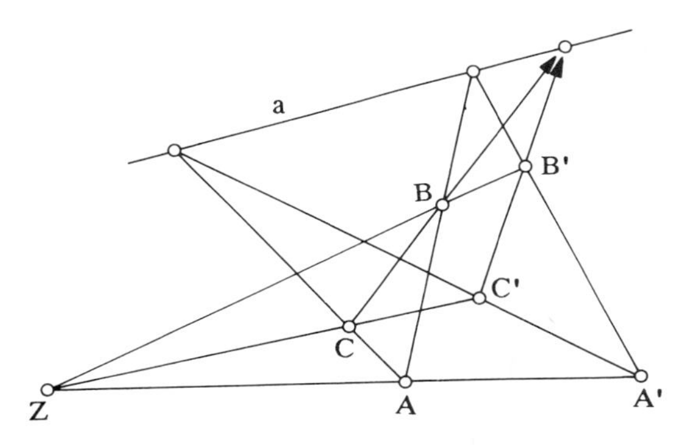


Fig. 48

Likewise, with Hilbert we can assert:

$(K, +, \cdot)$ is a skew field $\Leftrightarrow$ In $\mathbf{A}$ the theorem of Desargues is an additional axiom.

(iii) The 'Little' Closure Theorem of Desargues.
In this we make the additional assumption that the *axis* a already passes through the *centre*, so the 'Little' theorem is weaker than (ii) above.

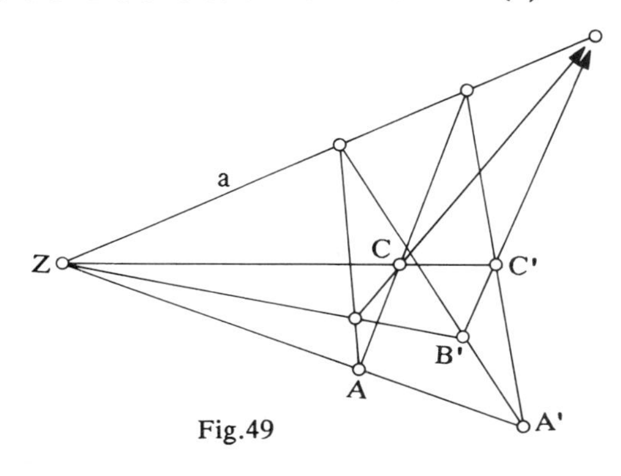


Fig.49

In a series of works (1931–1937), Ruth Moufang introduced the concept of an *alternative field* that we can express in the following way. An alternative field is a skew field with a weakened associativity law of the sort we find in the Cayley numbers. (We do not need greater precision here.)

Theorem of Moufang (1933/37):

$$(K, +, \cdot) \text{ is an alternative field} \Leftrightarrow \begin{cases} \text{In A the little theorem of} \\ \text{Desargues occurs as an} \\ \text{additional axiom.} \end{cases}$$

Around 1950 it was proved that one can obtain any alternative field through a construction similar to that of the Cayley numbers. The interested reader will find this and much further information about such connections between algebra and geometry in the book of G. Pickert: *Projective Planes* [55]. (Ruth Moufang, 1905–1977, came from Darmstadt and was a Professor in the University of Frankfurt.)

# 6

# Sets and numbers

INTRODUCTION.

With "pure sets", one can do little else than count their elements and compare their sizes; deliberately, by the very description "pure sets" we do not take cognisance of their other properties. If we were to regard the number systems in this way, only as 'pure' sets, we would be adopting a quite unnaturally abstract standpoint because the very concept of number includes the possibility of calculating as a constitutive feature. In spite of this, we obtain in this direction (following Cantor) some very interesting insights. With infinite sets we have already mentioned earlier the distinction, between the countable and the uncountable sets. This theme will be taken up once more in Section 6 A/B and treated systematically. The theory of sets opens up new possibilities of arithmetic, and with them access to new number– systems, e.g. with the help of the concept of cardinal number, to which Sections 6B, D and E are devoted. These new methods allow us in Section 6F to solve old problems – first we prove the isomorphism $(\mathbb{R}, +) \approx (\mathbb{R}^2, +)$ which has already been indicated, and then the Cauchy functional equation is treated.

Literature: Many of the newer books on sets are so detailed, that one can not see the wood for the trees. For learning more about this problem, one may refer to one of the following books. P. Halmos, *Naive Set Theory* [25]; Kamke: *Mengenlehre* [37]; I. Kaplansky : *Set Theory and Metric Spaces* [38].

In particular, with the application of Zorn's Lemma, the present text follows Kaplansky.

The volumes of Felscher [17], especially Vol.III, go considerably more deeply into the details. Felscher's general commentary is extraordinarily stimulating. His historical introduction III Section 6.1, on Cantor's discoveries, is recommended to every reader. For a survey of questions connected with the Axiom of Choice and the Continuum Hypothesis, one should read Jan Felix: "Freedom from contradiction, and Independence" [18].

## 6A EQUIPOTENT SETS.

Many sets have relatively few elements, for example the set of atoms in a pencil, and one can find an appropriate counting process to determine how many there are. This process can be described mathematically by means of bijective functions. First, let $\mathbb{N}_m$ denote the set $\{1, 2, \ldots, m\} \subseteq \mathbb{N}$.

We say: a set $M$ has $m$ elements if there is a bijective function

$$\varphi : M \to \mathbb{N}_m .$$

If we wish to see whether two sets $M, N$ have equally many elements, we count each set, using bijections

$$\varphi : M \to \mathbb{N}_m, \ \psi : N \to \mathbb{N}_n .$$

and then decide whether or not $m = n$. Underlying this procedure, we can really say : if we only want to know whether or not $M$ and $N$ have equally many elements, we could ignore the precise number $m$ and it would be simpler to approach the problem as follows. If there were bijections $\varphi : M \to \mathbb{N}_m$, $\psi : N \to \mathbb{N}_m$, then we would have a bijection $\psi^{-1} \circ \varphi : M \to N$ as suggested by diagram (i):

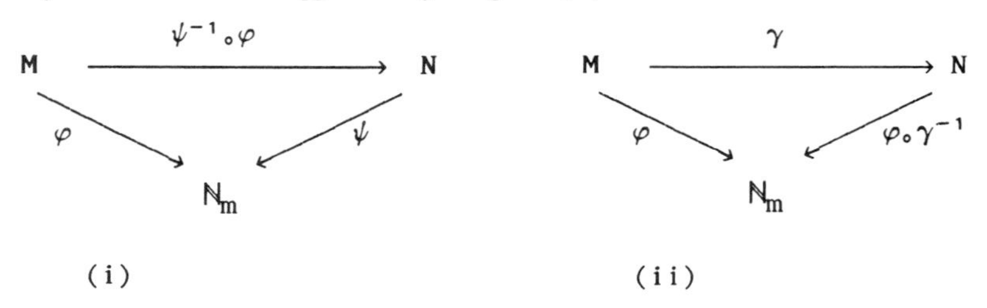


(i) (ii)

Conversely, if there are bijections $\gamma : M \to N$, and $\varphi : M \to \mathbb{N}_m$ then there is a bijection $\varphi \circ \gamma^{-1} : N \to \mathbb{N}_m$ as suggested by diagram (ii).

Thus one can say: two (finite) sets $M, N$ have equally many elements, if there is a bijection

$$\varphi : M \to N .$$

The existence of a bijection still makes sense for sets that are not finite. With this we have at hand a means to generalise the concept 'equally many elements' to arbitrary sets.

<u>Definition</u>. (Cantor 1878). Two sets M, N are called **equipotent** if there is a bijection

$$\varphi : M \to N .$$

For brevity we write: $M \approx N$.

If $\varphi$ is a bijection, it has an inverse $\varphi^{-1} : N \to M$ which is also a bijection. Therefore equipotency, like isomorphism, is a symmetric relation. One can say: equipotency is isomorphism of sets. Equipotency is also reflexive, because with the identity mapping

$$id_M : M \to M$$

every set is equipotent with itself. Further, we have transitivity, for if $\varphi : K \to M$, $\psi : M \to N$ are bijections, so also is $\psi \circ \varphi : K \to N$. In words : if $K \approx M$ and $M \approx N$, then $K \approx N$. Equipotency is therefore an *equivalence relation* between sets. Literally, "equipotent" means "of equal power" (in German, gleichmächtig = equally mighty) and "a power" is often used to mean a family of equipotent sets (see Section 6C).

*REMARK*. With our passage from finite to infinite sets, there is a vague analogy in elementary geomtry: the theorem of Thales tells us the precise size of certain angles, and in particular that angles in the same semi-circle are each equal to a right angle. The generalisation concerns angles on the same arc: it tells us that they are equally large, but not how large.

After having formalised the notion "equipotent", we must make sure that it does not have undesirable properties. For example, suppose $m, n \in \mathbb{N}$ and $m \neq n$. Can there be a bijection $\varphi : \mathbb{N}_m \to \mathbb{N}_n$? If so, that would be unpleasant, for then the concept would not meet our intentions. The following lemma reassures us.

<u>LEMMA</u>. *There is an injective mapping* $\alpha : \mathbb{N}_n \to \mathbb{N}_k$ *only when* $n \leq k$.

<u>*PROOF*</u>. We use induction on n. If $n = 1$, there certainly is such an injection when $1 \leq k$; and k cannot be $< 1$ in $\mathbb{N}$.

Now consider the inductive step. If $n \leq k$, there is an obvious injection $\alpha : \mathbb{N}_n \to \mathbb{N}_k$ given by

$$\alpha : \{1,\ldots,n\} \to \{1,\ldots,n,\ldots,k\} .$$

For the converse, suppose there is an injective mapping $\alpha : \mathbb{N}_n \to \mathbb{N}_k$. Let $\alpha(n) = i$. If by chance $i = k$, then the restriction

$$\alpha : \{1,\ldots,n-1\} \to \{1,\ldots,k-1\} .$$

is also an injection; so by an inductive hypothesis, $n-1 \leq k-1$, and hence $n \leq k$.

If, however, $\alpha(n) = i \neq k$, we apply a mapping $\beta$ as shown:

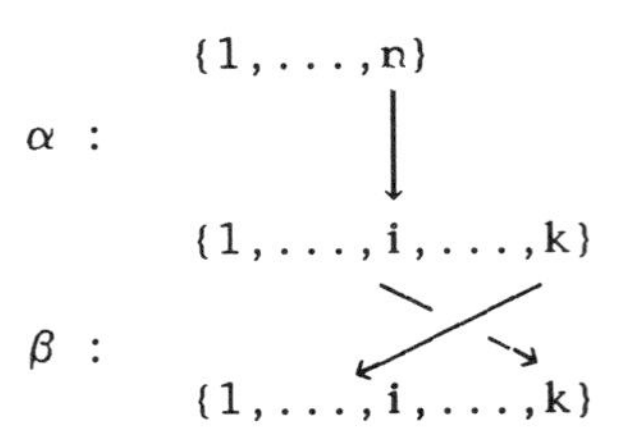


Then $\beta \circ \alpha : \{1,\ldots,n\} \to \{1,\ldots,k\}$ is an injection with $\beta \circ \alpha(n) = k$. Therefore by the earlier argument $n \leq k$. The lemma follows by induction. □

COROLLARY. *There is a bijective mapping* $\alpha : \{1,\ldots,n\} \to \{1,\ldots,k\}$ *when, and only when* $n = k$.

For if $\alpha$ is a bijection, then $\psi^{-1}$ exists and both $\psi$, $\psi^{-1}$ are injections. Hence $n \leq k$ and $k \leq n$. The rest of the proof is left to the reader. □

We can now give a formal definition of what is meant by a finite set.

Definition. A set M is said to be **finite** if either M is empty, or there is some $m \in \mathbb{N}$, with

$$M \approx \{1, 2, \ldots, m\}$$

The same content is naturally conveyed if one writes, say, $M = \{a_1,\ldots,a_m\}$ or a similar notation. The integer m is called the **cardinal number** of M, and by the preceding Corollary, this cardinal is uniquely defined.

In this definition, knowledge of an *infinite* set is used, for defining finiteness. This seems somewhat remarkable, and has led to philosophical discussion concerning the priority of "infinite" over "finite". Dedekind has shown how one can remove the dependence of "finite" from the pre-existence of the natural numbers, and we shall follow his line of thought. First, however, we still use the above definition.

THEOREM. *Every infinite set contains a subset that is equipotent with* $\mathbb{N}$.

*PROOF*. M is non-empty, so it contains some $a_1 \in M$. But then $M - \{a_1\}$ is non-empty, otherwise M would be the finite set $\{a_1\}$. Therefore there exists $a_2 \in M - \{a_1\}$. Again, $M - \{a_1,a_2\}$ cannot be empty, therefore there exists $a_3 \in M - \{a_1,a_2\}$. Proceeding in this way, one sees how we arrive at a subset $\{a_1,a_2,a_3,\ldots\} \approx \mathbb{N}$ in M. □

In a certain way then, $\mathbb{N}$ is the "smallest infinite set" — it can be inserted into any infinite set. (To within a change of notation; for example, in accord with our conventions, $\mathbb{N}$ can be inserted into the infinite set of squares 1, 4, 9, 16, 25,... .)

The proof of the theorem seems harmless enough, but the process used in it leads to far-reaching consequences. For, the *choice* of the elements $a_i$ from the infinitely many sets $M - \{a_1, _2, \ldots, a_{i-1}\}$ cannot follow from a process prescribed in advance, because one simply does not know enough about M. This, then raises the problem of the **Axiom of Choice,** (AC for short), and for further discussion the reader should look at the suggested Literature. We shall return to this again, later.

Exercise 1.

(a) (Revision) Find examples of mappings of the form $\varphi : \mathbb{N} \to X$, such that (a) $\varphi$ is an injection but not a surjection, (b) a surjection but not an injection.

(b) Prove that the composition of two injections, or two surjections, is respectively an injection or surjection. Give visual 'proofs' using arrows, and then symbolic proofs.

(c) Show how the Axiom of Choice arises in a proof of the statement

$$\exists\alpha : A \to B \ \text{ injective } \ \Leftrightarrow \ \exists\beta : B \to A \ \text{ surjective.}$$

Definition. A set M is called **Dedekind–finite** (briefly: $\text{fin}_D$) provided it satisfies:

For all $\psi : M \to M$, [$\psi$ is injective $\Leftrightarrow \psi$ is surjective].

To distinguish the previous notion of finiteness from this one, we write $\text{fin}_{\mathbb{N}}$ for the former, to indicate finiteness that refers to knowledge of $\mathbb{N}$.

THEOREM. (Dedekind). M *is* $\text{fin}_{\mathbb{N}} \Leftrightarrow$ M *is* $\text{fin}_D$.

*PROOF*. Suppose M is $\text{fin}_{\mathbb{N}}$; we may write briefly $M = \{a_1, \ldots, a_m\}$. If a given injective mapping

$$\alpha : \{a_1, \ldots, a_m\} \to \{a_1, \ldots, a_m\}$$

were not surjective, we could remove a suitable $a_i$ from the right, and after relabelling we would obtain an injection

$$\alpha : \{a_1, \ldots, a_m\} \to \{a_1, \ldots, a_{m-1}\}$$

By our earlier theorem, this would imply $m \leqq m-1$ which is impossible. Hence $\alpha$ must have been surjective.

Suppose, on the other hand, that M is not $\text{fin}_{\mathbb{N}}$. Then M contains a subset $N = \{a_1, a_2, a_3, \ldots\}$ equipotent with $\mathbb{N}$. If we define $\psi : M \to M$ by

$$\psi(x) = \begin{cases} x & \text{if } x \notin N \\ a_{i+1}; & \text{if } x = a_i \in N \end{cases}$$

then $\psi$ is injective but not surjective, since $a_1$ is not a value of $\psi$. □

Further possibilities for the so-called autonomous ($\text{fin}_D$) definition of finiteness can be found in Felscher [17] Vol.III.

So far, then, our state of knowledge concerning the number of elements of sets is, that we can distinguish those sets with 0, 1, 2,...,m elements, but the infinite sets lie before us, quite undifferentiated. In this direction, our intuition is quite unpractised, and we cannot make the distinctions inherent in the bare number of elements. At this stage, we resemble (pre) Stone Age people, who count "one, two, many".

## 6B THE NUMBER SYSTEMS AS UNSTRUCTURED SETS (COMPARISON OF CARDINALS)

The set $\mathbb{N}$ of natural numbers is not finite. For one reason, $\mathbb{N}$ is not equipotent with any of its sections $\{1,\ldots,m\}$ (so $\mathbb{N}$ is not $\text{fin}_{\mathbb{N}}$); for another, the correspondence $i \to 2i$ defines an injective mapping $\alpha : \mathbb{N} \to \mathbb{N}$ which is not surjective (so $\mathbb{N}$ is not $\text{fin}_D$). The bijective mapping of $\mathbb{N}$ on the even numbers shows that two sets can be equipotent, even though one has fewer eleemnts than the other. This is a considerable difficulty: is our concept "equipotent" defined correctly ÷ does it express what we intended or must we try to grasp that intention in a different way?

We shall, in fact, stay with the original definition on the following grounds: with respect to "numerosity", our intuition is accustomed only to small finite sets, and there we do have the assertion: "It cannot happen that a proper subset N of a finite set M is equipotent with M". Thus for sets, "fewer" and "equipotent" are mutually exclusive terms, provided we interpret the literary word "fewer" by the formal equivalence:

"N has fewer elements than M" $\Leftrightarrow_{def}$ $N \subseteq M$ and $N \neq M$.

It is not so with infinite sets, as we have seen: the concept "equipotent" does not have all the properties, for arbitrary sets, that we are used to with the concept "equal numbers of elements" applied to finite sets. Thus $\mathbb{N} \div \{1,2,3,4\}$ has equally many elements as $\mathbb{N}$, and yet fewer. With the statement $\mathbb{N} \approx 2\mathbb{N}$, one sees that we can remove from $\mathbb{N}$ as much as an infinite set (here, all odd numbers) and still be left with precisely as many as before!

This is the first "paradox of the infinite" from which Galileo (for example) stepped back in fright: "fewer and yet as many" was so hard to accept that the concept was allowed to lie fallow. Cantor was the first to cross this threshold, and it was his decisive, simple stroke of genius to substitute "equipotent" for "having equally many elements". People had already previously considered bijections, but had not recognised their fruitfulness. Further material concerning Cantor's discoveries can be found in Felscher [17], and elsewhere.

Let $\mathbf{P}$ denote the set of all prime numbers. Then $\mathbf{P} \approx \mathbb{N}$ because we can form a bijection $\mathbb{N} \to \mathbf{P}$ by enumerating the primes as a sequence $2 = p_1$, $3 = p_2$, $5 = p_3$. $7 = p_4$, $11 = p_5$, $13 = p_6, \ldots$ . As this sequence never terminates (by Euclid's theorem), the correspondence $i \to p_i$ gives a bijection $\varphi : \mathbb{N} \to \mathbf{P}$.

By a wholly similar argument, one derives:

<u>THEOREM</u>. *If* $M$ *is a subset of* $\mathbb{N}$, *then either* $M$ *is finite or* $M \approx \mathbb{N}$.

Here, the empty set is naturally regarded as finite. If $M$ is not finite, then it is not empty and must therefore contain a <u>smallest</u> element $m_1 \in M$. For $m_2$ we take the smallest element of $M - \{m_1\}$, and so on. As we always take the smallest element of $M - \{m_1,\ldots,m_k\}$, all elements of $M$ are eventually exhausted, and so a bijection of $\mathbb{N}$ on $M$ is found.

Within $\mathbb{N}$, therefore, there is only one type of "infinite", and the concept does not allow of any further refinement. We now consider sets that contain $\mathbb{N}$, and which are therefore "bigger" than $\mathbb{N}$.

<u>THEOREM</u>. $\mathbb{Z} \approx \mathbb{N}$.

<u>*PROOF*</u>. We write out $\mathbb{Z}$ in the following way:

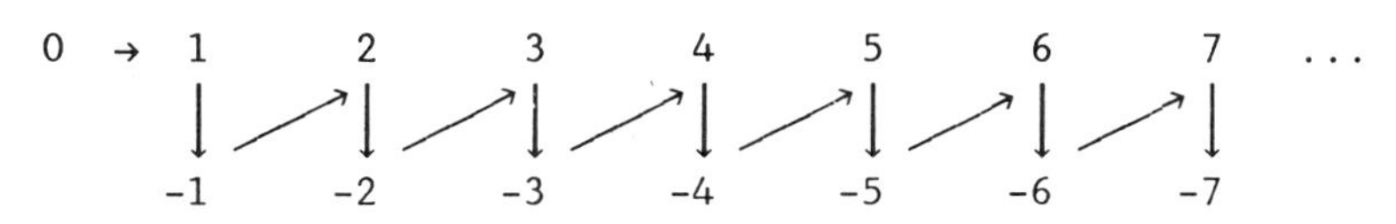


and obtain a bijection $\alpha : \mathbb{N} \to \mathbb{Z}$ by going along the top row and then down, to get

| $\mathbb{N}$ | 1 | 2 | 3 | 4 | 5 | 6 | 7 | ... |
|---|---|---|---|---|---|---|---|---|
| $\alpha \downarrow$ | ↓ | ↓ | ↓ | ↓ | ↓ | ↓ | | |
| $\mathbb{Z}$ | 0 | 1 | -1 | 2 | -2 | 3 | -3 | ... □ |

From now on, we shall usually present a bijection $\alpha : \mathbb{N} \to M$ in the form $M = (m_1, m_2, m_3, \ldots)$.

THEOREM. $\mathbb{Q} \approx \mathbb{N}$.

The proof results from the so-called first Cantor diagonal process, which is self-explanatory:

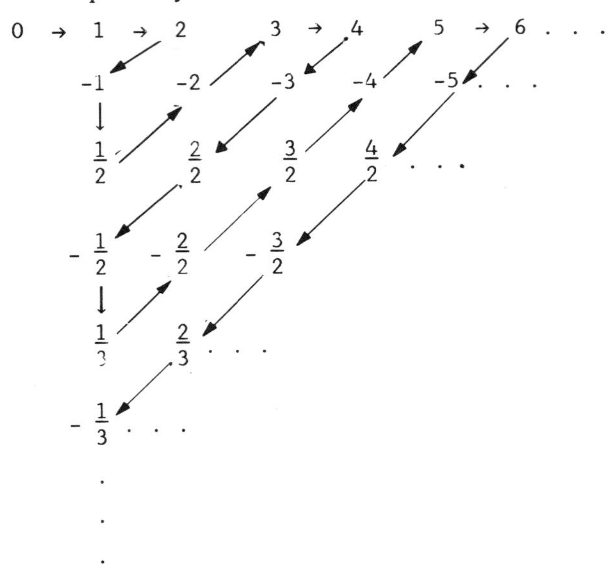


In this scheme, all rational numbers appear at least once. If we strike out those that have already appeared, (e.g. 2/2, because 1 already appears) then we obtain the enumeration of $\mathbb{Q}$:

0, 1, 2, -1, 1/2, -2, 3, 4, -3, -1/2, 1/3, 3/2, -4, ...

As the set $\mathbb{Q}^+$ of positive rational numbers, and Dez, are infinite subsets of $\mathbb{Q}$, then they are also countable (i.e. equipotent with $\mathbb{N}$).

Exercise 2.
(a) Show that $\mathbb{N} \times \mathbb{N} \approx \mathbb{N}$ by using the mapping $\alpha : (j, k) \to 3^j 5^k$.

(b) Show that $\{1, -1\} \times \mathbb{N} \times \mathbb{N} \approx \mathbb{N}$ using the mapping

$$\beta : (i, j, k) \to 2^{i+1} \cdot 3^j 5^k.$$

(c) With the help of these results conclude that $\mathbb{Q} \approx \mathbb{N}$.

THEOREM. *Let* **A** *denote the set of the algebraic numbers. Then*

$$\mathbf{A} \approx \mathbb{N}.$$

We have already given Cantor's proof in Section 3B.

THEOREM. $\mathbb{N}$ *and* $\mathbb{R}$ *are not equipotent.*

*PROOF*. In Section 1E we showed, using the so-called second Cantor diagonal process, that there is no surjective mapping $\beta$ of $\mathbb{N}$ on to the open real interval $I_o = \{x \mid 0 < x < 1\}$. If there were a bijective mapping $\alpha : \mathbb{N} \rightarrow \mathbb{R}$, then a subset of $\mathbb{N}$ would be surjectively mapped onto $I_o$, which is not possible. □

We have already given a second proof in Section 1F with the help of Cantor's characterization of $(\mathbb{Q}, \trianglelefteq)$.

The last theorem lifts our counting capabilities up to a new level. We have broken through the stone-age mentality "many is equally many". The concept "infinite" allows us to make distinctions, it organises, and allows contours to be recognised. In physics, there was a comparable insight when it was recognised that an atom is not an indivisible at all but has structure within itself.

As almost always in the natural sciences and mathematical research, such a discovery answers not only the old questions but poses immediately many new problems. In our case these are easy to name. Are there only two types of infinite sets (the countable ones, and those $\approx I_o$) or more? One would guess: many more. Which sets are $\approx I_o$? Obviously $I_o$ has countable subsets, e.g. $N = (1/2, 1/3, 1/4, 1/5,...\}$. Is there a set $M$ with $N \subset M \subset I_o$ which is neither countable nor $\approx I_o$? If two sets are equipotent, is there not some "potency" or "power" that each possesses in equal measure? Can one then calculate with "powers" in the same way as with the cardinals of finite sets? Can one arrange them in order of size? The reader will certainly be able to formulate many more questions.

*REMARK*. What would one say to the following objection, which was once made in a discussion? Surely, $\mathbb{R} \approx \mathbb{Q}$ — for, between any two irrational numbers there is a rational, and between any two rationals there is also an irrational. Therefore we have just as many irrationals as rationals, hence only countably many reals altogether.

*REPLY*. The opponent has hit on an illustrative idea which however only applies to ordered sets like $(\mathbb{Z}, \trianglelefteq)$ where each element has an upper and lower neighbour. If one takes, say, the even and odd numbers in $\mathbb{Z}$ then the "between each two" argument applies, and one obtains a bijection from the odd to the even numbers (and conversely) via the diagram

in which "x is mapped onto its upper neighbour". The unconscious transference of this idea onto the *dense* situation in $\mathbb{R}$ leads to wrong conclusions. There is no upper neighbour onto which one can map. Our intuition no longer extends this far: only logical argumentation helps us go further, and to do that, we need at least some elementary concepts concerning ordered sets.

We modify this objection a bit, to obtain a comparison of the (real) algebraic numbers **A** with the (real) transcendental numbers T. The set **A** is countable. If T were also countable, then one would have an enumeration of $\mathbb{R}$, which is not possible. Therefore T is not countable: there are more transcendentals than algebraic numbers in $\mathbb{R}$. This is Cantor's famous (non-constructive) proof of the existence of uncountably many transcendental numbers. As we have already mentioned, it is much harder to decide whether given, specific, numbers are transcendental or not.

What we still don't know is the exact size of T. Can we have, perhaps, $T \approx \mathbb{R}$? This question can be answered as follows, for example. Let $a_1, a_2, a_3,\ldots$ be an enumeration of the algebraic numbers. From the transendental numbers we choose a particular one, say e. Then 2e, 3e,...,ne,... are also transcendental for each $n \in \mathbb{N}$. Now define a mapping $\beta : \mathbb{R} \to T$ as follows:

let $$\beta(a_n) = 2ne \ , \quad \beta(je) = (2j-1)e$$

and if x is not of the forms $a_n$ or je, then we take $\beta(x) = x$. One sees immediately that $\beta$ is a bijection and hence that $\mathbb{R} \approx T$.

Researches on infinite subsets M of $\mathbb{R}$ always led to the conclusion that either $M \approx \mathbb{N}$, or $M \approx \mathbb{R}$. On the basis of this experience, Cantor formulated his so-called **Continuum Hypothesis**:

(CH) *Each (infinite) uncountable subset* M *of* $\mathbb{R}$ *is equipotent with* $\mathbb{R}$.

In other words, the "continuum" $\mathbb{R}$ represents the first new "power" after $\mathbb{N}$. The decision as to the truth or otherwise of (CH) proved to be extraordinarily difficult. In his famous lecture to the International Mathematical Congress, 1900, in Paris, Hilbert described (CH) as one of the most important problems for mathematics of the coming century. It soon turned out, that the treatment of this problem would require that the fundamental rules of Set-theory be made much more strictly precise than Cantor had done. The result was the so-called Axiomatic Set-Theory, with (say) the axiom system (ZF) of Zermelo and Frankel. In 1938, Gödel succeeded in proving that (CH) could not lead to any contradiction in (ZF); and in 1963, Cohen showed that (CH) is independent of (ZF) and of the Axiom of Choice, (AC). Thus one cannot decide the statement (CH) at all within the framework of a theory based on (ZF). The situation is quite similar to that of the discovery of non-euclidean geometry: the parallel axiom can neither be proved nor refuted on the basis of the remaining axioms of Euclid. In other words, it is *independent* of them.

<u>Exercise 3</u>. Construct simple bijections to prove that each open interval $\langle a, b \rangle$ in $\mathbb{R}$ is equipotent with $I_o = \langle 0, 1 \rangle$. Also prove:

$$I_o \approx \leqslant 0, 1 \rangle \approx \leqslant 0, 1 \geqslant \approx \mathbb{R}.$$

The search for new "powers" between those of $\mathbb{N}$ and $\mathbb{R}$ having remained unsuccessful, we will again go a step further and look beyond $\mathbb{R}$. A plausible argument for higher "powers" is suggested by the following idea:

Consider the half-open interval

$$J = \langle 0, 1\rangle = \{x \in \mathbb{R} \mid 0 \leqq x < 1\}$$

and the square

$$J^2 = J \times J$$

$$= \{(x, y) \mid 0 \leqq x, y < 1\} .$$

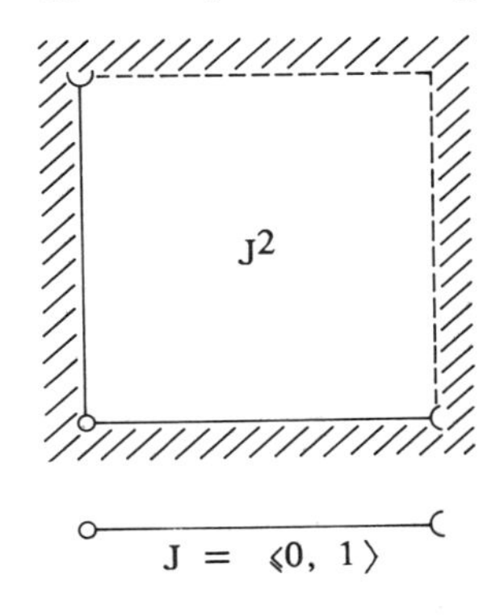


Fig. 50

The square has area 1 and the interval has area 0, so we expect more points to lie in the square than the interval. Cantor worked in this direction for three years until he discovered to his own surprise a simple proof that nevertheless $J \approx J^2$. We repeat his proof here.

<u>THEOREM</u>. *For the half−open interval* $J = \langle 0, 1\rangle$, *we have*

$$J \approx J^2 .$$

<u>*PROOF*</u>. In order to display the required bijection, we again use the infinite decimal expansion without recurring nines. First one tries to form a mapping rule for sending $(a, b) \to c$ through simple mixing of the subscripts, thus

$$(0.a_1a_2a_3,\ldots,\ 0.b_1b_2b_3,\ldots) \to 0.a_1b_1a_2b_2,\ldots$$

But with this, the inverse does not work, one would have e.g.

$$0.109090909,\ldots, \to (0.199,\ldots;\ 0.000,\ldots)$$

and therefore a decimal expansion with recurring nines would arise. In order to avoid this, one must try to think of a somewhat cleverer partition of the digits of a decimal expansion into two sequences, and we soon come quickly to one that works. The decimal expansion

$$x = 0.x_1x_2x_3 \ \ldots$$

is divided into sections in which each stops at the first number $\neq 9$, and then the next section begins.

*Example.* If $x = 0.39919629973899406\ldots$ , the partitioning looks like:

$$\underset{A_1}{3}\ \Big|\ \underset{A_2}{991}\ \Big|\ \underset{A_3}{96}\ \Big|\ \underset{A_4}{2}\ \Big|\ \underset{A_5}{997}\ \Big|\ \underset{A_6}{3}\ \Big|\ \underset{A_7}{8}\ \Big|\ \underset{A_8}{994}\ \Big|\ \underset{A_9}{0}\ \Big|\ \underset{A_{10}}{6}\ \Big|\ \ldots$$

The method results always in finite sections because no recurring nines are allowed.

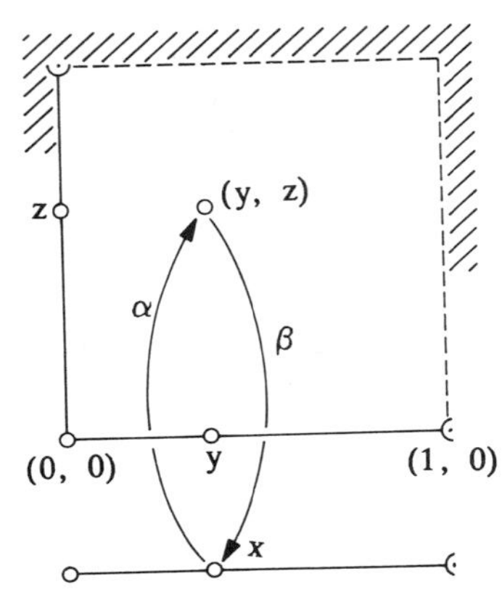


Fig. 51

Now form the decimal expansion

$$y = 0.A_1A_2A_3 \ldots$$

$$z = 0.A_2A_4A_6 \ldots$$

Each has no recurring nines because each $A_i$ ends on a number $\neq 9$. Using this method we have a correspondence $\alpha : x \to (y, z)$ which yields a mapping

$$\alpha : \langle 0, 1\rangle \to \langle 0, 1\rangle^2 ;$$

e.g. $\alpha(0.2745300 \ldots) = (0.24300 \ldots; 0.75000 \ldots)$

Next we must show that $\alpha$ is bijective, and to do this, we specify the inverse mapping

$$\beta : \langle 0, 1\rangle^2 \to \langle 0, 1\rangle .$$

It is simple enough to find $\beta$; given $(y,z) \in \langle 0, 1\rangle^2$; partition each of $y$ and $z$ by the above process, and obtain them in the forms

$$y = 0.B_1B_2B_3 \ldots, \quad z = 0.C_1C_2C_3 \ldots$$

Then we set

$$\beta(y,z) = 0 \cdot B_1C_1B_2C_2B_3C_3 \ldots$$

One sees immediately that $\beta$ is the inverse mapping of $\alpha$, and this completes the proof. □

What has all this got to do with area? Again one of the "paradoxes of the infinite" appears. The explanation is as follows. In the measurement of areas, one does not measure unstructured sets but sets which are structured by certain of their "situational properties", i.e. their relationships with other objects. A possibly inexact analogy: even though a sack of potatoes and a sack of table-tennis balls may contain just as many objects, they will have of course, completely different weights. This latter property comes from something which cannot be defined solely through counting.

Situational properties are respected e.g. by continuous functions. Our intuition is influenced much more strongly by geometrical (or rather, "topological") relationships within a structure, than by algebraic properties.

Every small child knows the topological difference between a line and a circle, for some time before it can even draw a straight line. Just because topological differences are so fundamental, they are very difficult to pin down mathematically. The appropriateness of topological notions will now be demonstrated using the comparison of the square $J^2$ with the segment $J = \langle 0,1 \rangle$.

***Definition***. Two subsets $M \subseteq R^m$ and $N \subseteq R^n$ are called topologically isomorphic (**homeomorphic**), whenever there is a bijective function $f : M \to N$ with inverse function $f^{-1} : N \to M$ such that both $f$ and $f^{-1}$ are continuous. (Continuity of $f$ alone does not suffice.)

THEOREM. $J_o$ *and* $J^2$ *are not topologically isomorphic.*

*PROOF*. Let $f : J^2 \to J$ be continuous. We show that $f$ can not be injective. To this end we choose a triangle with vertices P, Q and R in $J^2$ (see Fig.52). If f(P), f(Q), and f(R) are not all different then f is not injective. If they are different then in $J_o$ they are ordered by size, say $f(P) < f(Q) < f(R)$. On the segment PR, f is continuous, so by the intermediate value theorem there is a point C on PR with $f(C) = f(Q)$.

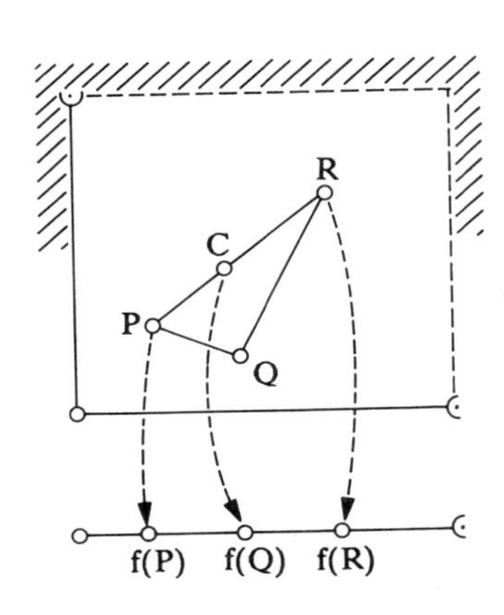


Fig. 52

Because P, Q and R are not collinear, $C \neq Q$; therefore f is not injective. □

By contrast with the pure Set-theory, Topology therefore keeps distinct what, in our intuition, should be distinct.. (The proofs however, become more difficult if one wants to show e.g. that for $n \neq m$ the spaces $R^m$ and $R^n$ are not topologically isomorphic).

We now return to Set-theory and will extend our result $J \approx J^2$ to the result that $R \approx R^2$. We first show that if I is the closed interval $\langle 0, 1 \rangle$ in R, and $I_o$ its interior $\langle 0,1 \rangle$ then $J \approx I_o$.

We have to insert the point 0 into the open interval by a suitable bijection, and Fig.53 illustrates the process, which shifts only the points of the sequence 0, 1/2, 2/3. ....

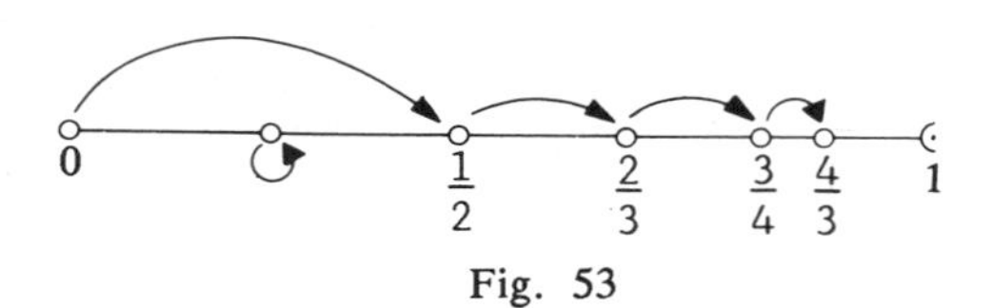


Fig. 53

Exercise 4. Show that $I_0$ and J are not topologically isomorphic, nor are $I_0$ and I, nor J and I, nor J and $\mathbb{R}$.

We can now put the different bijections together in the following way: From Exercise 3, there is a bijection $\alpha : \mathbb{R} \to J$ and with $\beta : J \to J^2$ we have $\mathbb{R} \approx J^2$. Also from the bijection $(x, y) \to (\alpha(x), \alpha(y))$ we have $\mathbb{R}^2 \approx J^2$. Consequently, using $\beta_0^{-1} \circ \alpha^{-1}$ we have $\mathbb{R}^2 \approx \mathbb{R}$. Since $\mathbb{R}^3 \approx \mathbb{R}^2 \times \mathbb{R}$, it immediately follows that $\mathbb{R}^3 \approx \mathbb{R}^2 \times \mathbb{R}$, etc.

Assembling the results of this section we obtain:

THEOREM. (Cantor). *For the various number systems, we have:*

$$\mathbb{N} \approx \mathbb{Z} \approx \mathbb{Q} \approx \mathbb{A} \not\approx \mathbb{R} \approx \mathbb{C} \approx \mathbb{R}^n \ . \qquad \square$$

This brings to light which fundamental difference is rooted in the completeness of $\mathbb{R}$. The most primitive aspect, the pure "power", is differentiated by the completeness (which is not to say, however that "uncountable $\Rightarrow$ complete" for all subsets of $\mathbb{R}^n$).

In the relationship between $\mathbb{R}$ and $\mathbb{R}^2$ we still have an old question: $\mathbb{R} \approx \mathbb{R}^2$ so they are "isomorphic as sets" but not topologically isomorphic. How do matters stand as between $(\mathbb{R}, +)$ and $(\mathbb{R}^2, +) = (\mathbb{C}, +)$? Are they algebraically isomorphic or not? The answer follows in Section 6F.

## 6C CARDINAL NUMBERS.

What we should take the "multiplicity" of a finite set to be is no problem: in that case $M \approx \{1,\ldots,m\}$ say , and so m represents the "power" of M; m is just the number of elements, or again the *cardinal number*, of M. However, just as with the first finiteness definition of $\mathrm{fin}_{\mathbb{N}}$, we have again fallen back upon the pre-existence of $\mathbb{N}$. With infinite sets, things are more difficult. We have no previously known "numbers" which correspond to the multiplicity of a set like $\mathbb{N}$ or $\mathbb{R}$. If we use the abbreviation card M for the cardinal number of M then we can indeed easily make a definition in terms of the relation $\approx$ as follows:

$$\text{card } M = \text{card } N \Leftrightarrow_{\text{def}} : M \approx N \ .$$

Thus, we do not thereby say what card M itself is, but only how we can *avoid* doing so. Early set-theorists like Bertrand Russell defined card M to be the S of all sets N with card M = card N, but the consideration of S has nowadays to be ruled out because such all-embracing classes are now known to involve possible contradictions.

Another approach is to fix a set from the class in question and define this as the cardinal number, – just as one works with remainders instead of remainder classes. For our purpose, it suffices that we use the above defined card M only in the sense of equipotence.

*REMARK.*

At the high point of the so-called modernising wave of mathematics teaching, at the beginning of the 1970's, it was opined that the natural numbers should be defined as cardinal numbers even for ten-year old pupils. Thus one finds, e.g. on p. 47 of the school text-book *Mathematik Heute*, (Vol.5, (1971) of Athen-Griesel) the explanation – presumably meant as a definition – *"A natural number is the common number property of equal-numbered* (gleichzähligen) *sets"*. How experienced teachers could write something like that in a school book is nowadays quite incomprehensible.

The Comparison of Sizes of Cardinal Numbers

The reason for using card M comes from trying to describe the multiplicity of M. In this spirit, we can say: if there is an injective mapping $\alpha : M \to N$, M will have fewer elements than N and we define

$$\text{card } M \leqq \text{card } N \; : \Leftrightarrow_{def} : \; \exists\, \alpha : M \to N \text{ injective.}$$

From an earlier exercise, this is equivalent to saying:

$$\exists\, \beta : N \to M \quad \text{surjective.}$$

For example, we have already seen that card $\mathbb{N} \leqq$ card $\mathbb{R}$. We can say more, however. It is natural to write (as in $\mathbb{N}$):

$$\text{card } M < \text{card } N \quad \text{iff} \quad \text{card } M \neq \text{card } N \text{ and card } M \leqq \text{card } N \; .$$

Then, since we proved in Section 1E that $\mathbb{R}$ is uncountable, we have

$$\text{card } \mathbb{N} < \text{card } \mathbb{R} \; .$$

From our intuitive idea of what multiplicity should mean, we would expect that our definition should imply a linear ordering. We test the corresponding properties.

1). *$\leqq$ is reflexive.* If card M = card K, then there is a bijective mapping $\alpha : M \to K$. This is certainly injective, so we have

$$\text{card } M = \text{card } K \Rightarrow \text{card } M \leqq \text{card } K \; .$$

(card M $\leqq$ card M does not suffice, since we do not usually handle only a single set.)

2). *$\leqq$ is transitive.* Suppose

$$\text{card } M \leqq \text{card } K \quad \text{and} \quad \text{card } K \leqq \text{card } L \; .$$

From the diagram, one can see that there is then also an injective function $\beta \circ \alpha : M \rightarrow L$ and therefore that card $M \leqq$ card L.

$$M \xrightarrow{\alpha} K \xrightarrow{\beta} L$$

$$\beta \circ \alpha$$

3). $\leqq$ *is antisymmetric.* We have card $A \leqq$ card B and card $B \leqq$ card A. From this we need to show that card $A =$ card B. But that task is not simple, and it is the content of the following theorem:

<u>THEOREM</u> of Schröder–Bernstein. *If there exist injective functions* $\alpha : A \rightarrow B$ *and* $\beta : B \rightarrow A$, *then there is also a bijective function* $\gamma : A \rightarrow B$.

*<u>PROOF</u>.* On the image $\alpha A \subseteq B$ of the injective function $\alpha : A \rightarrow B$, we define the inverse function $\alpha^{-1} : (\alpha A) \rightarrow A$, and correspondingly for B. It is in this sense that we understand $\alpha^{-1}$, and $\beta^{-1}$ in the following.

(a) *First proof–step*: partitioning the set A. For each element $a \in A$, we define its *Family tree* T(a) as a succession of elements specified by the following procedure:

if $a \in \beta B$, go to $\beta^{-1}a$, otherwise stop;
if $\beta^{-1} \in \alpha A$, go to $\alpha^{-1}(\beta^{-1}a)$, otherwise; stop
if $\alpha^{-1}(\beta^{-1}a) \in \beta B$, go to $\beta^{-1}(\alpha^{-1}\beta^{-1}a)$...
... etc.

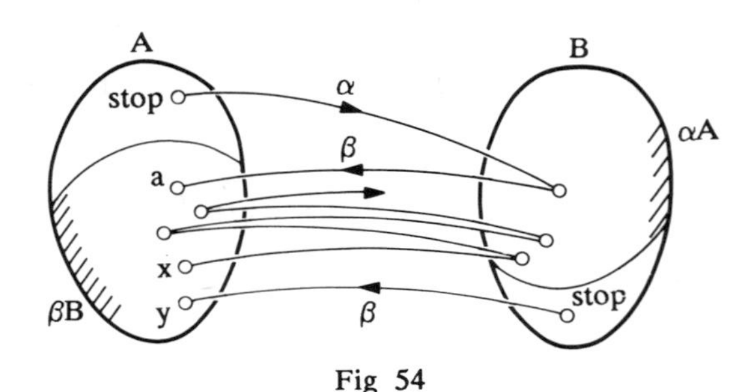


Fig 54

The ancestry of a then looks something like this:

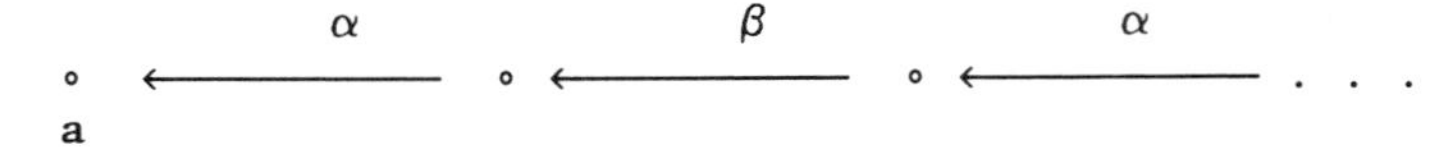


Fig.55

As Fig.55 indicates, there are three possible cases for the Family tree T(a):

(i) T(a) finishes after finitely many steps in A (and even with the first step, in the case $a \notin \beta B$).

(ii) T(a) finishes after finitely many steps in B.

(iii) T(a) does not end.

From these three possibilities, we can split the set A into three disjoint sets:

$$\begin{aligned} A_{(A)} &= \{a \in A \mid T(a) \text{ ends in } A\} \\ A_{(B)} &= \{a \in A \mid T(a) \text{ ends in } B\} \\ A_{(\infty)} &= \{a \in A \mid T(a) \text{ does not end.} \end{aligned}$$

The corresponding procedure is also followed through for the set B, and produces a splitting of B into the disjoint subsets $B_{(A)}$, $B_{(B)}$ and $B_{(\infty)}$.

(b) *Second proof–step*: construction of a bijective mapping from A to B.

We now look to see how these various parts of A and B will be mapped onto one another. Considering first $a \in A_{(A)}$, then the element $\alpha(a) \in B$ has the following ancestry:

$$\alpha a \xrightarrow{\alpha^{-1}} a \xrightarrow{\beta^{-1}} \ldots \xrightarrow{\alpha^{-1}} \circ \quad \text{stop in A},$$

for, T(a) is a portion of T($\alpha$a) and ends in A, so $A_{(A)}$ will be mapped by $\alpha$ into $B_{(A)}$. However, the mapping $\alpha : A_{(A)} \to B_{(A)}$ is also surjective, because if $b \in B_{(A)}$ then T(b) looks as follows:

$$b \xrightarrow{\alpha^{-1}} a \xrightarrow{\beta^{-1}} \circ \quad \ldots \quad \text{stop in A.}$$

Here T(b) ends after finitely many steps in A. Therefore it must return after the first step to some $a \in A$ for which T(a) ends in A. Hence $b = \alpha a$ with $a \in A_{(A)}$.

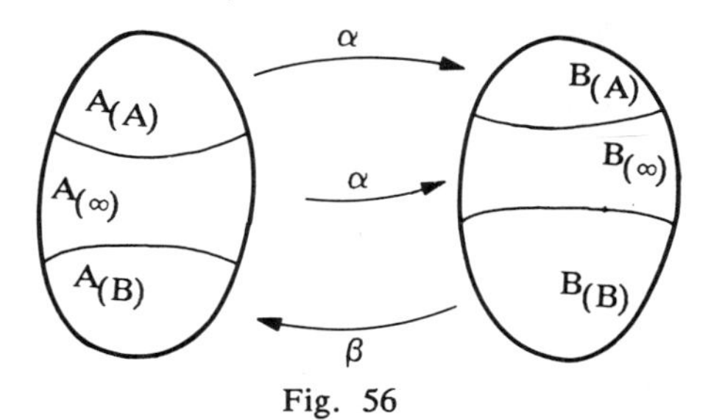


Fig. 56

Correspondingly one shows that the restrictions $\alpha : A_{(\infty)} \to B_{(\infty)}$ and $\beta : B_{(B)} \to A_{(B)}$ are bijective. Then $\beta^{-1} : A_{(B)} \to B_{(B)}$ is also bijective, and from the disjoint splitting we obtain a bijective function

$$\gamma : A \to B \text{ with } \gamma(a) = \begin{cases} \alpha(a) & \text{if } a \in A_{(A)} \text{ or } a \in A_{(\infty)} \\ \beta^{-1}(a) & \text{if } a \in A_{(B)} \end{cases}.$$

This completes the proof of the Schröder–Bernstein theorem. □

Our order-relation card M $\leq$ card N has now been shown to be reflexive, transitive and anti-symmetric but we still have to prove its linearity. For this we need a new method of proof to be introduced in the next section. Before doing this, we shall create for ourselves from card $\mathbb{N}$ and card $\mathbb{R}$ some bigger cardinal numbers. Otherwise, if we had but two infinite cardinal numbers, we would always be somewhat poverty-stricken.

Given a set M, we denote by $\wp M$ the set of all subsets of M – the so-called **power set** of M. For a finite set $M = \{a_1, \ldots a_m\}$ with m elements, the power set $\wp M$ has exactly $2^m$ elements. One can see this in different ways, e,.g, with the help of the characteristic function $\chi_T$ of a subset $T \subseteq M$, which is a mapping $\chi_T : M \to \{0, 1\}$, defined by

$$\chi_T(a) = \begin{cases} 1 & \text{if} \quad a \in T \\ 0 & \text{if} \quad a \notin T \end{cases}$$

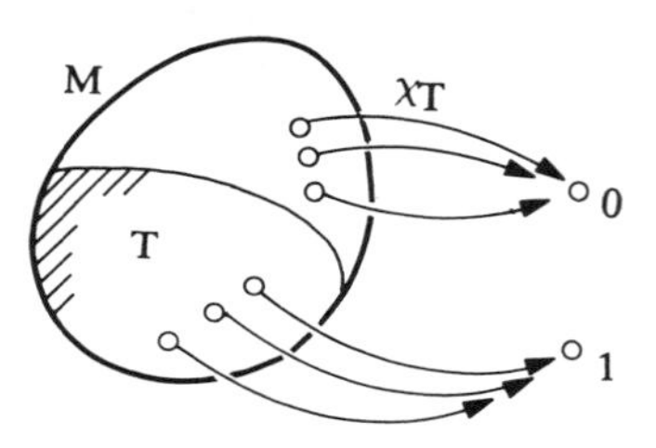


Fig. 57

The assignment $T \to \chi_T$ yields a bijective mapping from $\wp M$ onto the set C of the characteristic functions of the subsets of M. Usually, one denotes by $M^k$ the set of all functions $f : K \to M$; with the abbreviation $2 = \{0, 1\}$, we can write $C = 2^M$ . Therefore $\wp M$ is also commonly written as $2^M$ though the precise meaning refers to the characteristic functions. Cantor noticed that the increase from the multiplicity of M to that of $2^M$ is so strong that it also leads, in the infinite case, to different powers (Even when $M = \emptyset$, $\wp M$ is the singleton $\{\emptyset\}$, so card $\wp M$ is here $1 \neq 0$.)

<u>THEOREM</u>. (Cantor). *Always, card* M *is strictly smaller than card* $\wp M$, *so*:

$$\text{card } M < \text{card } \wp M \, .$$

<u>*PROOF*</u>. With the help of the injective mapping $\alpha : M \to \wp M$ given by $\alpha(m) = \{m\}$, we immediately obtain card $M \leq \wp M$.

To prove card $M \neq$ card $\wp M$, we need the fact that there is no bijective function $M \to \wp M$. Because injective functions do exist we claim: *there are no surjective functions* $\beta : M \to \wp M$.

To establish this, we must show that for any function $\beta : M \to \wp M$ there is always a subset T of $\wp M$ which is not the image under $\beta$ of some $m \in M$. To this end Cantor introduced the subset $A \subseteq M$ with

$$A = \{a \in M \mid a \notin \beta a\} \; .$$

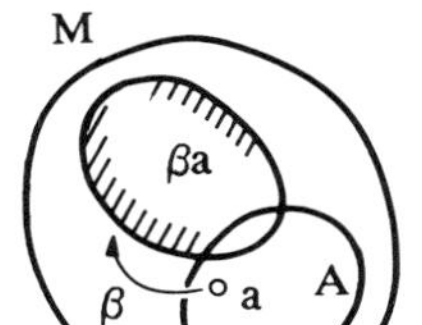


Fig. 58

It might happen that $A = \emptyset$ as with the function $\alpha$ above, but this does not matter because then $\emptyset \in$ ϸM.

*CLAIM*. There is no $b \in M$ with $\beta b = A$.

To prove this we distinguish two cases for elements x, y in M:

(i) Suppose $x \notin A$. Then x belongs to $\beta x$, but not to A. Therefore $\beta x \neq A$.

(ii) Suppose $y \in A$. Then y does not belong to $\beta y$, and hence $y \in A$. Therefore $\beta y \neq A$.

With these two cases, all elements of M are taken into account, so A does not occur as a value of $\beta$. This completes the proof. □

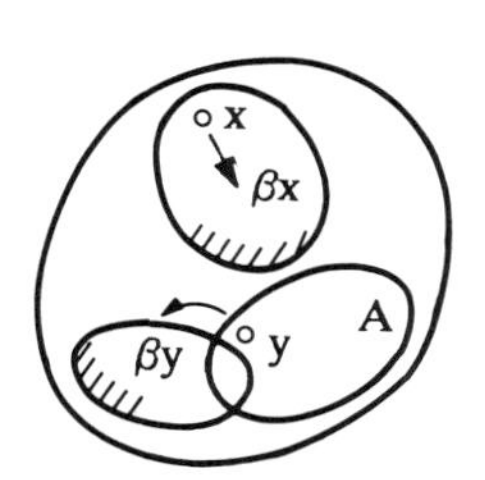


Fig. 59

*CONSEQUENCE*. Putting $K_1 = \mathbb{N}$ and ϸ$K_i = K_{i+1}$, we obtain the following chain of infinitely many, strictly increasing cardinalities:

$$\mathbb{N} = K_1, \; ϸ\mathbb{N} = K_2, \; ϸK_2 = K_3, \; \ldots$$

with

$$\text{card } \mathbb{N} < \text{card } ϸ\mathbb{N} < \text{card } K_3 < \text{card } K_4$$

contemplating this chain, we raise at once a number of questions:

1) Can we find card $\mathbb{R}$ somewhere in the chain, or alternatively insert it in the right gap?

2) Does the linearity of the ordering relation hold for the cardinal numbers? Is there perhaps still a set M, for which card M lies "alongside" (rather than in) the chain?

3) Between any members of this chain, can we find still more cardinal numbers, and are there beyond the "far end" of the chain, still bigger cardinal numbers?

The most immediate — and important — question concerning insertion of card $\mathbb{R}$ can be answered fairly easily.

THEOREM. *We have:* card $\mathbb{R}$ = card $\mathfrak{p}\mathbb{N}$ .

This means that there are just as many real numbers as there are subsets of $\mathbb{N}$.

*PROOF*. We use the characteristic functions of the subsets T of $\mathbb{N}$. These are nothing other than 0–1 sequences:

$$\tau_i = \begin{cases} 1 & \text{if } i \in T \\ 0 & \text{if } i \notin T \end{cases}$$

Let J denote the interval $\langle 0, 1\rangle$. With the help of $\tau_i$, we form the decimal expansion $0.\tau_1\tau_2\tau_2\ldots$, and so we obtain an ***injective*** mapping $\alpha : \mathfrak{p}\mathbb{N} \to J$ given by

$$\tau : T \to 0.\tau_1\tau_2\tau_3 \ldots = \Sigma \frac{\tau_i}{10^i}$$

If we also use the binary expansion, we obtain the following surjective mapping $\beta : \mathfrak{p}\mathbb{N} \to J$ :

$$\beta : T \to 0.\tau_1\tau_2\tau_3 \ldots = \Sigma \frac{\tau_i}{2^i} .$$

From $\alpha$ we obtain card $\mathfrak{p}\mathbb{N} \leq$ card J and from $\beta$, card I $\leq$ card $\mathfrak{p}\mathbb{N}$. Therefore by the Schröder–Bernstein theorem, we obtain: card $\mathfrak{p}\mathbb{N}$ = card J. We already know that $J \approx \mathbb{R}$. □

Note: The last two theorems together lead to a third proof for uncountability of $\mathbb{R}$, because we now have

$$\text{card } \mathbb{N} < \text{card } \mathfrak{p}\mathbb{N} = \text{card } \mathbb{R} .$$

*Exercise 5*. The completeness of $\mathbb{R}$ also enters into this argument. Find the place(s) where it is used implicitly.

In the next section, we shall use a new method of proof, to demonstrate the linearity of the ordering relation for cardinal numbers. We just add here a few words concerning the third question, that of the existence of a cardinality that lies between card $\mathbb{N}$ and card $\mathfrak{p}\mathbb{N}$ (i.e. between card $\mathbb{N}$ and card $\mathbb{R}$). This is again the already mentioned Continuum problem. If the question is generalised to cardinalities between M and $\mathfrak{p}M$ for infinite M, then we have the *generalised continuum hypothesis* (GCH). The situation here is of course, just as difficult as with the continuum hypothesis (CH).

*EXAMPLE.* We consider the set $E\mathbb{N}$ of all finite subsets of $\mathbb{N}$. Is card $E\mathbb{N}$ = card $\mathbb{N}$ or card $E\mathbb{N}$ = card $\mathbb{R}$? Again the characteristic functions will help us. If $T \subseteq \mathbb{N}$ is finite, and $\tau$ its characteristic function, then there are only finitely many $\tau_i$ with $\tau_i = 1$. Using the set $\mathbf{P} = \{p_1, p_2, p_3, \ldots\}$ of prime numbers we can form

$$\alpha : T \to p_1^{\tau 1} \; p_2^{\tau 2} \ldots = \Pi \, p_i^{\tau i} .$$

The theorem of the uniqueness of prime factorization in $\mathbb{N}$ implies the injectivity of $\alpha : E\mathbb{N} \to \mathbb{N}$, and therefore card $E\mathbb{N} \leq$ card $\mathbb{N}$. In other words: $E\mathbb{N}$ countable.

The second part of Question 3 asked about even bigger cardinal numbers. As there we have the sets $K_{i+1} = \mathfrak{P}K_i$, and $K_1 = N$; we form the set

$$K = \bigcup_i K_i$$

which has a colossal size!. Obviously there is an injective mapping $\alpha_i : K_i \to K$ for each $i$, so card $K_i \leq$ card $K$. But because also card $K_{i+1} \leq$ card $K$ and card $K_i <$ card $K_{i+1}$, then card $K$ is strictly greater than each card $K_i$. If now we put $L_1 = K$ and form $\mathfrak{P}L_1$ etc., then by iterating the process, even bigger monsters occur. The commited researcher can easily see how dangerous this might be. If one forms in a naive way the set $\mathbf{M}$ of all sets and its cardinal number card $\mathbf{M}$, we must have card $\mathbf{M} <$ card $\mathfrak{P}\mathbf{M}$. However, $\mathbf{M}$ is supposed to contain all sets, (including $\mathfrak{P}\mathbf{M}$ itself), so card $\mathbf{PM} \leq$ card $\mathbf{M}$. The formation of concepts like these, soon after Cantor had opened the paradise garden of Set-theory, led to contradictions that seemed to make the expulsion from it inescapable. With united efforts, mathematicians have succeeded up to now in keeping the contradictons out of the theory by restricting the permitted types of set-formation, and by similar expedients.

## 6D ZORN'S LEMMA AS A PROOF PRINCIPLE

We return again to the question: Is the ordering of cardinal numbers linear? For treating this question, we must get to know a method of working which we have already used – almost unconsciously. It depends on the use of the Axiom of choice or equivalent procedures. For an explanation of the problem, consider the use of the axiom of induction in the proof of statements about the natural numbers. In many circumstances, the inductive conclusions are so simple that one can be content with a simple "etc" instead of the formal proof, and with that the matter can be regarded as settled. Or, instead of induction, one can work with the so-called *well-ordering* of $\mathbb{N}$, – i.e. with the fact that each non-empty sub-set of $\mathbb{N}$ has a smallest element.

As a type of substitute for induction with natural numbers, conclusions can be made about finite and (countably) infinite domains of Set–theory that use the choice axiom in its different, mutually equivalent versions. The most important of these versions are:

(AC) **Axiom of Choice** (Zermelo). *Let* $\mathbf{M}$ *be a set of subsets of a set* A. *Then there exists a function* $f : \mathbf{M} \to A$ *with* $f(M) \in M$ *for all* $M \in \mathbf{M}$.

This function f chooses from each subset M of the system $\mathbf{M}$ an element $f(M) = m \in M$, and is therefore called a *choice function*.

(WO) **Well ordering axiom** (Cantor). *Each set can be well–ordered.*

In this, one calls a chain $(K, \trianglelefteq)$ **well–ordered**, whenever each non–empty subset $M \subseteq K$ has a smallest element. We have already referred to the well–ordering of $(\mathbb{N}, \trianglelefteq)$. We can immediately prove the choice axiom with the help of the well–ordering axiom: simply well–order the set A and take f(M) to be the smallest element of the subset $M \subseteq A$.

(ZL) **Zorn's Lemma**. *Let* $(\mathbf{M}, \trianglelefteq)$ *be an ordered set with the property:*

*each chain* K *in* $\mathbf{M}$ *has an upper bound* S *in* $\mathbf{M}$.

*Then there is a maximal element in* $\mathbf{M}$.

($M \in \mathbf{M}$ is ***maximal*** whenever the statement $M < N$ is false for every $N \in \mathbf{M}$.

This last statement looks somewhat more long–winded in its formulation than the previous ones, but it is, however, wholly analogous to the formulation of the induction axiom when compared with the well–ordering in $\mathbb{N}$. The transition from the chain K to the upper bound S can be seen to be analogous to the inductive step (seen, in the following formulation, say: if $1,\ldots,k$ belongs to the subset $T \subseteq \mathbb{N}$, then $s = k + 1$ does too.) Because of the formation of S, one usually must go beyond K, as with the transition from k to k + 1. In most cases of the use of Zorn's lemma, one can easily form the bound S, as described below. The harder problem is however, to show that S belongs to the set $\mathbf{M}$.

In many applications, as we have already seen in Section 6E/F, the ordered set $\mathbf{M}$ consists of subsets of a set A and the order relation is simply set–inclusion. For a chain $\mathbf{K}$ of subsets, it is then easy to construct an upper bound S by taking their union $S = \bigcup \{K \mid K \in \mathbf{K}\}$. In this situation, Zorn's lemma takes the following form.

(ZL') *Let* $\mathbf{M}$ *be a set of subsets of a set* A *with the property: For each chain* $\mathbf{K}$ *in* $\mathbf{M}$, *its union* $S = \bigcup \{K \mid K \in \mathbf{K}\}$ *belongs also to* $\mathbf{M}$.

*Then there is a maximal element in* **M**.

One can read in Felscher (Vol.III), Kaplansky, or in other books on Set–theory that on the currently acceptable foundations of mathematics – say the axiom system (ZF) of Zermelo-Fraenkel – the statements (AC), (WO), (ZL) are mutually equivalent. Also, however, these statements are, like the continuum hypothesis (CH), independent of (ZF).

Zorn's lemma has proved itself to be the most flexible of the three equivalent formulations, and it has rather eclipsed the other two. We work in this section with (ZL'), and the work will enable us to become familiar with it. With our earlier reference to the analogy with induction, we hope to avert the danger, which Felscher (Vol.III, p.124) describes as follows: "In its completely unexplained effectiveness and motivated only by the consequences derived from its applications, Zorn's lemma seems like a true *deus ex machina*, which by eliminating the transfinite ordinal numbers, also immediately succeeds in the elimination of the understanding of the set–theoretic mode of proof". The analogy between the use of Zorn's lemma and induction becomes clearly visible in the proof of the following theorem.

THEOREM. *The cardinal numbers are linearly ordered by* $\leqq$.

In other words: *given sets* A, B *there is always an injective function* $\alpha : A \to B$ *or an injective function* $\beta : B \to A$.

*PROOF*. If $A = \emptyset$ then we have the (injective) empty function $\emptyset \to B$. If A and B are both non-empty, then there exist $a_1 \in A$ and $b_1 \in B$ and an injective partial mapping

$$\alpha_1 : \{a_1\} \to \{b_1\} .$$

We will now try – as with the simultaneous counting of two sets – to construct further injective mappings by adding in further elements, say

$$\alpha_2 : \{a_1, a_2\} \to \{b_1, b_2\}$$

$$\alpha_3 : \{a_1, a_2, a_3\} \to \{b_1, b_2, b_3\} .$$

Each time, we replace the previous mapping with the succeeding one, and we continue like this until either A or B is used up. This idea can be modified in such a way that it works also in the infinite (particularly uncountable) case.

Using some index set I, we index all bijective mappings of the form

$$a_i : A_i \to B_i \quad \text{with} \quad A_i \subseteq A \quad \text{and} \quad B_i \subseteq B$$

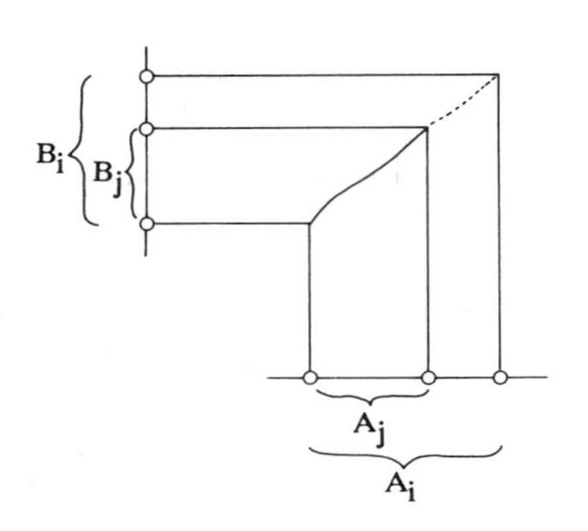


Fig. 60

On this set $\mathbf{M}$ of bijections, we shall define an order-relation $\leqslant$, in order to have the necessary ordered set $(\mathbf{M}, \leqslant)$ for applying Zorn's lemma. Thus, given mappings $\alpha_i$, $\alpha_j \in \mathbf{M}$ we write $a_j \leqslant \alpha_i$ whenever $\alpha_i$ continuation of $\alpha_j$, i.e. whenever $A_j \subseteq A_i$ and $B_j \subseteq B_i$ and $\alpha_i(a) = \alpha_j(a)$ for all $a \in A_j$. The sketch indicates that $\alpha_i$ can be described simply as having a larger domain and codomain than does $\alpha_j$. The way in which these sets are nested within each other shows immediately that our relation $\leqslant$ really is an order relation on $\mathbf{M}$. Now let $\mathbf{K}$ be a chain of nested functions $\kappa_j : A_j \to B_j$ in $\mathbf{M}$. Then we form

$$\kappa_o = \bigcup \{\kappa_j | \kappa_j \in K\} \quad \text{with} \quad \kappa_o : A_o \to B_o$$

where $A_o = \cup A_j$ and $B_o = \cup B_j$ are the corresponding unions of the subsets. We now claim that $\kappa_o$ is an upper bound of $\mathbf{K}$ that lies in $\mathbf{M}$.

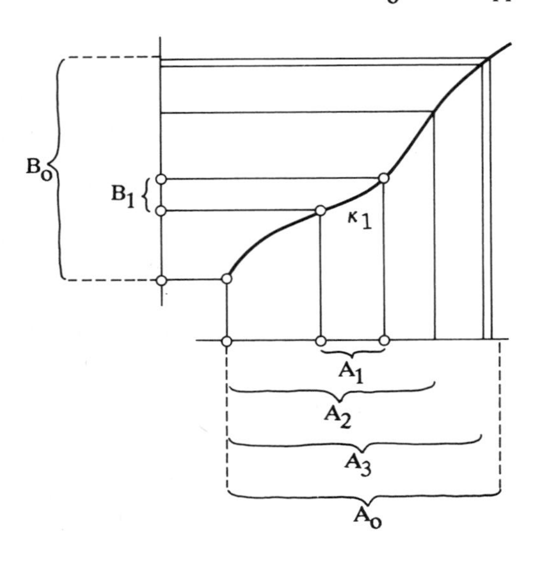


Fig. 61

It is intuitively clear from Fig.61 that in taking the union of all domains and codomains, a bijective mapping arises as claimed. The formal argument runs as follows. Certainly $A_o$ and $B_o$ are subsets of $A$ and $B$. If $a_i \neq a_j$ we must show that $\kappa_o(a_i) \neq \kappa_o(a_j)$ to establish that $\kappa_o$ is injective. Now $a_i$ and $a_j$ lie in subsets which belong to the chain of domains – say $a_i \in A_i$ and $a_j \in A_j$ with $A_j \subseteq A_i$. Then $\kappa_o(a_i) = \kappa_i(a_i)$ and $\kappa_o(a_j) = \kappa_j(a_j) = \kappa_i(a_j)$ because $\kappa_i$ is the continuation of $\kappa_j$. Since $\kappa_i$ is injective then

$$\kappa_o(a_i) = \kappa_i(a_i) \neq \kappa_i(j) = \kappa_o(a_j);$$

thus $\kappa_o$ is also injective. Correspondingly, by considering a given $b_i \in B$ one can show that $\kappa_o$ is surjective. Therefore $\kappa_o : A_o \to B_o$ is bijective and belongs to **M**. Further, from its definition, $\kappa_o$ is a continuation of each $\kappa_i \in$ **K** so $\kappa_i \leqslant \kappa_o$ for each $\kappa_i \in$ **K**; hence, $\kappa_o$ is an upper bound of the chain **K** and it lies in **M**.

We may now use Zorn's lemma, to obtain in **M** a maximal element which is a bijection of the form

$$\gamma : A^* \to B^*$$

where $A^* \subseteq A$ and $B^* \subseteq B$. This bijection is almost what we want, which is to find either an injection of A in B, or of B in A. We claim that either $A = A^*$ or $B = B^*$. Otherwise, we would have both $A^* \neq A$ and $B^* \neq B$; so there would exist elements $a \in A$ and $b \in B$ with $a \notin A^*$ and $b \notin B^*$. We could then enlarge the bijective function $\gamma$ to an even bigger bijection $\delta: A^* \cup \{a\} \to B^* \cup \{b\}$, where on $A^*$, $\delta$ is $\gamma : A^* \to B^*$, and $\delta(a) = b$.

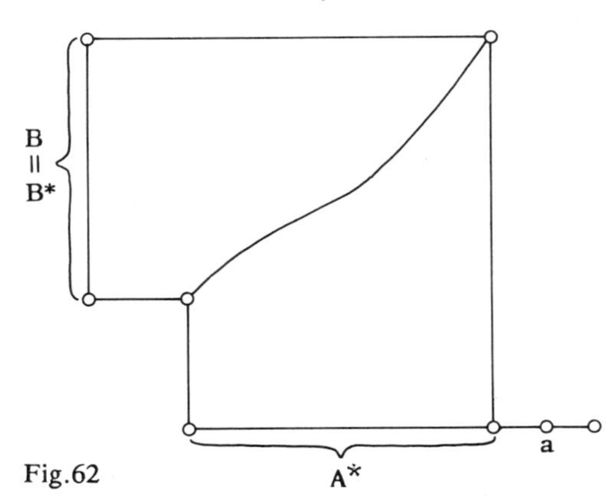


Fig.62

But this is not possible since $\gamma$ was maximal, so either $A = A^*$ or $B = B^*$ as claimed. If $A = A^*$, then we have the injective function $\gamma : A \to B^* \subseteq B$ from A to B. If $B = B^*$ then we have the injective function $\gamma^{-1} : B \to B^* \subseteq A^* \subseteq A$ from B to A. In either case however, we have an injective function as desired. □

With this, the question as to the linear ordering of the cardinal numbers is answered affirmatively. (One can also read in Felscher [17] Vol.III, pp. 165,66, that the theorem just proved is logically equivalent to (AC)).

## 6E THE ARITHMETIC OF CARDINAL NUMBERS.

For finite sets, the following description of addition, using the union of sets, is familiar. Suppose $M = \{m_1,\ldots,m_K\}$ and $N = \{n_1,\ldots,n_\ell\}$ are disjoint sets, then $M \cup N = \{m_1,\ldots,m_k, n_1,\ldots,n_\ell\}$ has exactly $k+\ell$ elements. The only restriction we need for this is that $M \cap N = \emptyset$, but otherwise M and N can be any finite sets. If one wishes also to add infinite cardinal numbers also, it is necessary only to drop the condition of finiteness.

Thus, suppose sets K and L are given. If K and L are not disjoint one can pass to disjoint sets $K^* \approx K$ and $L^* \approx L$, — say $K^* = K \times \{1\}$ and $L^* \times \{2\}$.

The power of the set $K^* \cup L^*$ does not depend on the special choice of $K^*$, $L^*$ as one can verify from the following diagram in which each arrow denotes a bijective mapping.

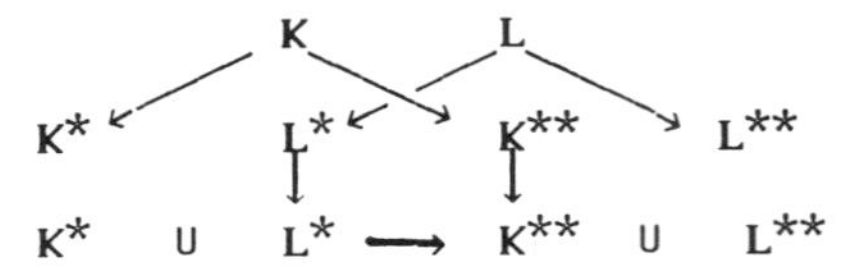


This clearly shows that one can replace K and L themselves by sets of equal cardinal. It is therefore justified to <u>define</u>

$$\text{card } K + \text{card } L =_{def}: \text{card } (K^* \cup L^*) \ .$$

<u>Exercise 6</u>. Using simple injective mappings, verify the following inequalities for infinite M when card K ≦ card M:

$$\text{card } M \leqq \text{card } K + \text{card } M \leqq \text{card } M + \text{card } M.$$

The sum, card M + card M, can easily be worked out after we have proved the following lemma.

<u>LEMMA</u>. *Let* B, R *and* T *be infinite sets of equal cardinal, and* $R \cap T = \emptyset$. Then $B \approx (R \cup T)$.

<u>*PROOF*</u>. As a key to the proof, we construct a disjoint splitting of B into countably infinite subsets which we obtain by use of Zorn's lemma. For the elements of the set **M**, we take those subsets X of B which are disjoint unions of countably infinite subsets $A_i$. If **K** is a chain in **M**, we again form

$$S = \bigcup \{K \mid K \in \mathbf{K}\} \ .$$

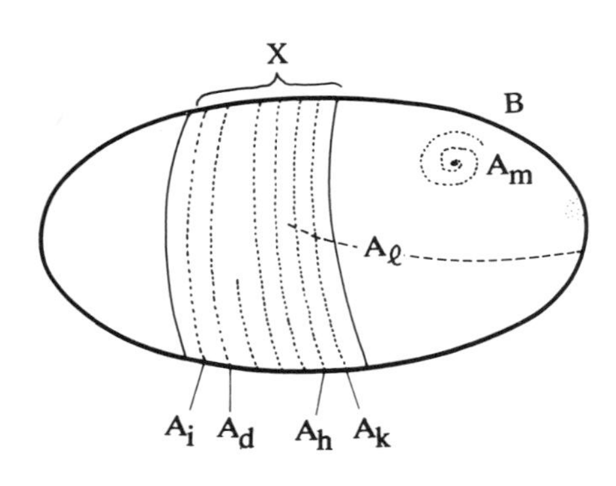


Fig. 63

Thus S is itself a disjoint union of countable infinite sets, so S belongs to **M** and is an upper bound of **K**. Using Zorn's lemma, we obtain a maximal element M in **M**. The set B\M cannot be infinite, otherwise M could be enlarged further by adding a countably infinite set. Therefore, there are, at most, finitely many elements in B\M. We add this remainder to one of the $A_i$'s, and this leaves us with the splitting of B that we want.

(Incidentally, it is necessary that there be at least one $A_i$ in B, but that follows from the infiniteness of B.)

With the equipotent sets B, R and T we have bijections

$$\rho : B \to R \quad \text{and} \quad \tau : B \to T \ .$$

These bijections carry over the disjoint splitting of B into corresponding splittings of R and T into countable sets $Y_i$, $Z_i$ respectively. We index the sets $Y_i$ with the even natural numbers and $Z_i$ with the odd ones, in order to obtain a bijection $\alpha_i : Y_i \cup Z_i \to A_i$. Collecting together all the $\alpha_i$'s, we obtain the required bijection $\alpha : R \cup T \to B$. □

*EXAMPLE.* A splitting of $\mathbb{R}$ into disjoint countable sets can be given by taking $A_x = x + \mathbb{Z}$ with $x \in J = \langle 0, 1 \rangle$.

Exercise 7. Write down corresponding splittings for $I_o$, I, $\$^1$, $\mathbb{R}$, $\$^2$, $\mathbb{R}^2$.

COROLLARY. If M is *infinite*, then card M + card M = card M. For, one forms the sum with the help of a disjoint union of two sets each equipotent with M, so the assertion follows directly from the Lemma. □

Special cases of the Lemma are already familiar to us, — e.g.

$$\mathbb{Z} \approx \mathbb{N}, \quad \text{or} \quad (\langle 0,1 \rangle \cup \langle 2, 3 \rangle) \approx \langle 0, 1 \rangle .$$

If we now apply the Schröder–Bernstein theorem to the inequalities given in Exercise 6, viz:

$$\text{card } M \leq \text{card } K + \text{card } M \leq \text{card } M$$

then we obtain at once:

THEOREM. *If* M *is infinite and* card K ≤ card M, *then*

$$\text{card } K + \text{card } M = \text{card } M . \qquad \square$$

Therefore the addition of two infinite cardinal numbers always gives us the greater of the two.

With multiplication, we get something even simpler. We use the Cartesian product and define

$$\text{card } K \cdot \text{card } L =_{def}: \text{card } (K \times L)$$

where again the replacement of K, L by equipotent sets is allowed. We already know some products of infinite cardinal numbers — for example

$$\text{card } \mathbb{N} \cdot \text{card } \mathbb{N} = \text{card } (\mathbb{N} \times \mathbb{N}) = \text{card } \mathbb{N}$$

by the first Cantor diagonal method, and

$$\text{card } \mathbb{R} \cdot \text{card } \mathbb{R} = \text{card } (\mathbb{R} \times \mathbb{R}) = \text{card } \mathbb{R}$$

from Cantor's Theorem that $\mathbb{R}^2 \approx \mathbb{R}$. Using a lemma analogous to that for addition, it can be shown (cf. Kaplansky):

THEOREM. *If* M *is infinite and* $0 \neq$ card K ≤ card M *then*

$$\text{card } K \cdot \text{card } M = \text{card } M.$$

The narrow connection between set operations and arithmetic has already been used for ages in the Primary school for illustrating addition and multiplication. Also the so-called Pythagorean "arithmetic of the calculating pebbles" (cf. Van der Waerden [71] Vol.1, p.47) can be interpreted in this way. The explicit introduction of set-theoretical concepts into schools has however, not led to the educational consequences that were hoped for, and has not resolved the controversy about* "set theory".

*Exercise 8*.

(a) Use addition of cardinal numbers to show that the set T of transcendental numbers is equipotent with $\mathbb{R}$.

(b) Prove that the set of irrational real numbers has the same cardinal as $\mathbb{R}$.

(c) State the cardinal numbers of the following sets, giving brief justifications for your answers:
  (i) the set of all polynomials with rational coefficients,
  (ii) the set of all non-degenerate triangles in $\mathbb{R}^2$,
  (iii) the set of all sequences $(a_n)$, where $a_n \in \mathbb{N}$
  (iv) continuous monotonic functions $f:\mathbb{R} \to \mathbb{R}$
  (v) points of discontinuity of a given monotonic function $f : \mathbb{R} \to \mathbb{R}$. (Southampton University, various years).

(d) Explain the terms partially ordered set, well ordered set.

  (i) Let A be a well-ordered set and B a subset of A, and let $f:A \to B$ be a similarity. Prove that $a \leqq f(a)$ for every $a \in A$.

  (ii) A countable set T is partially ordered in such a way that every element has only finitely many predecessors. Prove that T is similar to a subset of $\mathbb{N}$ ordered by divisibility. [It may be helpful to observe that via the equivalences $P \sim \mathbb{N} \sim \mathbb{N} \times \mathbb{N}$ the primes P may be written as a denumerable disjoint union of denumerable sets.]

(Southampton 1983).

---

*Translator's Note: The author is referring to a political wrangle that occurred in Germany, when parents brought a court-case involving the efficacy of teaching set-theory in schools.

## 6F VECTOR SPACES OF INFINITE DIMENSION, AND THE CAUCHY FUNCTIONAL EQUATION.

From Section 4E we still have the unproved, and intuitively surprising, claim that there is an isomorphism between the additive group $(\mathbb{R}, +)$ of real numbers and the vector group of the plane $(\mathbb{R}^2, +)$. We had already seen there, with the example $(\mathbb{Q}^+, \cdot)$ that to prove such a claim, it would be vital to use something like an "infinite basis". For such use we shall, more generally, prove the existence for any vector space of a finite or infinite basis — again with the help of Zorn's lemma.

Instead of regarding $(\mathbb{R}, +)$ and $(\mathbb{R}^2, +)$ as additive groups only, it is a good plan to permit also the multiplication of the elements $x \in \mathbb{R}$ or $\binom{x}{y} \in \mathbb{R}^2$ by *rational* scalars. The advantage of this point of view lies in the fact that in place of the additive group we can then work with *vector spaces* $(\mathbb{R}, +, \mathbb{Q})$, $(\mathbb{R}^2, +, \mathbb{Q})$ respectively, over $\mathbb{Q}$. Immediately we can use the fruitful and effective concepts of linear algebra. (If you don't believe that these are indeed vector spaces, then you should check the corresponding axioms yourself.) It is quite significant for our problem that *only rationals*, and not arbitrary real, scalars are allowed.

To begin, let us recall the basic concept and its most important property: Given a set B of vectors of a vector space $(V, +, K)$ with scalar field K, (in brief, V), we call B a **basis** of V whenever B is linearly independent, and spanning. The linear combinations of vectors that appear in the definition of "linearly independent" and "spanning" are always taken as *finite* in number, even when B itself is infinite. Now for the important basis property: If V and W are vector spaces with the same scalar field, if B is a basis of V and $\beta : V \to W$ is any mapping, then there is exactly one *linear* mapping $\varphi : V \to W$ which extends $\beta$, which is to say that $\varphi b = \beta b$ for all $b \in B$.

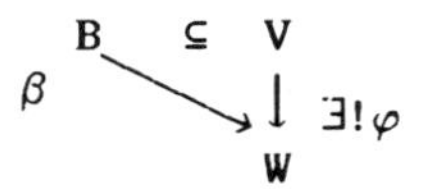


(Of course, one has $\varphi(x_1 b_1 + \ldots + x_n b_n) = x_1 \beta b_1 + \ldots + x_n \beta b_n$, for all choices of $b_i$ from the basis B).

The fundamental theorem we need is:

THEOREM. *Each vector space has a basis.*

*PROOF.* Let $V = (V, +, K)$ be a vector space with the scalar field K. First, instead of bases we consider only linearly independent subsets of V. To apply Zorn's Lemma, the set **M** will consist of all such subsets, and is ordered by inclusion; thus

$$(\mathbf{M} = \{T \mid T \text{ is a linearly independent subset of } V\} .$$

Next, suppose we have a chain **K** of such subsets. (Example: Two chains of linearly independent subsets in $\mathbb{R}^3$ are

$K_1 : \{a\} \subsetneq \{a, b, c\}$ , and $K_2 : \{b\} \subsetneq \{b, a\} \subsetneq (b, a, d\}$ .)

In the chain **K** we find an upper bound S by forming the union

$$S = \bigcup \{K \mid K \in \mathbf{K}\}$$

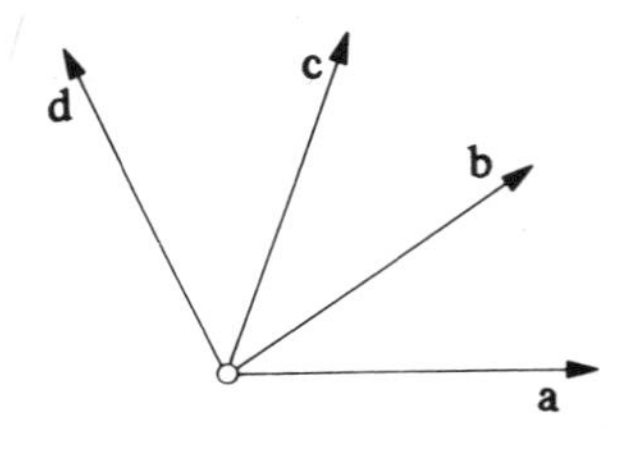


Fig.64

We next prove S $\in$ **M**. Let $s_1,\ldots,s_r$ be vectors of S. Then $s_1$ lies in $K_1,\ldots,s_r$ in $K_r$, (say). Of the (finitely many) sets $K_i$ in the chain **K**, one is largest, say $K_j$. Then $s_1,\ldots,s_r$ lie in the linearly independent set $K_j$ and are therefore themselves linearly independent. Hence, so is S, and S $\in$ **M**.

By definition of S, K $\subseteq$ S for all K $\in$ **K**, and therefore S is an upper bound of **K**. Each chain in **M** possesses therefore an upper bound, so we can apply Zorn's lemma. Thus there is a maximal linearly independent subset of **M**, and we denote it by B.

*CLAIM*. B is a basis of V. We already have the linear independence of B because B belongs to **M**. It remains to show that B is also a spanning set. Then let **v** $\in$ V be a vector. If by chance **v** $\in$ B, then **v** = 1.**v** is a representation of **v** as a linear combination of finitely many vectors from B. If **v** $\notin$ B, then the vector set B $\cup$ {**v**} must be linearly dependent because B is a maximal linearly *in*dependent subset of V. Therefore there exist $b_1,\ldots,b_r \in B$, and scalars $x_1,\ldots,x_r$, y, which are not all zero with

$$0 = x_1b_1+\ldots+x_rb_r + yv \quad .$$

In this linear combination, y cannot be zero; otherwise, the $b_i$'s would be linearly dependent. Therefore, $y^{-1}$ exists in the field K and we have

$$v = - y^{-1}\cdot(x_1b_1+\ldots+x_rb_r)$$

so B is also spanning. Hence B is a basis of V. □

We apply this to the vector spaces $\mathbb{R}$ and $\mathbb{R}^2$ over $\mathbb{Q}$. From our theorem, $\mathbb{R}$ has a basis B over $\mathbb{Q}$. (Such a basis is often called a *Hamel–Basis* after G. Hamel; and his existence proof and first application of such a basis will be discussed later.) The basis B cannot be finite. (Why?) From B we get the following basis C for $\mathbb{R}^2$:

$$C = \{\begin{bmatrix} b \\ 0 \end{bmatrix} \mid b \in B\} \cup \{\begin{bmatrix} 0 \\ b \end{bmatrix} \mid b \in B\} \ .$$

Here, C is the disjoint union of two (infinite) sets which are equipotent with B, and therefore C $\approx$ B by the Lemma from Section 6E.

Therefore there exists a bijection $\varphi : B \to C$. From this we obtain, as an extension, the vector space isomorphism $\varphi : (\mathbb{R}, +, \mathbb{Q}) \to (\mathbb{R}^2, +, \mathbb{Q})$. If we consider $\varphi$ only with respect to addition, then we have an isomorphism

$$\varphi : (\mathbb{R}, +) \cong (\mathbb{R}^2, +)$$

of the additive groups. Iteration of this process leads to $(\mathbb{R}^3, +) \cong (\mathbb{R}, +)$ etc. As a result, we have:

THEOREM. *For the additive groups* $(\mathbb{R}^n, +)$ *of the n-dimensional real spaces, we have*

$$(\mathbb{R}^n, +) \cong (\mathbb{R}, +) \quad \text{for all} \quad n \in \mathbb{N}.$$

*In particular therefore,*

$$(\mathbb{R}, +) \cong (\mathbb{C}, +).$$

From this theorem a large number of interesting structure statements can also be derived for other groups of numbers like ($, ·) for example. These can be found in Salzmann [65], Vol.I.

We can equally well become acquainted with a further consequence of the existence of the basis B of $\mathbb{R}$ over $\mathbb{Q}$. This was given by G. Hamel in his 1905 paper "*A Basis for all Numbers and the Discontinuous Solutions of the Functional Equation* $f(x+y) = f(x) + f(y)$"[26]. Already we have seen his proof for the existence of B, the basis mentioned in his title. Zermelo had proved the well-ordering theorem in the year 1905 with the help of the choice axiom (AC). Hamel immediately picked up this theorem and solved an old problem with the help of the new method. The reader should consider whether he/she finds Hamel's argument using (WO) or the above argument using (ZL) more sympathetic to the taste. Here is Hamel's proof:

"We think of the continuum as well ordered in some way, which is possible from the theorem of Mr. Zermelo. If a is the first number in this ordering, we take a as a member of the basis we intend to construct, and strike out all numbers which are rational multiples of a. The remaining part of the continuum is still a subset of a well-ordered set, and has again a first element b (say). We will take this b as the second number of our basis and now strike out all numbers of the form

$$\alpha a + \beta b$$

where $\alpha$ and $\beta (\neq 0)$ are rational. In this way, we imagine the process as continually repeated.

This agreed, we can now decide, for each number x, whether it belongs to the basis or not. In so doing, we meet the following proposition:

Let x be the section determined by x in the original well-ordered set; then either x can be represented in the form (1) below by a finite number of elements which belong to x, or this is not possible. In the last case, x will be added to the basis, but in the first case, it is not.

From this supposition, there never can exist a relation of the form

$$\alpha a + \beta b + \gamma c + \ldots = 0 \tag{1}$$

with rational coefficients $\alpha, \beta, \gamma, \ldots$ between a finite set of basis elements a, b, c, ... . Otherwise one of the elements a, b, c, ... — say c — would be the last in the original ordering; it could then be expressed in the form (1) by means of a finite number of earlier elements, which contradicts the above proposition.

But then, it immediately results that numbers which can be represented in the form (1), can moreover be so represented in only one way.

If x is not a basis element, then it can always be brought into the form (1) where the a, b, c, ... are basis elements.

Suppose there were still numbers which were neither basis elements nor representable in the form (1); then there would exist amongst these, as a subset of the well-ordered set, a first, say x. By the above proposition,

$x$ can be represented by a finite number of earlier elements in the form (1). These earlier elements are, however, either basis elements or they can be represented in the form (1) by *basis* elements. Since they are only finite in number, a similar basis representation is possible for $x$ itself.

With this the theorem is completely proved".

## The Cauchy Functional Equation

Through Zermelo's proof of the well-ordering theorem and the related work that came from it which had as content such noteworthy results as $(\mathbb{R}, +) \cong (\mathbb{R}^2, +)$, a lively debate was set in train concerning the admissibility of the proof principle (WO) (or alternatively, (AC); (ZL) had still not been formulated by then). It is worth getting to know yet another of the stones of the attack.

In his work, Hamel cited the problem set by Cauchy as an exercise, to give all real functions $f$ which satisfy the condition

$$f(x+y) = f(x) + f(y) \quad \text{for all} \quad x, y \in \mathbb{R} .$$

As one can immediately verify, the linear functions $x \to kx$ fulfill this so-called Cauchy functional equation. Cauchy himself had also shown that the linear functions are the only *continuous* solutions one can have. The decision as to the eventual existence of discontinuous solutions remained open until the work of Hamel, cited above. He argued in the following way:

Let $B$ be a basis of the vector space $(\mathbb{R}, +, \mathbb{Q})$. Fix a constant $k \in \mathbb{Q}$, and define a mapping $\beta : B \to \mathbb{R}$ as follows. For a basis element $b_1$ let

$$\beta b_1 = k \cdot b_1$$

and for a second, choose $v_2 \in \mathbb{R}$ to define

$$\beta b_2 = v_2 \neq k \cdot b_2 \ ;$$

for all other basis elements $b_i$ let

$$\beta b_i = v_i$$

where $v_i$ are any real numbers (here = vectors).

The resulting basis mapping $\beta : B \to \mathbb{R}$ can be extended to a $\mathbb{Q}$-linear mapping

$$f : (\mathbb{R}, +, \mathbb{Q}) \to (\mathbb{R}, +, \mathbb{Q}) \ .$$

As a mapping of a linear vector spaces, $f$ satisfies the condition

$$f(x+y) = f(x) + f(y) \quad \text{for all vectors} \quad x, y.$$

The vectors **x, y** are however, nothing other than real numbers, so f satisfies the Cauchy functional equation. Since $\beta b_1 = fb_1 = kb_1$ and $\beta b_2 = fb_2 \neq kb_2$, then f is certainly not a function of the form $x \to kx$. (It is here that linear mappings of the vector space $(\mathbb{R}, +, \mathbb{Q})$ are different from the usual real linear functions!) From Cauchy's result, f can certainly not be continuous. Therefore, we have found further solutions of the Cauchy functional equation, which are different from the continuous ones.

From this result, we can also get interesting information about the exponential function. The function exp and its relatives $x \to a^x$ satisfy the functional equation

$$h(x+y) = h(x) \cdot h(y)$$

This equation however, does <u>not</u> suffice to characterize these functions (all of them continuous). For, if we take a discontinuous solution f (say) of the Cauchy functional equation and we put

$$h(x) = \exp f(x))$$

then we have a discontinuous function which satisfies the required equation $h(x+y) = h(x) \cdot h(y)$.

Can we make an intuitive picture of a discontinuous solution f of the Cauchy functional equation in the coordinate plane? It can certainly not be a "curve", because f is not continuous. The following idea, which was already pointed out by Hamel, stems from the essay of A. Kirsch: "The functional equation $f(x+x') = f(x)+f(x')$ as a theme for lessons in the upper school" [40], pp.172–187. As the first basis vectors, we choose $b_1 = 1$ and $b_2 = e$ ($b_2$ must be irrational); the remaining basis vectors are of little interest. Further, following the lines of Hamel's proof above, let

$$\beta(1) = \tfrac{1}{2} \cdot 1 = \tfrac{1}{2} \quad \text{(therefore } k = \tfrac{1}{2}\text{)}$$

$$\beta(e) = 1 \cdot e \quad \text{and}$$

$$\beta(b_i) = \tfrac{1}{2}\, b_1 \quad \text{for all other } b_i$$

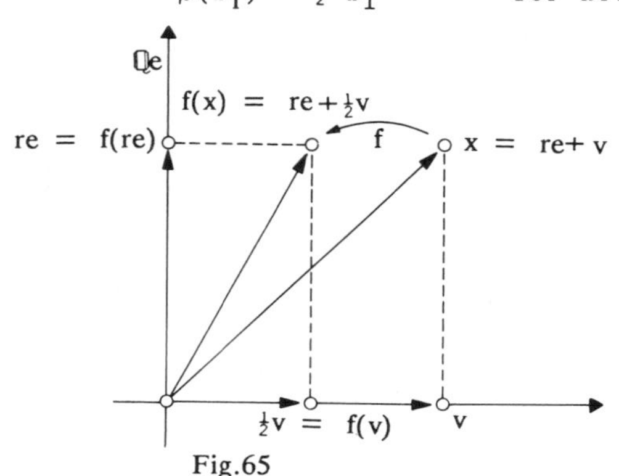

Fig.65

On the "axis" $\mathbb{Q}b_i$ the equations $f(rb_i) = rf(b_i)$ hold because of the $\mathbb{Q}$-linearity of f.

If we collapse together all the axes, apart from $\mathbb{Q}e$, then in Fig.65 we can illustrate f quite harmlessly in the usual way, as a linear mapping of a two dimensional space.

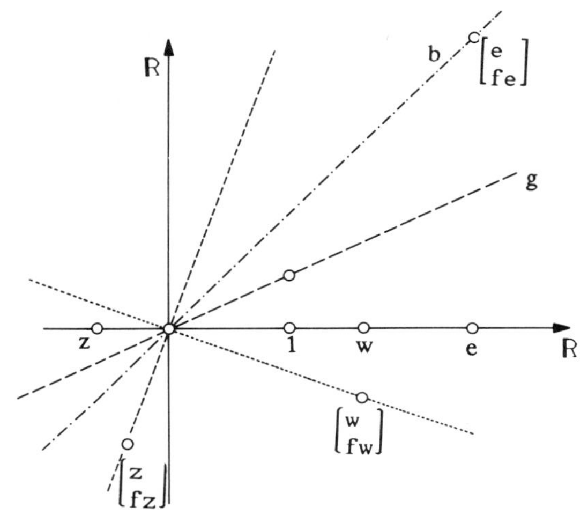


Fig. 66

The situation becomes different, if all the axes in $\mathbb{R}$ are put together as in Fig.66, so that $f$ should be illustrated as the graph of a real function. To deal with this, we first consider the function value $f(r)$ for rational $r$. Here, $f(r) = rf(1) = r \cdot 1/2$, because we may take out rational scalar factors. Therefore, for rational $r$, the points $(r, f(r))$ of the graph of $f$ all lie on the line $g$ through $(0, 0)$ with the gradient $1/2$. (Other points of the graph lie on this line, but they do not interest us here.) For the rational multiples $re$ of $e$, we have $f(re) = rf(e) = re$ and therefore the points $(x, f(x))$, with $x = re$, all lie on the line $h$ through $(0, 0)$ with gradient 1. Now consider, say, $z = 2 - e$. Then

$$f(z) = f(2 \cdot 1 - 1 \cdot e) = 2f(1) - 1f(e) = 1 - e.$$

The points $(x, f(x))$ with $x = rz$, all lie on the line $k$ through $(0, 0)$ and $(2-e, 1-e)$. Correspondingly, for $w = 4-e$ (see Fig.66). From the pictures, it becomes plausible that the points $(x, f(x))$ of the graph of $f$ are scattered over the whole plane and lie densely. (Try to prove this yourself, sometime.) The graph of a bijective function can therefore look like this – it need not always be a "thread-like" graph like that of the elementary functions in Calculus. Fortunately, as the theorems of Analysis tell us, it is just the lack of continuity that allows such "wild" behaviour.

# 7

# Non-standard numbers

## INTRODUCTION.

In the 1960's, a new branch of Mathematics began, with A. Robinson's book *Non–Standard Analysis*. It is called 'non–standard analysis', and uses novel number domains that contain 'infinitesimal' as well as infinite, numbers. These therefore also serve to elucidate the conception of the infinite. They are adapted to the purposes of Analysis and permit us to take up certain procedures previously followed by Euler and his contemporaries which cannot be comprehended within ordinary (standard) Analysis. Whereas Cantor was mainly interested in enumeration and ordering of a set, one now wants rather to be able to express limit processes and similar phenomena by using infinitely small (infinitesimal) and infinitely big (infinite) numbers. This is therefore more a concern with calculating than with enumerating. A typical example is the treatment of the so called Dirac function: this function is zero except on an infinitesimal interval in which it takes a (constant) value among the infinite numbers, but is such that the integral of the function (here, simply the length of the interval times the function–value) is exactly unity.

We will neither go into the historical assessment of non–standard Analysis, nor into its applications in Analysis, nor into the important relations with logic; we shall introduce only the underlying domain of numbers. Therefore we comment quickly on the wider connections. The infinitesimal numbers are designed to follow Euler's usual practice of replacing limit considerations by arithmetic. For example, in non–standard analysis one can define the term "infinitesimally neighbouring"; then one defines a function $f$ to be continuous whenever for infinitesimal by neighbouring $x$ and $y$, the function values $f(x)$ and $f(y)$ are also infinitesimally neighbouring If something like this is to have a pay-off for real Analysis, then one needs a theorem of the form: a statement of a certain type about a function is correct in real Analysis, precisely when the "corresponding" statement of non-standard analysis for "the same" function is correct. There are two aspects of interest here: the first is that as many functions on the reals as possible must be continuable to the non-standard numbers, and the second is that the relevant assertions must apply to as large a domain as possible. Both these demands are only fulfilled to a satisfactory extent when the field $^*\mathbb{R}$ has been constructed with the help of ultrafilters at the end of the chapter. The function field $\mathbb{R}(x)$ considered in Section 7A does not permit e.g. a square root function $\sqrt{\ }$, and with the ring $^\Omega\mathbb{R}$ from Section 7B one has the same problem with the continuation of the function $x \to 1/x$.

Taken altogether, one has a sort of accumulation of difficulty, first with the construction of $^*\mathbb{R}$ (where $\mathbb{R}$ is already assumed), then with the decision as to the "permissible" sets, functions and statements for which the transfer principle holds, and finally with the proof of the transfer principle itself. Beautiful as the theory is, some enthusiasts have advocated its use in the education of beginners (at school and High school). Surely the Arithmetisation of the concept of limit that it accomplishes is too dearly bought to be practicable for that particular purpose.

Concerning the continuation problem and the relevant theorems of logic, one obtains a good overview in A.H. Lightstone: 'Infinitesimals" [46]. In the newer work of W.S. Hatcher: "Calculus is algebra" [28], the problems treated by Lightstone are somewhat played down. The biggest part of this work is concerned with the construction of $^*\mathbb{R}$, which we ourselves also carry out here. Whoever wishes to study non-standard analysis more deeply should consult one of the following books: *Infinitesimal Calculus* by D. Laugwitz [43], or *Ideal points, Monads and non-standard methods* by N. Richter [63]. There, the aspects introduced above are worked out and many literature citations can be found, which can help further with special questions. For the text that follows, we have used, besides the book quoted, some notes of W.A.J. Luxemburg (California Institute of Technology, Pasadena 1962). The author also wishes to thank Professor Laugwitz for useful advice on our theme. We make use of the elementary properties of ideals and ring homomorphisms (Lang III [42], Chapter 2/3) without further explanations. All the rings appearing in these paragraphs are commutative.

## 7A PREPARATION: THE NON-ARCHIMEDEAN ORDERED FIELD $\mathbb{R}(X)$ OF RATIONAL FUNCTIONS.

A field which contains, as well as the real numbers, both infinitesimally small and infinitely big numbers must be non-Archimedean ordered, as we know from Chapter I. We shall first use an example we have already met in Chapter I, to help us become familiar with concepts of non-standard analysis.

The field $(\mathbb{R}(x), +, \cdot)$, – briefly $\mathbb{R}(x)$ – of rational functions is the quotient field of the ring $\mathbb{R}[x]$ of polynomials with real coefficients, i.e. for each $f \in \mathbb{R}(x)$ there are polynomials $p, q$ such that $q \neq$ zero polynomial and

$$f(x) = p(x)/q(x) \ .$$

In Chapter I C we defined orderings of $\mathbb{Q}(t)$ and $\mathbb{Q}[t]$ which can similarly be introduced into $\mathbb{R}(x)$. We briefly repeat the description: a polynomial $p \neq 0$ in $\mathbb{R}[x]$ has only finitely many zeros and therefore takes from a certain point onwards always positive or always negative values (so to speak, "completely to the right or to the left (respectively) of infinity"). In the first case the polynomial is called *positive*, and in the second, *negative*. This ordering of polynomials can be transferred in the usual way to the quotients. In this way we obtain the linearly ordered field $(\mathbb{R}(x), +, \cdot, \trianglelefteq)$. The field $(\mathbb{R}, +, \cdot, \trianglelefteq)$ is embedded isomorphically in $\mathbb{R}(x)$ by identifying $r \in \mathbb{R}$ with the constant rational function $x \to r$. In this ordering, the identity function $x \to x$ is infinitely big, because it finally becomes ("completely to the right") bigger than each $r \in \mathbb{R}$ (i.e. bigger than each constant function): see Fig. 67.

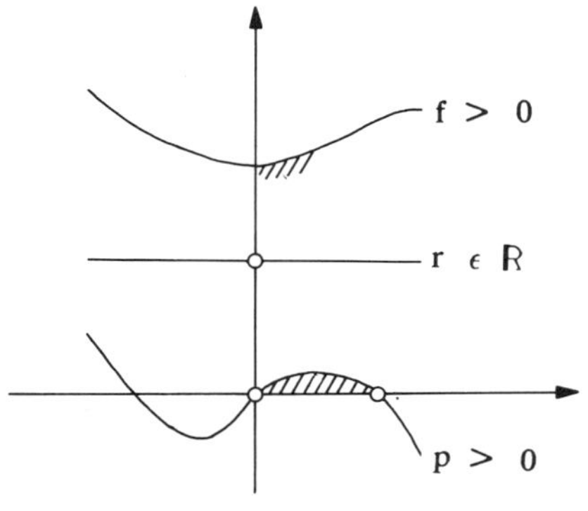
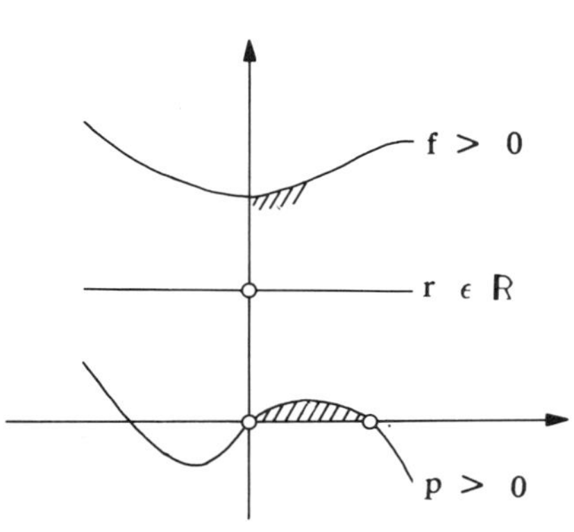


Fig. 67

Correspondingly $x \to 1/x$ is infinitely small, because it is smaller than every constant $1/n$ with $n \in \mathbb{N}$: see Fig.68. To help understand this even better, we consider another ordering of $\mathbb{R}(x)$. Thus, for $p \in \mathbb{R}[x]$ we define :

$$p > 0 \text{ in } \mathbb{R}(x),$$

provided that in $\mathbb{R}$,

$$\exists\ \varepsilon > 0 : p(x) > 0 \text{ for all } x \text{ with } 0 < x < \varepsilon \ .$$

Fig. 68

**Exercise 1.**

(a) Check that, with this definition of $>$ all requirements of a linearly ordered ring, and then of its quotient field, are fulfilled. (It is essential for the proof that polynomials have only finitely many zeros).

(b) Show that if each $r \in \mathbb{R}$ is regarded as a member of $\mathbb{R}(x)$, then $r < s$ in $\mathbb{R} \Leftrightarrow r < s$ in $\mathbb{R}(x)$.

This new ordering comes from the behaviour of functions "to the right of zero", and in this ordering we now find that $x \to x$ is infinitely small, because to the right of zero it is smaller than each constant $1/n$. On the other hand $1/x$ is infinitely big. Also $x^2$ is smaller than $x$, and $x^3$ is even smaller etc. All functions $f \in \mathbb{R}(x)$ with $f(0) = 0$ are "infinitesimally small" regardless of whether they are positive or negative to the right of 0. If one imagines $\mathbb{R}$ enlarged, by adding the set of these $f$ to each $r \in \mathbb{R}$, then one gets the "infinitesimally neighbouring elements" of $r$ in $\mathbb{R}(x)$. In the "right of zero" ordering of $(\mathbb{R}(x), +, \cdot, <)$ they are, so to speak, squashed between the real numbers. Borrowing an expression from the philosophy of Leibniz, we call the set of all infinitesimally neighbouring elements to r, the **Monad** of r, without thereby going further into the philosophical significance of the concept.

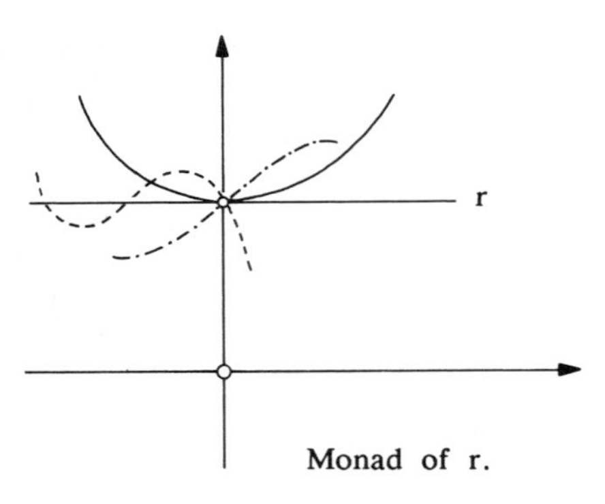


Fig. 69

In order to get back from the Monad of r, to the real number r, one only needs to find the common intersection point at which all the functions cut the y-axis. We now describe this intuitively simple process using the language and concepts of algebra, in order to enable it to carry over to situations where the intuition is no longer adequate.

## The Annihilation of Monads.

Let $\mathbb{R}(x)$ again be ordered by the 'right of zero' process. In that process, we have so far not mentioned the infinitely big elements of $\mathbb{R}(x)$. Therefore let

$$S = (f \in \mathbb{R}(x) \mid \exists\, n \in \mathbb{N}: -n < f < n)$$

be the set of elements which are not infinitely big. These are simply those functions of $\mathbb{R}(x)$ which have no pole at zero. S is a sub-ring of $\mathbb{R}(x)$ which contains the real numbers (check: if $f, g \in S$ then $f \pm g$, $fg \in S$, and $r \in S$ for each $r \in \mathbb{R}$). S also contains infinitesimal elements, the set of which is

$$J = \{f \in S \mid \forall\, n \in \mathbb{N}: -\frac{1}{n} < f < \frac{1}{n}\} .$$

<u>**Exercise 2**</u>

(a) Show that $x, -x^2, x^3, x^4, \ldots$ belong to $J$.

(b) Show that $J$ can also be described by:

$$J = \{f \in S \mid f \text{ has no multiplicative inverse in } S\} .$$

(c) Prove that $J$ is an ideal in $S$, by verifying the ideal properties: if $f, g \in J$ and $h \in S$, then $f \pm g$ and $hf$ lie in $J$.

(d) Prove that this ideal $J$ is even maximal in $S$, that is to say, there is no ideal $I$ with $J \subseteq I \subseteq S$, yet $J \neq I \neq S$.

If we now form the factor ring $S/J$, then infinitesimal neighbouring elements of $S$ will be identified (i.e. regarded as indistinguishable), as we have already discussed intuitively. The mapping

$$st : S \to S/J$$

is called the **transition to the standard part** of a non-standard number. The monad at $u \in S$ is the coset $u + J$, and st annihilates $J$, the monad of $0$. Note that *infinitely large numbers have no standard part.*

## 7B THE RING $^{\Omega}\mathbb{R}$ OF SCHMIEDEN AND LAUGWITZ.

Instead of polynomials and rational functions, we now consider sequences

$$a : \mathbb{N} \to \mathbb{R} \quad \text{with} \quad a : n \to a_n$$

which we denote as usual by $(a_n)$. We denote by $\mathbb{R}^{\mathbb{N}}$ the set of these functions; with operations defined term by term, $(\mathbb{R}^{\mathbb{N}}, +, \cdot)$ becomes a ring, which can also be ordered via the rule.

$$(a_n) \geqq 0 \quad \text{provided} \quad \forall\, i : a_i \geqq 0 .$$

Here the sum of positive elements, $(a_n) + (b_n) = (a_n + b_n)$, is again positive, and similarly for multiplication etc.; but the order is not linear, because e.g. the sequence $(-1, 1, -1, \ldots) = ((-1)^i)$ is neither positive nor negative. The ordered ring $(\mathbb{R}^{\mathbb{N}}, +, \cdot, \leqq)$ is nothing other than the infinite direct product of $(\mathbb{R}, +, \cdot, \leqq)$, so all rules of calculation are very simple to verify. By the usual assignment $r \to (r, r, r, \ldots)$, of $r \in \mathbb{R}$ to the constant sequence $(r)$ we obtain an isomorphic embedding of $(\mathbb{R}, +, \cdot, \leqq)$ in $(\mathbb{R}^{\mathbb{N}}, +, \cdot, \leqq)$. An important difference from $\mathbb{R}(x)$ is that here we have sequences with infinitely many zero terms which are not the zero sequence. Such sequences are $(1, 0, 1, 0, 1, 0, \ldots)$ say, or $(a_1, a_2, \ldots, a_n, 0, 0, 0, \ldots)$. One would like to eliminate sequences of the second type, i.e. those which are zero from a certain point on (equivalently: "for almost all $i$", or "finally") should not be distinguished from zero. The formal construction of this identification again comes from a suitable ideal. Let

$$J = \{(g_n) \mid g_i = 0 \text{ for almost all } i\} ;$$

then $J$ is an ideal in $\mathbb{R}^{\mathbb{N}}$. For, $0 \in J$, and if $(f_n), (g_n) \in J$, then $(f_n) + g_n) \in J$, while if $(h_n) \in \mathbb{R}^{\mathbb{N}}$, then $(h_n f_n) \in J$. The factor ring

$$^{\Omega}\mathbb{R} = \mathbb{R}^{\mathbb{N}}/J$$

is the "ring of omega-numbers" introduced by Schmieden and Laugwitz, and we study this ring in more detail. Its elements are equivalence classes of sequences modulo $J$, using the equivalence relation

$$f \sim g \Leftrightarrow f - g \in J$$

i.e. two sequences from $\mathbb{R}^{\mathbb{N}}$ are identified in $\mathbb{R}^{\mathbb{N}}/J$ whenever they agree for almost all $i$. Again $r \in \mathbb{R}$ embeds in $^{\Omega}\mathbb{R}$, now as the class of those sequences which have all but finitely many terms equal to $r$. The ring homomorphism $\mathbb{R}^{\mathbb{N}} \to {}^{\Omega}\mathbb{R}$ preserves only addition and multiplication, but when we take factor-rings it also respects (or rather in our case, transfers) the ordering. We see this as follows. If $a_i > 0$ for almost all $i$ in a sequence $(a_n)$, then this must hold also for any sequence equivalent to $(a_n)$ modulo $J$, because $(b_n)$ agrees with $(a_n)$ from a certain index onwards. Using this, one can define the whole class $(a_n) + J$ as positive, independently of representatives. It is again nothing more than a tedious calculation, to verify the monotonic laws (properties of an ordered ring) for $(^{\Omega}\mathbb{R}, +, \cdot, \trianglelefteq)$. Again too, the ordering is non-linear, since the class of the sequence $((-1)^n)$ is neither positive nor negative. Besides, $^{\Omega}\mathbb{R}$ has zero-divisors: for neither of the sequences $(1, 0, 1, 0, 1, 0,\ldots)$ and $(0, 1, 0, 1, 0, 1,\ldots)$ is equivalent to zero yet their product is the sequence $(0, 0, 0,\ldots)$, and similarly for their equivalence classes in $^{\Omega}\mathbb{R}$.

Exercise 3. Let $f : \mathbb{R} \to \mathbb{R}$ be a function, with $\mathbb{R}$ embedded in $^{\Omega}\mathbb{R}$. For each coset $(a_n) + J = X \in {}^{\Omega}\mathbb{R}$, define a function $F : {}^{\Omega}\mathbb{R} \to {}^{\Omega}\mathbb{R}$ by $F(X) = (b_n) + J$, where $b_n = f(a_n)$ for each $n \in I$. Show that $F(X)$ is independent of the choice of representative $(a_n)$ from $X$; and that the restriction of $F$ to $\mathbb{R}$ is $f$ (i.e. if $X \in \mathbb{R}$ then $F(X) = f(X)$).

Although we can therefore obtain neither a linear ordering nor a field, still $^{\Omega}\mathbb{R}$ is well suited for many purposes of non-standard analysis – in particular, because the elements of $^{\Omega}\mathbb{R}$ are good tor inspecting quickly. Whenever we describe related homomorphic images of $^{\Omega}\mathbb{R}$, which are fields, we lose this clarity. Concerning analysis with $^{\Omega}\mathbb{R}$, see Chapter I of Laugwitz [43].)

The existence of divisors, and the non-linear nature of the ordering in $^{\Omega}\mathbb{R}$, might be supposed to have two causes. It is possible that we have permitted too many sequences in $\mathbb{R}^{\mathbb{N}}$, or that we have too weak an equivalence relation, i.e. too few sequences are identified with zero. A comparison with Cantor's construction of the real numbers might make the situation clearer.

With Cantor, a) not all the sequences are admitted from $\mathbb{Q}^{\mathbb{N}}$, only the Cauchy sequences; and b) not only do we identify those sequences which are almost everywhere the same, but all those which differ only by a null-sequence. Therefore, e.g. the class $\omega$ in $^{\Omega}\mathbb{R}$ of the sequence $(1, 1/2, 1/3, 1/4, 1/5,...)$, is not zero, whereas with Cantor, this sequence is a null-sequence, hence equivalent to zero. Cantor takes, therefore, fewer sequences and then also even bigger equivalence classes. We shall in the next sections continue to permit all sequences from $\mathbb{R}^{\mathbb{N}}$ but we shall form larger equivalence classes.

To conclude, we want to retrieve the function field $\mathbb{R}(x)$ again in $^{\Omega}\mathbb{R}$. To do this let $\Omega$ be the class of $(1, 2, 3, 4, 5,...)$. In $^{\Omega}\mathbb{R}$, $\Omega$ is larger than every constant sequence and therefore infinitely large. Thus $\Omega$ is analogous to $x$ in $\mathbb{R}(x)$, and therefore we consider $\mathbb{R}(x)$ with the "left of infinity" ordering, because there, $x$ is infinitely large. In the construction of $\mathbb{R}(x)$ one first forms a polynomial ring $\mathbb{R}[x]$ (in which $x$ appears). Here, we show that $\Omega$ already behaves in $^{\Omega}\mathbb{R}$ just as $x$ does in $\mathbb{R}(x)$. Certainly, we have already seen this as far as the ordering goes. We will now show that $\Omega$ cannot be a zero of a polynomial with real coefficients and thus that $\Omega$ is *transcendental* over $\mathbb{R}$. (Everything takes place in $^{\Omega}\mathbb{R}$, in which the real numbers are embedded). The following conclusion is typical for work in $^{\Omega}\mathbb{R}$: Let $p$ be a polynomial with real coefficients. What is $p(\Omega)$? Because addition and multiplication are defined term by term, then

$$p(\Omega) = \text{class of } (p(1), p(2), p(3), p(4),...,p(i),...) \ .$$

The polynomial $p$ has only finitely many zeros; in particular $p(i) \neq 0$ for almost all $i$. Therefore $p(\Omega) \neq 0$ for each polynomial $p$, and in $^{\Omega}\mathbb{R}$, $\Omega$ is transcendental over $\mathbb{R}$. (Analogy: $\pi$ is transcendental over $\mathbb{Q}$ in $\mathbb{R}$.) By means of the mapping

$$p \to p\ (\Omega)$$

we obtain an injective homomorphism, i.e. an embedding of $\mathbb{R}[x]$ in $^{\Omega}\mathbb{R}$, in which the image of $\mathbb{R}[x]$ is $\mathbb{R}[\Omega]$ .

As we said earlier, this embedding respects the ordering. The image of the polynomial $p \in \mathbb{R}[x]$ is the sequence $(p(i))$ of the function values at the points $i \in \mathbb{N}$. Correspondingly this holds for the quotient field $\mathbb{R}(x)$. The function $1/x$ is mapped on to $\omega$ like $x$ on $\Omega$, because $\omega\Omega = 1$ is valid in $^{\Omega}\mathbb{R}$. The embedding

$$\mathbb{R}(x) \to {}^{\Omega}\mathbb{R}$$

is naturally not surjective, because sequences such as $(1, 0, 1, 0, 1, 0,...)$ cannot appear as sequences of any function $f \in \mathbb{R}(x)$.

## 7C FILTERS AND ULTRAFILTERS.

We want to enlarge the equivalence classes of sequences of $\mathbb{R}^{\mathbb{N}}$, and we now prepare the tools for the task. Let $I$ be an infinite set which we will use as an index set. Usually, in the examples, $I = \mathbb{N}$ or $I = \mathbb{R}$.

<u>Definition</u>. A set $F$ of subsets of $I$ is called a **filter** on $I$, whenever the following conditions are fulfilled:

(F0) $\emptyset \notin F$ and $I \in F$ ,

(F1) $(A \in F$ and $A \subseteq B) \Rightarrow B \in F$ ,

(F2) $(A \in F$ and $B \in F) \Rightarrow A \cap B \in F$ .

The filter $F$ is called **free**, whenever

(F3) $$\bigcap \{A \mid A \in F\} = \emptyset$$

*EXAMPLES*. The set consisting only of the set $I$ itself is a (rather trivial) filter.

(a) *The filter* Cofin *on* $\mathbb{N}$. A set $A \subseteq \mathbb{N}$ belongs to Cofin whenever it has a finite complement (we call it "complement finite"). Therefore:

$$A \in \text{Cofin} \Leftrightarrow \mathbb{N} \setminus A \text{ is finite}$$

$$\Leftrightarrow n \in A \text{ for almost all } n \in \mathbb{N} .$$

The properties (F0) – (F2) are quickly verified. ($\mathbb{N}$ belongs to Cofin, because $\mathbb{N} \setminus \mathbb{N} = \emptyset$ is finite). This filter is free, because for each $a \in \mathbb{N}$, the complement $A = \mathbb{N} \setminus \{a\}$ is a set belonging to Cofin and therefore

$$a \notin \bigcap_{A \in \text{Cofin}} A,$$

from which $\bigcap A = \emptyset$ follows.

The free filter Cofin is also often called the *Fréchet–filter*; one can form it from any infinite set, just as one does from $\mathbb{N}$.

<u>Exercise 4</u>.

(a) Consider the ideal $J \subseteq \mathbb{R}^{\mathbb{N}}$, that was used in Section 7C to construct ${}^{\Omega}\mathbb{R} = \mathbb{R}^{\mathbb{N}}/J$. Show that for each $g \in J$, the set of zeros of $g$ is an element of Cofin.

(b) (The **neighbourhood filter** on $\mathbb{R}$.) Let $r$ be a given real number. If the set $U \subseteq \mathbb{R}$ is an open interval surrounding $r$ (say $\langle r-\varepsilon, r+\varepsilon\rangle$), then $U$ is called a neighbourhood of $r$. The set $V_r$ of all neighbourhoods of $r$ is a filter on $\mathbb{R}$, the so-called neighbourhood filter of $\mathbb{R}$. This filter is not free, because

$$\bigcap_{U \in V_r} U = \{r\} ,$$

it "filters" the number $r$ out of $\mathbb{R}$. (The designation "filter" presumably comes from this way of looking at it).

(c) The **principal filter** $\mathbf{W}_r$ of $r \in \mathbb{R}$ consists of *all* the subsets of $\mathbb{R}$ that contain $r$. Hence, if $A \subseteq \mathbb{R}$ then either $A$, or the complement $\mathbb{R} \setminus A$ belongs to $\mathbf{W}_r$; for, $r$ lies in exactly one of these two sets. Therefore this filter has a *maximal* property: indeed, we cannot add to $\mathbf{W}_r$ a single extra set $X$ that does not already belong to $\mathbf{W}_r$, without violating the condition (F0). For, if $X \notin \mathbf{W}_r$ then its complement $\mathbb{R}\setminus X$ must contain $r$, and hence belongs to $\mathbf{W}_r$; so in $\mathbf{W}' = \mathbf{W}_r \cup \{X\}$ then we have $X \cap (\mathbb{R} \setminus X) = \emptyset \in \mathbf{W}'$ by (F2), which is forbidden by (F0). Thus $\mathbf{W}'$ is not a filter, so $\mathbf{W}_r$ is maximal.

(d) $\mathbf{V}_r$ and $\mathbf{W}_r$ already show that one can sometimes not enlarge a filter. Each set of $\mathbf{V}_r$ belongs also to $\mathbf{W}_r$, but there are many more sets in $\mathbf{W}_r$ (e.g. $\langle r, r+1 \rangle$) than in $\mathbf{V}_r$. A similar situation arises with the filter Cofin in $\mathbb{R}$. We call $A \subseteq \mathbb{R}$ **co–countable** whenever $\mathbb{R} \setminus A$ is countable or finite and so the resulting filter Co-count contains Cofin. On $\mathbb{N}$, Cofin can also be enlarged, say to the filter of sets that contain almost all even numbers.

<u>Definition</u>. We write $\mathbf{F} \subseteq \mathbf{G}$ for two filters on $I$, whenever each set of $\mathbf{F}$ also belongs to $\mathbf{G}$. (One also says: $\mathbf{F}$ is coarser than $\mathbf{G}$, or $\mathbf{G}$ is finer than $\mathbf{F}$,) A filter $\mathbf{U}$ is called an **Ultrafilter** whenever there is no filter which contains $\mathbf{U}$; thus $\mathbf{U}$ is maximal in the ordered set of all filters on $I$ (ordered by $\subseteq$).

We have already seen examples in (d). In particular, the free filter Cofin is not an ultrafilter. In (c) it turns out that the principal filter $\mathbf{W}_r$ is an Ultrafilter. This type of Ultrafilter is however too crude for our use, as we will soon see. (In particular, $\mathbf{W}_r$ is not free.) Some theorems on free filters and Ultrafilters now follow.

<u>THEOREM</u>. *(Characterization of the free filters.)*

$$\mathbf{F} \text{ is free on } I \iff \text{Cofin} \subseteq \mathbf{F} .$$

*PROOF*.

(a) Let $\mathbf{F}$ be free. Then by definition $\bigcap \{F \mid F \in \mathbf{F}\} = \emptyset$. We must show that given any set $I \setminus \{a_1, \ldots, a_n\} \in$ Cofin, then it also belongs to $\mathbf{F}$. Now $a_1$ cannot belong to all the sets $F \in \mathbf{F}$ otherwise

$\bigcap \{F \mid F \in \mathbf{F}\} = \{a_1\}$. In particular then, there is a set $A \in \mathbf{F}$ with $a_1 \notin A$. Thus, $A \subseteq I \setminus \{a_1\}, = B_1$ say, and by (F1), $B_1$ belongs to $\mathbf{F}$. Correspondingly for $B_2 = I \setminus (a_2)$. By (F3), we have $B_1 \cap B_2 = I \setminus \{a_1, a_2\} \in \mathbf{F}$. Repeating this with $a_3, \ldots, a_n$ we get $I \setminus \{a_1, \ldots, a_n\} \in \mathbf{F}$ as required.

(b) Conversely, suppose Cofin $\subseteq \mathbf{F}$. Then

$$\bigcap \{F \mid F \in \mathbf{F}\} \subseteq \bigcap \{A \mid A \in \text{Cofin}\} .$$

So the condition (F3) for $\mathbf{F}$ is fulfilled, and $\mathbf{F}$ is free. □

THEOREM. *(The existence of Ultrafilters.) Each filter* $\mathbf{F}$ *on* I *is contained in some Ultrafilter.*

*PROOF.* We invoke Zorn's Lemma. Let $\mathbf{M}$ be the set of all those filters on I, which contain $\mathbf{F}$. Then $\mathbf{M}$ is ordered by inclusion ($\subseteq$). Suppose that $\mathbf{K}$ is a chain of filters in $\mathbf{M}$. As usual, we form the union $\mathbf{S}$ of all filters $\mathbf{G}$ from the chain $\mathbf{K}$: $\cup\{G \mid G \in \mathbf{K}\}$. The conditions (F0) – (F2) are easily verified for $\mathbf{S}$, so $\mathbf{S}$ is also a filter. Because $\mathbf{F}$ is contained in each of the filters $\mathbf{G}$, $\mathbf{F}$ also lies in $\mathbf{S}$. The filter $\mathbf{S}$ therefore belongs to $\mathbf{M}$ and is an upper bound of the chain $\mathbf{K}$.

This proves that each chain in $\mathbf{M}$ has an upper bound in $\mathbf{M}$. Therefore Zorn's Lemma guarantees for us a maximal element $\mathbf{U}$ in $\mathbf{M}$. That is, $\mathbf{U}$ is an ultrafilter which contains $\mathbf{F}$. □

*CONSEQUENCE.* This theorem also produces for us free ultrafilters, namely, those that contain Cofin.

THEOREM. *(Prime Property of Ultrafilters.) If* $\mathbf{U}$ *is an ultrafilter on* I *and* $A, B \subseteq I$ *are two sets with* $A \cup B \in \mathbf{U}$ *then either* $A \in \mathbf{U}$ *or* $B \in \mathbf{U}$.

After the proof, we shall explain how this is exactly analogous to the prime factor condition in Algebra ($p \mid ab \Rightarrow p \mid a$ or $p \mid b$ for each ordinary prime number $p$).

*PROOF of the Theorem.* Assume that $A \cup B \in \mathbf{U}$ and $A \notin \mathbf{U}$. We must show that $B \in \mathbf{U}$, and begin by forming an auxiliary filter

$$\mathbf{G} = \{G \subseteq I \mid A \cup G \in \mathbf{U}\} .$$

That $\mathbf{G}$ is indeed a filter which includes $\mathbf{U}$ can be seen thus: if $U \in \mathbf{U}$, then $A \cup U \supseteq U$ therefore $A \cup U \in \mathbf{U}$ and $U \in \mathbf{G}$. Further $\emptyset \notin \mathbf{G}$ because $A = A \cup \emptyset \notin \mathbf{U}$. Hence $\mathbf{U} \subseteq \mathbf{G}$ and (F0) is satisfied. One can just as easily verify (F1). For (F2) let $G, H \in \mathbf{G}$; then $G \cup A, H \cup A \in \mathbf{U}$ and

$$(G \cap H) \cup A = (G \cup A) \cap (H \cup A) \in \mathbf{U} ;$$

therefore $G \cap H$ is also in $\mathbf{G}$.

The set $B$ lies in $\mathbf{G}$ because of the requirement $A \cup B \in \mathbf{U}$. Because $\mathbf{U}$ is maximal, $\mathbf{G} = \mathbf{U}$ because $\mathbf{G} \supseteq \mathbf{U}$ and no filter is larger than $\mathbf{U}$. Therefore it follows that $B \in \mathbf{U} = \mathbf{G}$, which we wanted to show. □

*CONCLUSION.* *(Analogy with the prime factor condition.)* A number $p \in \mathbb{Z}$ is called prime whenever $p$ has no proper factors. That is exactly equivalent to saying that the ideals $p\mathbb{Z}$ are maximal in the ring

$\mathbb{Z}$. Assume that $p|ab$; then $ab \in p\mathbb{Z}$ because $ab$ is a multiple of $p$. If we assume further that $p|a$, then $a \notin p\mathbb{Z}$. Now analogously to the previous proof, we form the subset

$$G = \{g \in \mathbb{Z} \mid ag \in p\mathbb{Z} .$$

Using the conditions for $p\mathbb{Z}$ to be an ideal, one shows that $G$ is an ideal in $\mathbb{Z}$. By hypothesis, $G$ contains the number $b$. Further, $p\mathbb{Z} \subseteq G$; hence, since $a = a \cdot 1 \notin p\mathbb{Z}$, then $1 \notin G$ and $G$ is a proper ideal that contains $p\mathbb{Z}$. Because $p\mathbb{Z}$ is maximal, it follows that $G = p\mathbb{Z}$, and $b \in p\mathbb{Z} = G$, so $b$ is a multiple of $p$, and $p|b$ as required. (Behind all this stands the general theorem: "maximal" implies "prime".) The prime factor condition belongs to the oldest theorems of mathematics. It appears in Euclid's book VII, Theorem 30.

Exercise 5. Can the analogy be developed further to give an abstract theorem concerning ordered sets with a binary operation?

THEOREM. *(Characterization of the Ultrafilter by the complement property). Let* $U$ *be a filter on* $I$. *Then*:

$U$ *is an ultrafilter* $\Leftrightarrow$: *for all* $A \subseteq I$, *either* $A \in U$ *or* $I \setminus A \in U$.

We have already mentioned this property in connection with the filter $W_r$.

*PROOF*.
(a) The filter $U$ always contains $I = A \cup (I\setminus A)$. If $U$ is an ultrafilter, then it has the prime property; therefore $A \in U$ or $I\setminus A \in U$.
(b) Suppose the filter $U$ has the complement property, and yet $U \subseteq F$ for some filter $F$. If $B$ lies in $F$ then $I\setminus B$ cannot lie in $U$ otherwise we would have in $F$ that $B \cap (I\setminus B) = \emptyset \in F$. By the complement property, $B \in U$ must hold, and it follows that $F = U$. Thus $U$ is maximal. □

## 7D THE FIELDS $^*\mathbb{R}(I,U)$ AS AN ULTRAPRODUCTS.

### 1. The Function ring $\mathbb{R}^I$.

As in the previous section, let $I$ be an infinite set. Instead of $\mathbb{R}$, one could consider any (ordered) field but we have no need to do this. Let

$$\mathbb{R}^I = \text{set of all functions } f : I \to \mathbb{R} .$$

For the value of $f$ at the point $i \in I$, we shall write $f_i$ or $f(i)$ as convenient.

As an example, one can take the set of all sequences $\mathbb{R}^{\mathbb{N}}$, or the set of all real functions $\mathbb{R}^{\mathbb{R}}$, or also the set $\mathbb{R}^{\{1,2\}} = \{(f_1, f_2)\} = \mathbb{R}^2$ (here of course, and exceptionally, $I = \{1, 2\}$ is finite.) The operations $+, \cdot$ and the ordering $\mathbb{R}^I$ are defined "component-wise":

$$f + g : i \to f(i) + g(i)$$

$$fg : i \to f(i) \cdot g(i)$$

$$f \leqq g \Leftrightarrow \forall i \left[f(i) \leqq g(i)\right] .$$

One can verify once more by direct calculations, that we have here an ordered ring. As the example $\mathbb{R}^2$ shows, this contruction is a generalization of the usual direct product. Therefore, one also calls $\mathbb{R}^I$ the *product* of $\mathbb{R}$ over I. The unity element of the ring $\mathbb{R}^I$ is the constant function

$$i \to 1 \quad \text{for all} \quad i \in I .$$

More generally, an embedding $\eta$ of $\mathbb{R}$ in $\mathbb{R}^I$ is obtained, by assigning to each $r \in \mathbb{R}$ the constant function $\eta(r)$ on I that is given by

$$i \to r \ (\text{or } r_i = r \text{ for all } i) .$$

As we have already done before, we usually make no distinction between $\mathbb{R}$ and its embedding $\eta(\mathbb{R})$ in $\mathbb{R}^I$.

Exercise 6. Verify that $\eta:\mathbb{R} \to \mathbb{R}^I$ is an injective homomorphism.

## 2. Filters on I and Ideals in $\mathbb{R}^I$.

In Exercise 4 we saw how the construction of the ring $^\Omega\mathbb{R}$ was related to the filter Cofin. We now copy the construction, using any filter in place of Cofin. Let F be a filter on I. With this filter we associate an ideal in $\mathbb{R}^I$ in the following way: let $\text{Zero}_f$ be the set of all the zeros of the function f, so

$$\text{Zero}_f = \{i \in I \mid f(i) = 0\}$$

Now let

$$J_F = \{f \mid \text{Zero}_f \in F\}$$

be the family of those functions whose zero-sets belong to F. If one knows only that a proper subset of $\text{Zero}_f$ belongs to F, then that is already enough because (F1) implies that $\text{Zero}_f$ also lies in F.

*CLAIM.* $J_F$ is a proper ideal in $\mathbb{R}^I$.
We must check the usual conditions for an ideal.

(Id 1). $0 \in J_F$. For, the constant function 0 has the whole of I as a zero-set, and by (F0), $I \in F$ for every filter.

Here we can already see that $J_F = \mathbb{R}^I$ because the constant function 1 has $\emptyset$ as zero-set, and $\emptyset \notin F$ so $1 \notin J$.

(Id 2). $f, g \in J_F \Rightarrow f + g \in J_F$. For, we have

$$\text{Zero}_{f+g} \supseteq \text{Zero}_f \cap \text{Zero}_g.$$

Therefore by (F1) and (F2), $\text{Zero}_{f+g}$ belongs to **F**.

(Id 3). $f \in J_{\mathbf{F}}$ and $h \in \mathbb{R}^I \Rightarrow hf \in J_{\mathbf{F}}$. This holds by (F1), because $\text{Zero}_{hf} \supseteq \text{Zero}_f$.

This establishes the Claim. □

Starting with the filter **F** on I, one forms the ideal $J_{\mathbf{F}}$ and with it the homomorphic image $\mathbb{R}^I/J_{\mathbf{F}}$ of the ring $\mathbb{R}^I$:

$$F \mapsto \mathbb{R}^I/J_F$$

(which is not a mapping with domain **F**, but with a domain in which **F** is an element).

The homomorphic image is also called the **reduced product modulo F**. In $\mathbb{R}^I/J_{\mathbf{F}}$, all those functions are treated as zero if their zero sets belong to **F**.

The construction of $\mathbb{R}^I/J_{\mathbf{F}}$ involves at first only addition and multiplication. We shall deal with the ordering subsequently.

<u>Exercise 7</u>.

(a) Using the embedding $\eta : \mathbb{R} \to \mathbb{R}^I$ of Exercise 6, show that $\eta(\mathbb{R}) \cap J_{\mathbf{F}} = \{0\}$, and hence that the natural surjection $\nu : \mathbb{R}^I \to \mathbb{R}^I/J_{\mathbf{F}}$ gives an injective embedding $\nu \circ \eta : \mathbb{R} \to \mathbb{R}^I/J_{\mathbf{F}}$. (Thus $\mathbb{R}^I/J_{\mathbf{F}}$ always contains this copy of $\mathbb{R}$).

(b) Now carry out the construction used in Exercise 3, to show that each function $f : \mathbb{R} \to \mathbb{R}$ has an extension $F : \mathbb{R}^I/J_{\mathbf{F}} \to \mathbb{R}^I/J_{\mathbf{F}}$ (so the restriction of $F$ to $\mathbb{R}$ is $f$).

*EXAMPLES*.

(a) Let **F** = Cofin on $\mathbb{N}$. We have already studied the associated homomorphic image:

$$\Omega\mathbb{R} = \mathbb{R}^{\mathbb{N}}/J_{\text{Cofin}} .$$

For simplicity of notation one often leaves out J and only indicates the filter while meaning, however, the ideal – as in $\Omega\mathbb{R} = \mathbb{R}^{\mathbb{N}}/\text{Cofin}$ or more generally, $\mathbb{R}^I/\mathbf{F}$ instead of $\mathbb{R}^I/J_{\mathbf{F}}$.

(b) Let $\mathbf{F} = \mathbf{W}_r$ on $\mathbb{R}^{\mathbb{R}}$. The filter $\mathbf{W}_r$ consists of all sets which contain r. Therefore the ideal $J_{\mathbf{W}_r}$ consists of all functions f which have a zero at r, that is, for which $\{r\} \subseteq \text{Zero}_f$.

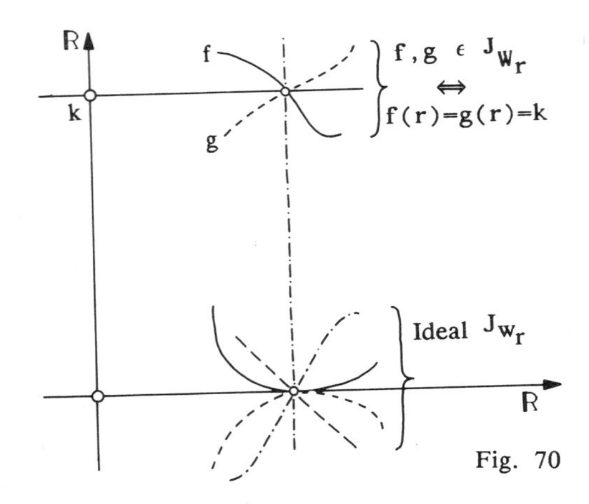


Fig. 70

In this reduced product, those functions become identified, that take the same value r (say $k \in \mathbb{R}$) at the point r. They are therefore identified with the constant k, so we have

$$\mathbb{R}^{\mathbb{R}}/J_{W_r} \cong \mathbb{R}$$

in this case, nothing comes out of the entire construction except $\mathbb{R}$ itself again. Filters of the type of $\mathbf{W}_r$ are therefore not fruitful for producing new objects.

<u>Exercise 8</u>. Show that the last argument is saying that the homomorphism $\nu \circ \eta : \mathbb{R} \to \mathbb{R}^{\mathbb{R}}/ J_{\mathbf{W}_r}$, described in Exercise 7(a), is an isomorphism.

Now to the way in which the homomorphism $\varphi : \mathbb{R}^I \to \mathbb{R}^I/J_{\mathbf{F}}$ transports the ordering. We can define $\varphi f \geqq 0$ in the same way as we did with ${}^{\Omega}\mathbb{R}$, so that one first requires $f \geqq 0$ and then shows that the definition is independent of the representative of the class. To do this assume that $f_i \geqq 0$ for all i and $\varphi f = \varphi g$, so $f - g \in J_{\mathbf{F}}$. Hence $F = Zero_{-g}$ lies in $J_{\mathbf{F}}$, which means that f and g agree everywhere on a filter set F. Thus $g_i \geqq 0$ for all $i \in F$. We therefore define, in the ring $\mathbb{R}^I/J_{\mathbf{F}}$:

$$\varphi g \geqq 0 \quad \text{iff} \quad \{ i \,|\, g_i \geqq 0 \} \in \mathbf{F} \ .$$

An easy modification of the previous idea shows that this definition is independent of the representative g of the class $\varphi g$ (consider h with $\varphi h = \varphi g$). The monotonic laws must again be verified here, which one can easily do by using the filter properties of $\mathbf{F}$. One cannot expect linearity of the ordering ÷ as we already know of ${}^{\Omega}\mathbb{R}$.

<u>Exercise 9</u>.

(a) Show that the above definition of $\geqq$ really is independent of the choice of representative g.

(b) Show that if $\varphi g \geqq 0$ and $\varphi g \leqq 0$ then $\varphi g = 0$.

(c) Verify that $\geqq$ satisfies the monotonic laws.

From now on we use only the abbreviated notation $\mathbb{R}^I/\mathbf{F}$.

THEOREM. *(The smaller homomorphic image is associated with the bigger filter.) Let* **F** *and* **G** *be filters on* I *with* $\mathbf{F} \subseteq \mathbf{G}$; *then there is a surjective homomorphism* $\psi : \mathbb{R}^I/\mathbf{F} \to \mathbb{R}^I/\mathbf{G}$.

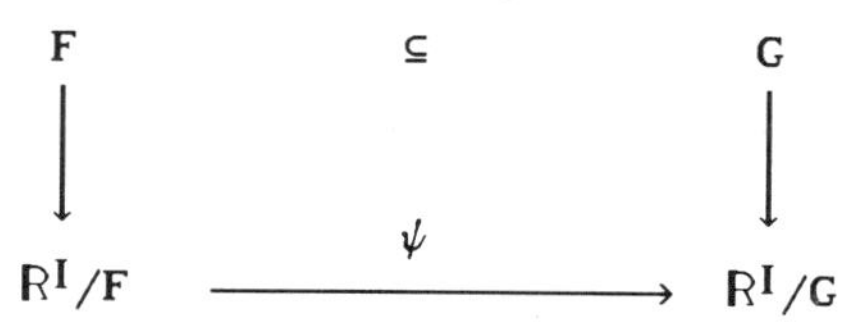


*COMMENT*: If one considers the embedding of **F** in **G** as an injective mapping of filters, then this process makes a surjective homomorphism between the quotient rings correspond to the injective mapping between the filters.

Altogether, one therefore has a correspondence $\mathbf{F} \mapsto \mathbb{R}^I/\mathbf{F}$ between objects of two families (here filters, and rings) and on top of that, a correspondence between mappings which is compatible with the corresponding objects. We have here an example of what is called a *Functorial correspondence*, or a *Functor* but we do not need here to go into the exact details of the formal definition.

*PROOF of the Theorem*. From $\mathbf{F} \subseteq \mathbf{G}$ it follows immediately that for the associated ideal $J_\mathbf{F} \subseteq J_\mathbf{G}$. One has $\alpha : \mathbb{R}^I \to \mathbb{R}^I/\mathbf{F}$ with kernel($\alpha$) $= J_\mathbf{F}$, and $\beta : \mathbb{R}^I \to \mathbb{R}^I/\mathbf{G}$ with kernel($\beta$) $= J_\mathbf{G}$ as in the diagram

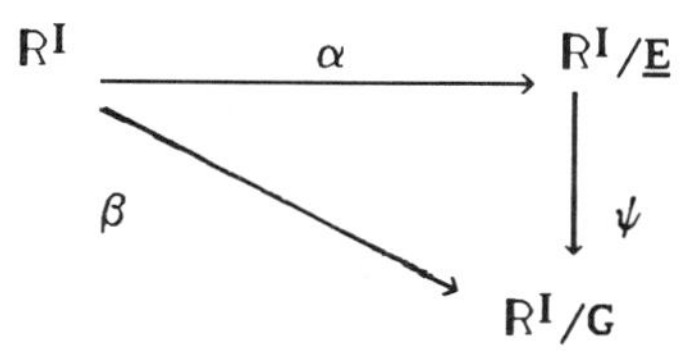


in which $\psi$ is obtained from a variation of the homomorphism theorem for rings (see Lang [42], III, Section 3, Exercise 11). Indeed, that theory tells us that $\psi$ is a surjective homomorphism and $\psi \circ \alpha = \beta$, which is more than required to prove our theorem. □

## 3. Ultrafilters and Fields.

Earlier, we objected to the presence of zero divisors and non-linear orderings in $^\Omega\mathbb{R}$. These disappear whenever we use ultrafilters. This we have already seen in the example of the ultrafilter $\mathbf{W}_r$, but we now show that it is always valid.

THEOREM. **U** *is an ultrafilter on* I $\Leftrightarrow$ $\mathbb{R}^I/\mathbf{U}$ *is a field.*

*PROOF*. (a) Let **U** be an ultrafilter. We already know that $\mathbb{R}^I/\mathbf{U}$ is a commutative ring and we now show that multiplicative inverses exist.

Let $\alpha$ be the homomorphism $\alpha : \mathbb{R}^I \to \mathbb{R}^I/U$ and suppose $\alpha f \neq 0$ in $\mathbb{R}^I/U$. Because $\alpha f \neq 0$, then the set $A = \mathrm{Zero}_f \notin U$. Because of the complement property of the ultrafilter, therefore, $I\backslash A \in U$; and on this set $f(i) \neq 0$ for all $i \in I\backslash A$. We now define

$$g(i) = \begin{cases} 1 & \text{if} \quad i \in A = \mathrm{Zero}_f \\ 1/f(i) & \text{if} \quad i \in I\backslash A \,. \end{cases}$$

From this definition we have

$$g(i) \cdot f(i) = 1 \text{ for all } i \in I\backslash A \in U \,.$$

Because $gf$ agrees on the filter set $I\backslash A$ with the constant function $1$ then $gf$ is identified with $1$ in $\mathbb{R}^I/U$, which is to say

$$\alpha(gf) = \alpha g \cdot \alpha f = 1 \quad \text{in } \mathbb{R}^I/U \,,$$

and hence we have $\alpha g = (\alpha f)^{-1}$, as the inverse that we sought.

(b) For the converse, we assume that $F$ is <u>not</u> an ultrafilter. Then — again because of the complement property of ultrafilters — there exists a set $A \subseteq I$ with $A \notin F$ and $I\backslash A \notin F$. The images of the characteristic functions of $A$, and $I\backslash A$ respectively are then zero divisors in $\mathbb{R}^I/F$, which therefore cannot be a field. □

<u>THEOREM</u>. *If* $U$ *is an ultrafilter, then the ordering of* $\mathbb{R}^I/U$ *is linear.*

<u>*PROOF*</u>. It suffices to show: given an element $\alpha h \in \mathbb{R}^I/U$, then $\alpha h \geq 0$ or $\alpha h < 0$. We form

$$A = \{i \in I \mid h(i) \geq 0\}$$

$$B = \{i \in I \mid h(i) < 0\}$$

Because the function values $h(i)$ lie in $\mathbb{R}$ then either $i \in A$ or $i \in B$, (and not both) so $B = A\backslash I$. By the complement property then, $A \in U$ or $B \in U$. In the case $A \in U$, we have $\alpha h \geq 0$, and in the other case, $\alpha h < 0$. This proves that the ordering is linear. □

The example of the ultrafilter $W_r$ shows that we should not expect, without doing anything further, that the construction of $\mathbb{R}^I/U$ should yield a new "number system".

Before discussing this further, we examine how the land lies with free ultrafilters.

<u>THEOREM</u>. *If* $U$ *is a free ultrafilter on* $\mathbb{N}$ *then* $\mathbb{R}^{\mathbb{N}}/U$ *is not isomorphic to* $\mathbb{R}$.

*PROOF*. As in $^{\Omega}\mathbb{R}$ we let $\Omega$ denote the class of $(1, 2, 3, 4,\ldots)$ in $\mathbb{R}^{\mathbb{N}}/U$, and $\omega$ will denote the class of $(1/2, 1/3, 2/4,\ldots)$. Both classes are different from zero in $\mathbb{R}^{\mathbb{N}}/U$ because the zero-set of each sequence is empty, and the empty set does not belong to U. As in $^{\Omega}\mathbb{R}$, one can see that $\Omega$ is infinitely large and $\omega$ is infinitely small in the ordering of $\mathbb{R}^{\mathbb{N}}/U$. Again, therefore, we have in $\mathbb{R}^{\mathbb{N}}/U$ the sub-field $\mathbb{R}(\Omega)$, which is isomorphic to $\mathbb{R}(x)$ with the non-Archimedean ordering.

From this we shall now conclude that also, even without the ordering, the fields $(\mathbb{R}, +, \cdot)$ and $(\mathbb{R}^{\mathbb{N}}/U, +, \cdot)$ cannot be isomorphic. For, we know from Section 1C that $\mathbb{R}$ permits only one single ordering, namely the usual Archimedean ordering. The reason lay in the fact that in $\mathbb{R}$, the positive elements are characterised as squares. This characterization transfers itself through any isomorphism of fields and therefore cannot allow a field isomorphic to $\mathbb{R}$ to have a non-Archimedean ordering. But $\mathbb{R}^{\mathbb{N}}/U$ is non-Archimedean ordered, and can therefore not be isomorphic to $\mathbb{R}$. □

With the help of ultrafilters, we have therefore obtained genuinely new fields. In non-standard analysis, it is usual to denote these fields by $^*\mathbb{R}(\mathbb{N}, U)$, instead of $\mathbb{R}^{\mathbb{N}}/U$ or simply by $^*\mathbb{R}$ whenever the precise construction is not in question. We know from the previous section that every free ultrafilter contains the filter Cofin. From the theorem on filters and homomorphic images, we have for free ultrafilters U, V the following situation:

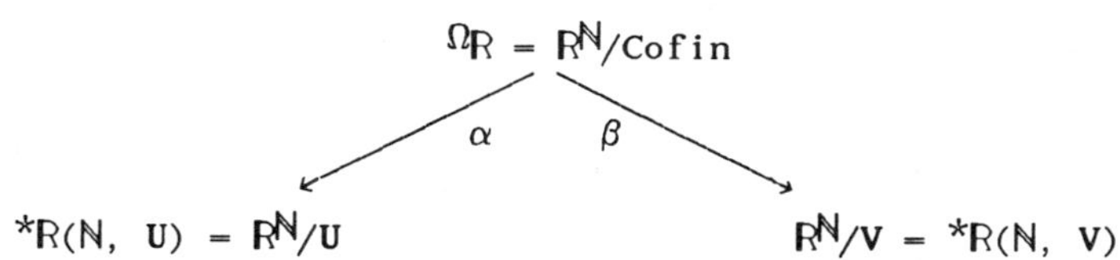


with surjective order-preserving homomorphisms $\alpha$ and $\beta$. One would obviously want to know whether $^*\mathbb{R}(\mathbb{N}, U)$ and $^*\mathbb{R}(\mathbb{N}, V)$ are isomorphic to each other.

Before we go into the isomorphism question, we fix the cardinal numbers of $^{\Omega}\mathbb{R}$ and $^*\mathbb{R}(\mathbb{N}, U)$. The set of sequences $\mathbb{R}^{\mathbb{N}}$ has the same cardinal number as $\mathbb{R}$ (see e.g. Kaplansky [38], Section 2.6). Because sequences are identified in $^{\Omega}\mathbb{R}$ and $^*\mathbb{R}$, then card $^*\mathbb{R} \leq$ card $^{\Omega}\mathbb{R} \leq$ card $\mathbb{R}$. On the other hand, $\mathbb{R}$ is contained in $^*\mathbb{R}$, so card $\mathbb{R} \leq$ card $^*\mathbb{R}$. Therefore $^*\mathbb{R} \approx {}^{\Omega}\mathbb{R} \approx \mathbb{R}$ as sets independently of which ultrafilter is used in the construction of $^*\mathbb{R}$. The power therefore, does not stand in the way of the isomorphism. One can now cite a work of P. Erdös, L. Gillmann and M. Henriksen [16] in Corollary 3.9 of which it is asserted: If the continuum hypothesis (CH) holds, then the fields in question are isomorphic. The author did not understand the quite involved proof that is given there; however, there is another proof in the book *Ordered structures: groups, fields, projective planes* by Priess-Crampe [60].

Even without knowledge of these isomorphisms, one can still learn a little about the relationships in the different $^*\mathbb{R}$'s with the help of the homomorphisms $\alpha$, and $\beta$ above. Because the ultrafilter has not been constructively produced, working directly in $^*\mathbb{R}$ is not easily possible. For example, we know of the image in $^*\mathbb{R}(\mathbb{N}, \mathbb{U})$ :

$$\alpha(1, 0, 1, 0, 1, 0, \ldots) = \begin{cases} 0, & \text{if the set of even numbers belongs to } \mathbb{U}. \\ 1, & \text{if the set of odd numbers belongs to } \mathbb{U}. \end{cases}$$

However we do not know which of the two sets does lie in $\mathbb{U}$. One can certainly produce an ultrafilter which contains, besides Cofin, the set of all even numbers; but with any arbitrary ultrafilter, the situation is unclear. Because of this uncertainty, one must first carry out calculations in $^\Omega\mathbb{R}$. One then asks oneself which of the calculations is preserved by the homomorphism $\alpha : {}^\Omega\mathbb{R} \to {}^*\mathbb{R}$ (using any free ultrafilter for $^*\mathbb{R}$), and thus derives statements about $^*\mathbb{R}$ from it. The logicians have developed extensive theories on this in which yet other fields than those described here as $^*\mathbb{R}$ play a role (e.g. if one iterates our construction process with $^*\mathbb{R}$ in place of $\mathbb{R}$.)

If we compare $^\Omega\mathbb{R}$ with $^*\mathbb{R}$, we see advantages and disadvantages: $^\Omega\mathbb{R}$ is clearly fixed, described constructively and permits specific calculations. Among the disadvantages are the presence of zero divisors and the non-linear ordering relation. With $^*\mathbb{R}$ one has no direct grasp of the elements and indeed no isomorphic exemplars, but in compensation however, we have linear ordering and the field properties.

In a discussion of this problem, Laughwitz quotes a phrase of H. Weyl, on the Continuum as "Medium of free development", which allows us to justify the choice of a suitable $^*\mathbb{R}$ for each particular problem. That may well be. But the author cannot agree to the claim (p.18) that the restriction to the real numbers is "dogmatic". In this exposition we have recognised in many places the exceptional role which $\mathbb{R}$ (and $\mathbb{C}$) play in mathematics because of their structural properties – in particular because of the completeness of $\mathbb{R}$. We cannot call it "dogmatic" to accept these foundations and to recognize the real numbers as an essential foundation for Analysis and the whole of mathematics. For a detailed discussion of this whole problem-area the reader is referred to the introduction of Laugwitz's book.

## 4. The Standard Part of a Non–Standard Number.

Often with calculations that use complex numbers, one is interested at the end in results concerning $\mathbb{R}$. Similarly, it is desirable to have a convenient transition from $^*\mathbb{R}$ to $\mathbb{R}$. This can be arranged as with $\mathbb{R}(x)$. With the embedding $\mathbb{R} \to {}^*\mathbb{R}$ we also have the natural numbers in $^*\mathbb{R}$. Thus we can define

$$S = \text{set of non-infinitely large elements of } {}^*\mathbb{R}$$

$$= \{s \in {}^*\mathbb{R} \mid \exists\, n \in \mathbb{N} : -n < s < n\} .$$

One can quickly verify that $S$ is a sub*ring* of $^*\mathbb{R}$. In $S$ we find

$$J = \text{set of the infinitesimally small elements}$$

$$= (t \in S \mid \forall n \in \mathbb{N} : -\frac{1}{n} < t < \frac{1}{n}\} .$$

<u>Exercise 10</u>.
(a) Prove that $J$ can also be described as the set

$$J = \{t \in S \mid t \text{ has no inverses in } S\} .$$

(b) Show that $\mathbb{R} \cap J = \{0\}$.

We justify briefly that $J$ is an ideal in $S$: $0 \in J$ is clear. If $t, r \in J$, then each lies between $-1/2n$ and $1/2n$ for all $n$, whence $-1/n < r + t < 1/n$, so $r + t \in J$. If also $s \in S$ with $-k < s < k$, then $-k/n < st < k/n$ for all $n$; hence $-1/m < st < 1/m$ for all $m \in \mathbb{N}$, so $st \in J$. Thus $J$ is an ideal in $S$.

Using st for "standard", the homomorphism

$$\text{st} : S \to S/J$$

is called the transition to the *standard part* of $s \in S$, and the coset $s + J$ is called the **monad** of $s$. Those elements of $^*\mathbb{R}$, that do not lie in $S$, have no standard part, even though one can define a monad for them.) Now the fact that the standard part of $s \in S$ is the coset $s + J$ — the monad of $s$ — does not help us very much, because we really wanted a number in $\mathbb{R}$. Therefore, we now show that in each monad, there lies exactly one real number, and this yields $S/J \cong \mathbb{R}$. This will justify the description of the real number belonging to the monad of $s$, as the standard part of $s$.

If $x$ and $y$ are two real numbers (embedded in $^*\mathbb{R}$) with $x \neq y$, then $x$ and $y$ lie in $S$. Because we may assume that $x - y > 0$ in $\mathbb{R}$, $x$ and $y$ are not infinitesimally neighbouring in $S$; therefore $\text{st}\, x \neq \text{st}\, y$ in $S/J$.

The correspondence $x \to x + J$ therefore gives an (injective) embedding of $\mathbb{R}$ in $S/J$. We shall show that in each monad $s + J$ a real number can be found, so this embedding is a surjection, and $S/J$ is therefore isomorphic to $\mathbb{R}$.

We find the real number in $s + J$ in the following way: knowing $s$ we form

$$D_s = \{x \in R \mid x \leq s\} .$$

Because there is a natural number $n$ with $-n < s < n$, $D_s$ is a non-empty subset of $R$ bounded above; and because of the completeness of $R$, $D_s$ possesses a least upper bound $r = \sup D_s$. From the definition of a least upper bound we have $s - 1/n < r \leq s$ for all $n \in N$, and therefore $r$ and $s$ are infinitesimally neighbouring in $S$ and therefore belong to the same monad. Therefore $r + J = s + J$ and the real number $r$ can be taken as the standard part of $s$. Hence we have derived $S/J \cong R$, and by altering the notation we obtain the homomorphism

$$st : S \to R$$

as a surjective (and order-preserving) homomorphism.

*NOTE*. If in the above proof, $s$ is already in $R$, then $r = s$. Hence we have: *the standard part of a real number* $r$ *is* $r$ *itself*. Let us once more point out that here the completeness of $R$ has been used in an essential way.

It is useful to summarize matters now in the following diagram (in which $\simeq$ denotes isomorphism of ordered fields). If $^*R$ is constructed by using a fixed ultrafilter $U$ on $N$, then we have:

$$(\dagger) \qquad R \subseteq R\,(x) \simeq R(\Omega) \subseteq {}^{\Omega}R \xrightarrow{\alpha} {}^*R \supseteq S \xrightarrow{st} R$$

where $\alpha$ is the surjection from

$$^{\Omega}R = R^N/\text{Cofin}$$

to

$$^*R = R^N/U$$

arising from the inclusion of Cofin in $U$.

## 7E AN AXIOMATIC APPROACH TO NON-STANDARD ANALYSIS.

The 'analysis' part of non-standard analysis concerns the last three terms in ($\dagger$) above. We now give a brief taste of how one can work with $^*R$ in practice; it is based on notes by the Translator's colleague C.L. Thompson. He is concerned to give students a quick introduction to a system of 'hyperreal' numbers, and describes them by means of five axioms, as follows..

Axiom A. There is a complete linearly ordered field $R$.

Axiom B. There is a linearly ordered field $^*R$ which properly contains $R$ as an ordered subfield. ($^*R$ is called a hyperreal number system.)

By definition, an **n-ary real relation** is any subset of $R \times R \times \ldots \times R$. Similarly an **n-ary hyperreal relation** is any subset of $^*R \times {}^*R \times \ldots \times {}^*R$. Familiar examples on $R$ are the binary relations $E$ of equality and $L$ of being less than; by definition $E$ is the set

$\{(x,x) | x \in \mathbb{R}\}$ and L is the set $\{(x,x+u^2) | x \in \mathbb{R}, 0 \neq u \in \mathbb{R}\}$. Instead of writing $(x,y) \in E$ or $(x,y) \in L$ we usually write $x = y$ or $x < y$, but we may in this context wish to write $E(x,y)$, or $L(x,y)$.

Familiar unary relations $(r=1)$ on $\mathbb{R}$ are just the subsets of $\mathbb{R}$, and we have the choice of writing either $x \in \mathbb{N}$, $x \in \mathbb{Z}, \ldots$, or $\mathbb{N}(x)$, $\mathbb{Z}(x), \ldots$ . Similarly, unary relations on $^*\mathbb{R}$ are just subsets of $^*\mathbb{R}$. A familiar example of a ternary relation on $\mathbb{R}$ is $S = \{(x,y,z) \in \mathbb{R}^3 | z = x+y\}$. Then $(x,y,z) \in S \Leftrightarrow x$ is the sum of x and y.

A real function f of n variables is an example of an $(n+1)$-ary real relation S, with the property that if $(x_1,x_2,\ldots,x_n,y) \in S$ and $(x_1,x_2,\ldots,x_n,z) \in S$ then $y = z$; here it is usual to denote y by $f(x_1,x_2,\ldots,x_n)$.

We now state two further axioms.

Axiom C. (*Relation Axiom*) To any subset S of $\mathbb{R}^n$ corresponds a subset $^*S$ of $(^*\mathbb{R})^n$ whose real elements are precisely the elements of the original set S: i.e.

$$^*S \cap \mathbb{R}^n = S \ .$$

In particular, to the subsets $\mathbb{N}$, $\mathbb{Z}$, $\mathbb{Q}$ of $\mathbb{R}$ there correspond subsets $^*\mathbb{N}$, $^*\mathbb{Z}$, $^*\mathbb{Q}$ of $^*\mathbb{R}$, called respectively the **hypernaturals,** the **hyperintegers,** the **hyperrationals.** Since $^*\mathbb{N} \cap \mathbb{R} = \mathbb{N}$ it follows that $\mathbb{N} \subseteq {^*\mathbb{N}}$. Similarly $\mathbb{Z} \subseteq {^*\mathbb{Z}}$, and $\mathbb{Q} \subseteq {^*\mathbb{Q}}$.

For relations which are functions we have the expected:

Axiom D. (*Function Axiom*). If in Axiom C, S is a real function, then $^*S$ is a hyperreal function.

The really unusual axiom is the following, which has a 'metamathematical' flavour, and will be explained in greater detail after we have quoted it.

Axiom E. (*Transfer Axiom*). If **S** is any sentence in the language of $\mathbb{R}$, then **S** is true in $\mathbb{R}$ if and only if **S** is true in $^*\mathbb{R}$.

This axiom allows us to deduce facts about $^*\mathbb{R}$ from corresponding ones about $\mathbb{R}$. It does so by means of a formal language $L_{\mathbb{R}}$ called "*the language of* $\mathbb{R}$" which consists of the following symbols.

1) Constant symbols: a constant symbol $\hat{c}$ for each real number c.
2) Variables: $x_1, x_2, x_3, x_4, \ldots$
3) Grammatical symbols: the brackets "(" and ")".
   Connectives: $\vee, \wedge, \rightarrow, \neg$ [to be read: or, and, implies, not].
   Quantifiers: $\forall$, $\exists$.

4) Function symbols: a function symbol $\hat{f}$ for every real function of any number of variables. For the functions of addition, subtraction, multiplication and division, $L_R$ has the symbols $+$, $-$, $\times$, $\div$.

5) Relation symbols: a relation symbol $\hat{\rho}$ for every n-ary real relation $\rho$. For the real relations of equality and being less than, $L_R$ has the symbols $=$ and $<$.

(In particular, constants and functions like 1 or cos appear in $L_R$ as $\hat{1}$ and $\hat{\cos}$.)

These symbols can be assembled into formulae. More precisely, a (well formed) **formula** of $L_R$ is a finite string of symbols taken from the above list (1) – (5), the symbols being juxtaposed according to certain rules which are to be found in books on logic. Since the reader will have little difficulty in either recognising or constructing well-formed formulae, we do not give these rules here. However, an important point is that according to the rules, only *variables* can be quantified in $L_R$; function symbols and relation symbols may not be quantified in $L_R$.

A **sentence** of $L_R$ is a formula in which every variable is governed by a quantifier. Thus

$\forall x_1 \forall x_2 \ (x_1+x_2 = x_2 + x_1)$ is a sentence of $L_R$

$\forall x_1 \ x_1 < x_1+x_2$ is not a sentence of L

Here are some statements about real numbers which can be expressed by sentences in $L_R$.

1. Every number is equal to itself: $\forall x_1(x_1 = x_2)$
2. Addition is commutative: $\forall x_1 \forall x_2(x_1 \hat{+} x_2 = x_2 + x_1)$
3. $\mathbb{R}$ is Archimedean: $\forall x_1 \forall x_2[x_1 > 0 \rightarrow \exists x_3(\hat{N}(x_3) \wedge (x_3 x_1 > x_2))]$

However, consider the completeness axiom: "for all nonempty sets of real numbers E, if E has an upper bound then E has a least upper bound". This cannot be written in $L_R$ because the phrase "for all nonempty sets of real numbers" cannot be translated. We can quantify *numbers* in $L_R$ but not sets of numbers.

Interpretation. A sentence S of $L_R$ can be interpreted as a statement about $\mathbb{R}$ by removing all hats from symbols in S and replacing $\forall x_i$ and $\exists x_i$ by $\forall x_i \in \mathbb{R}$ and $\exists x_i \in \mathbb{R}$. If the resulting statement is a true one then we say that S is true in $\mathbb{R}$.

*EXAMPLES* 1. $\forall x_1(x_1 + \hat{1} = x_1)$ becomes $(\forall x_1 \in \mathbb{R})(x_1 + 1 = x_1)$, which is false.

2. $\forall x_1(\hat{\cos}\ \hat{2}x_1 = \hat{2}\hat{\cos}\ x_1 . \hat{\cos}\ x_1 - \hat{1})$ becomes the true $(\forall x_1 \in \mathbb{R})\ (\cos 2x_1 = 2\cos^2 x_1 - 1)$;

Similarly, a sentence S of $L_{\mathbb{R}}$ can be interpreted as a statement about $^*\mathbb{R}$ by replacing each $\hat{c}$ by c, each $\hat{f}$ by $^*f$, each $\hat{\rho}$ by $^*\rho$, and replacing $\forall x_i$ and $\exists x_i$ by $\forall x_i \in {}^*\mathbb{R}$, $\exists x_i \in {}^*\mathbb{R}$. Further examples are now:

3. $\forall x_1(x_1 + \hat{1} = x_1)$ becomes $(\forall x_1 \in {}^*\mathbb{R})(x_1 + 1 = x_1)$, which is false because $^*\mathbb{R}$ is a field.

4. $\forall x_1(\hat{\sin}\ x_1.\hat{\sin}\ x_1 + \hat{\cos}\ x_1.\hat{\cos}\ x_1 = \hat{1})$ becomes $(\forall x_1 \in {}^*\mathbb{R})$ $({}^*\sin\ x_1.{}^*\sin\ x_1 + {}^*\cos\ x_1.{}^*\cos\ x_1 = 1)$. Is this last statement true or false? It is certainly true for *real* $x_1$ (i.e. $x_1 \in \mathbb{R}$) that ${}^*\sin\ x_1.{}^*\sin\ x_1 + {}^*\cos\ x_1{}^*\cos x_1 = 1$, because ${}^*\sin\ x_1 = \sin\ x_1$ and ${}^*\cos\ x_1 = \cos\ x_1$ by the function axiom. But what about values of $x_1$ that are hyparreal and not real i.e. $x_1 \in ({}^*\mathbb{R}) \setminus \mathbb{R}$? This is where the transfer axiom comes in. Because the sentence

$$\forall x_1(\hat{\sin}\ x_1.\hat{\sin}\ x_1 + \hat{\cos}\ x_1.\hat{\cos}\ x_1 = \hat{1})$$

is a sentence of $L_{\mathbb{R}}$ that is true in $\mathbb{R}$ it must also be true in $^*\mathbb{R}$.

It is a basic theorem of the subject that there exists a system (indeed many such systems) $^*\mathbb{R}$ of hyperreals that satisfies Axioms A÷E. The proof uses the ultrafilter construction of Section 3; for example Axiom D (the function axiom) is verified for $^*\mathbb{R}(I,U)$ by the method of Exercise 7(b). However, the verification of Axiom E in particular requires a good deal of detail concerning the logical notions involved. Meanwhile, we now show some of the mathematics that can be done on the basis of the Axioms, by simply assuming the existence theorem.

Simple properties of $^*\mathbb{R}$ (using only Axioms A and B).

In (4) of 7D we used only Axioms A and B (without explicitly referring to them) in order to deduce the existence of the order-preserving surjection

$$st : S \rightarrow S/J \simeq \mathbb{R}$$

where S, J consist respectively of the finite, and the infinitesimal, elements of $^*\mathbb{R}$. The point is that $^*\mathbb{R}$ contains the complete ordered subfield $\mathbb{R}$. Thus the existence of $\mathbb{R}$ allows us to define various items in $^*\mathbb{R}$, and we can summarise the theory in the following way.

Definition.

$x \in {}^*\mathbb{R}$ is

| | | | |
|---|---|---|---|
| **infinitesimal** | if | $\lvert x\rvert < r$ | for all $r \in \mathbb{R}$, $r > 0$ |
| **finite** | if | $\lvert x\rvert < r$ | for some $r \in \mathbb{R}$, $r > 0$ |
| **infinite** | if | $\lvert x\rvert > r$ | for all $r \in \mathbb{R}$. |

We call $x, y \in {}^*\mathbb{R}$ **infinitely close** if $x-y$ is infinitesimal; we then write $x \approx y$.

The **monad** of x is: monad $(x) = \{y \in {}^*\mathbb{R}: x \approx y\}$

The **galaxy** of x is: galaxy $(x) = \{y \in {}^*\mathbb{R}: x - y$ is finite$\}$.

For example, galaxy (0) is S, the sub-ring of finite elements of $^*\mathbb{R}$.

Since the relation 'x−y ϵ S' is an equivalence relation, we have:

*Any two galaxies are equal or disjoint.*

We saw in the last section that J, which is monad(0) (= all infinitesimals), is a subring of $^*\mathbb{R}$ and a maximal ideal of galaxy(0).

Since ≈ is an equivalence relation on $^*\mathbb{R}$ then :

*Any two monads are equal or disjoint.*

Exercise 11. Prove that x is infinite ⟺ $x^{-1}$ is a non-zero infinitesimal.

As explained above, we already have the surjective homomorphism st:S → ℝ of ordered rings. Thus for every finite $x \in {}^*\mathbb{R}$, the **standard part** st(x) is defined; it is a real number infinitely close to x. The proof showed that the following properties hold.

*If* x,y *are finite elements of* $^*\mathbb{R}$ *then*

(i) $x \approx y \Leftrightarrow st(x) = st(y)$
(ii) $x \approx st(x)$
(iii) $r \in \mathbb{R} \Rightarrow st(r) = r$ . □

Exercise 12.
(a) Show that $^*\mathbb{R}$ contains both positive and negative infinitesimal elements.
(b) Show that $^*\mathbb{R}$ has positive and negative infinite elements.
Let $x,y \in {}^*\mathbb{R}$ be finite. Show that:
(c) if $st(y) \neq 0$ then $st(x/y) = st(x)/st(y)$,
(d) if $y = x^{1/n}$ then $st(y) = (st(x))^{1/n}$.

We come now to a more subtle matter: How to use the transfer axiom. We give some theorems of which the proofs make use of the transfer axiom.

THEOREM. $^*\mathbb{R}$ *contains infinite hypernatural numbers.* i.e. $^*\mathbb{N}$ *has infinitely large elements.*

*PROOF*. Because ℝ is Archimedean, the following is true:

$$(\forall y \in \mathbb{R})\ (\exists\, n \in \mathbb{N})\ (n > y)\ .$$

this can be written in $L_{\mathbb{R}}$, hence by transfer, the following is true:

$$(\forall y \in {}^*\mathbb{R})\ (\exists\, n \in {}^*\mathbb{N})\ (n > y)\ .$$

By Exercise 12(b), $^*\mathbb{R}$ contains a positive infinite element, say $y_0$. Thus there is an infinite hypernatural number $n > y_0$. □

Let $f : \mathbb{R} \to \mathbb{R}$ be a real function and let c be any real number. Consider the two definitions of continuity:

*Standard Definition*: f is continuous at c iff

$$(\forall \varepsilon \in \mathbb{R}^+)\ (\exists\ \delta \in \mathbb{R}^+)\ (\forall x \in \mathbb{R})\ (|x-c| < \delta \Rightarrow |f(x) - f(c)| < \varepsilon).$$

This has three quantifiers, whereas the Non-standard definition has only one:

*Non-standard Definition*: f is continuous at c iff

$$(\forall\ x \in {}^*\mathbb{R}^+)\ \left[(x \approx c) \Rightarrow ({}^*f(x) \approx {}^*f(c))\right]$$

<u>*EXAMPLES*</u>.
1. Let $f(x) = x^2$, $c = 1$. Let $x = 1+h$ where $h \approx 0$, i.e. $h \in J$. Then ${}^*f(x) = {}^*f(1+h) = (1+h)^2 = 1+2h+h^2 \approx 1$ (since J is a ring). So ${}^*f(x) - {}^*f(1) \approx 0$, and ${}^*f$ is continuous at 1.
2. Let $f(x) = 1(x>0)$, and $0(x \leq 0)$, and let $c = 0$. Let h be a positive infinitesimal. Then ${}^*f(h) = 1$, so ${}^*f(h) - {}^*f(0) = 1-0 = 1$ which is not infinitesimal. Thus f is discontinuous at 0.

<u>THEOREM</u>. *The standard and non-standard definitions of continuity are equivalent.*

<u>*PROOF*</u>. Suppose that f is continuous at c by the standard definition, and suppose that y is a hyperreal such that $y \approx c$. To show that f is continuous at c by the non-standard definition we must show that ${}^*f(y) \approx {}^*f(c)$. But if $\varepsilon$ is any positive real number, then there is another real number $\delta > 0$ such that

$$(\dagger) \qquad (\forall x \in \mathbb{R})\ (|x-c| < \delta \Rightarrow |f(x) - f(c)| < \varepsilon).$$

By transfer, (†) must be true in ${}^*\mathbb{R}$. Thus

$$(\forall x \in {}^*\mathbb{R})\ (|x-c| < \delta \Rightarrow |{}^*f(x) - {}^*f(c)| < \varepsilon).$$

Now $y \approx c$ so $|y-c| < \delta$ (because $\delta$ is real) hence $|{}^*f(y) - {}^*f(c)| < \varepsilon$. This is true for any real $\varepsilon > 0$, so ${}^*f(y) \approx {}^*f(c)$.

Conversely, suppose that f is continuous at c by the non-standard definition and suppose we are given any real $\varepsilon > 0$. We must show that

$$(\exists\ \delta \in \mathbb{R}^+)\ (\forall x \in \mathbb{R})\ (|x-c| < \delta \Rightarrow |f(x) - f(c)| < \varepsilon)$$

is true. By transfer, it suffices to show the truth of

$$(\exists \delta \in {}^*\mathbb{R}^+)\ (\forall x \in {}^*\mathbb{R})\ (|x-c| < \delta \Rightarrow |{}^*f(x) - {}^*f(c)| < \varepsilon)$$

But there does exist such a $\delta$ in ${}^*\mathbb{R}$, namely any positive infinitesimal $\eta$. For if $|x-c| < \eta$ then $x \approx c$ and so ${}^*f(x) \approx {}^*f(c)$ by hypothesis; hence $|{}^*f(x) - {}^*f(c)| < \varepsilon$. □

We conclude with two harder theorems, each of which asserts that a certain real number exists.

THEOREM. *Let* $f : \mathbb{R} \to \mathbb{R}$ *be continuous at every real number in the interval* $I = \langle a,b \rangle$. *Then* f *is bounded on* I.

*PROOF.* We want to prove that there exists a real number M such that $-M < f(x) < M$ for every real x in I. In the language of $\mathbb{R}$, we must prove: $(\exists y \in \mathbb{R})(\forall x \in \mathbb{R})((a \leq x \leq b) \Rightarrow (-y \leq f(x) \leq y))$. It is sufficient (by transfer) to prove

$$(\exists y \in {}^*\mathbb{R})(\forall x \in {}^*\mathbb{R})(a \leq x \leq b \Rightarrow -y \leq {}^*f(x) \leq y).$$

But this is true, for take y to be any positive infinite hyperreal: if $a \leq x \leq b$ with $x \in {}^*\mathbb{R}$ then since st preserves order, $a \leq st(x) \leq b$ and ${}^*f(x) \approx {}^*f(st(x)) = f(st(x))$ which is real, hence finite. Thus $-y \leq {}^*f(x) \leq y$. □

THEOREM. (*Intermediate Value Theorem*). *Let* $f : \mathbb{R} \to \mathbb{R}$ *be a real function which is continuous on* $I = \langle a,b \rangle$ *and suppose that* $f(a) < 0 < f(b)$. *Then there is a real number* c, *such that* $a < c < b$ *and* $f(c) = 0$.

*PROOF.* If we divide I into n equal subintervals $\langle x_m, x_{m+1} \rangle$ by division points $x_m = a + m(b-a)/n$, $(0 \leq m \leq n-1)$, then there must be two adjacent points $x_m$ and $x_{m+1}$ such that $f(x_m) < 0 < f(x_m)$.

Thus the following is true:

$$(\forall n \in \mathbb{N})(\exists m \in \mathbb{N})\left[(0 \leq m \leq n-1) \wedge \left[f\left[a + \frac{m(b-a)}{n}\right] \leq 0 \leq f\left[a + \frac{(m+1)(b-a)}{n}\right]\right]\right].$$

This statement is translatable into $L_{\mathbb{R}}$, so by transfer the following is true:

$$(\forall n \in {}^*\mathbb{N})(\exists m \in {}^*\mathbb{N})\left[(0 \leq m \leq n-1) \wedge \left[{}^*f\left[a + \frac{m(b-a)}{n}\right] \leq 0 \leq {}^*f\left[a + \frac{(m+1)(b-a)}{n}\right]\right]\right]$$

Suppose N is any infinite hypernatural. The last formula tells us that there is a hypernatural M, $0 \leq M \leq N-1$, such that ${}^*f(x_M) \leq 0 \leq {}^*f(x_{M+1})$, where $x_M = a + M(b-a)/N$, $x_{M+1} = a + (M+1)(b-a)/N$. Clearly $a \leq x_M < x_{M+1} \leq b$, and $x_M \approx x_{M+1}$. Let $c = st(x_M)$. Then c is real, and $a \leq c \leq b$; also $c \approx x_M$ and $c \approx x_{M+1}$. Because f is continuous and c is real we have

$${}^*f(x_M) \approx f(c) \approx {}^*f(x_{M+1}).$$

Since ${}^*f(x_M) \leq 0$ and ${}^*f(x_{M+1}) \geq 0$, and also ${}^*f(c)$ is real $(= f(c))$ it follows that $f(c) = 0$. Obviously $c \neq a,b$, so $a < c < b$. □

For further material in this direction see the books of Keisler [39], Henle and Kleinberg [30] and of course the original pioneering book of Robinson who shows how to use non-standard analysis in various areas of pure and applied mathematics.

We conclude with some exercises that have a revision content, to help readers to organise the preceding theory in their minds. They are taken from examination papers of Southampton University from 1982 onwards.

Exercise 13.

(a) Let $\mathbb{R}(t)$ be the ordered field of all formal rational expressions $x$ with real coefficients

$$x = \frac{a_0 + a_1 t + \ldots + a_m t^m}{b_0 + b_1 t + \ldots + b_n t^n} \qquad (b_n > 0),$$

where $x$ is positive if and only if $a_m b_n > 0$. Suppose that $x$ is positive. What conditions on $m$ and $n$ ensure that $x$ is (i) infinite, (ii) infinitesimal, (iii) finite and not infinitesimal?

Prove that there is no $x$ in $\mathbb{R}(t)$ such that

$$k \leq x \leq \frac{t}{k} \qquad \text{for all } k \text{ in } \mathbb{N}.$$

Discuss the convergence or divergence in $\mathbb{R}(t)$ of the following sequences.

$$\left[\frac{t^n}{n!}\right], \quad (n^{-1}), \quad \left[\sum_{j=0}^{n} t^{-2j}\right].$$

(b) Explain what is meant by "the language of $\mathbb{R}$", and state axioms for a hyperreal number system $\mathbb{R}^*$. Explain why the equation $x + x = x$ is true for some infinite cardinal numbers but is false for all infinite hyperreal numbers.

(c) Explain what is meant by saying that ${}^*\mathbb{R}$ is a hyperreal number system. Prove that every finite element $x$ of ${}^*\mathbb{R}$ is infinitely close to a unique real number, $st(x)$ say. Prove that

$$st(xy) = st(x)\, st(y)$$

whenever $x, y$ are finite elements of ${}^*\mathbb{R}$.

Let $A$ be a subset of $\mathbb{R}$. Prove that $A$ is bounded if and only if every element of ${}^*A$ is finite. Hence give a nonstandard proof that every continuous real function $f : \langle 0,1 \rangle \to \mathbb{R}$ is bounded.

(d) Let ${}^*\mathbb{R} = \mathbb{R}^{\mathbb{N}}/U$ be the non-standard real numbers, where $U$ is a suitable ultrafilter on $\mathbb{N}$.

Prove that, in a sense to be made precise, every element $x$ of ${}^*\mathbb{R}$ which is not infinitely large is infinitely close to a unique real number, called the standard part of $x$.

Explain what is meant by the non-standard extensions $^*X$, $^*\alpha$ of a subset $X \subseteq \mathbb{R}$ and of a function $\alpha: X \to \mathbb{R}$.

If $n$ is an infinitely large element of $^*\mathbb{N}$, what is the

standard part of $\sum_{k=1}^{n} \frac{1}{2^k}$ ?

# 8

# Pontrjagin's topological characterization of $\mathbb{R}$, $\mathbb{C}$ and $\mathbb{H}$

INTRODUCTION.

It is often through topological concepts that ideas of a geometrical nature come into their own for describing mathematical objects. We have already had a foretaste of this with $\mathbb{R}$ and $\mathbb{R}^2$, which we compared them using successively the criteria of set theory, of group theory, and of topology. In this, only with the topology were our intuitive ideas confirmed.

The goal of this chapter is the famous theorem of the Soviet mathematician Pontrjagin, which characterizes $\mathbb{R}$, $\mathbb{C}$ and $\mathbb{H}$ as the only locally compact connected topological skew fields. We shall see that for $\mathbb{R}$, "locally compact" is nothing other than a further variant of completeness; so that this basic property, which has already described connectivity, also has a decisive effect in this last chapter. For the proof of the theorem, we need at the beginning some painstakingly detailed work for which however, one is compensated at the end by a beautiful compactness argument which guarantees the finite dimension (over $\mathbb{R}$) of the skew field in question. With this one displays the connection with the theorem of Frobenius that we met in Chapter 5.

In the first paragraph a few basic concepts of topology will be recalled which however, will be assumed essentially as known. What is important is to know something of the many examples which can make the reader more familiar with the material. In order to fix the terminology, we use that of the book *Topology I* of W. Franz [20]. Our proof of Pontrjagin's theorem follows essentially the exposition of Pontrjagin [59], with variations at the beginning from Salzmann [65] I, Para 13, and at the end from Fuhrer [21] *General Topology with Applications.* Pontrjagin simplified his original proof of 1932, using ideas of Kowalsky (1953) to which he explicitly refers. A useful introduction into the basic concepts of topological algebra can be found in van der Waerden: [73] Vol. II, but his material is not a necessary pre-requisite for understanding the present chapter. In the early stages of the Chapter, some exercises are provided – largely for revision, or to introduce general topology to students who have met only metric spaces. It is not necessary to be able to do all the exercises before reading further!

## 8A TOPOLOGICAL GROUPS.

A topology on a set $X$ can be defined by means of neighbourhoods or by means of open sets (or even in other ways). We will switch freely between both possibilities (assumed as known) and assume also that the reader understands the concept of the basis of a topology even though a definition is repeated below, in Exercise 1. We write $(X, \text{top})$ or $(X, \mathcal{O})$ for a topological space (with system $\mathcal{O}$ of open sets), later mostly just $X$, whenever the relationships are clear.

There are two trivial possibilities for a topology on any set X. One obtains the discrete (or finest) topology on $X$, by taking as the open sets *all* the subsets of $X$; therefore here $\mathcal{O} = \mathfrak{P}X$. In this case, in particular the singleton sets $\{x\}$ are open. The other extreme is the coarsest topology on $X$, where we take $\mathcal{O} = \{\emptyset, X\}$ and here there are only two open sets.

The explicit description of all open sets for a space $(X,\text{top})$ is often complicated. It is much more convenient to manage with the help of the concept of a basis for a topology.

__Definition__. In a topological space $(X, \text{top})$ a family $B$ of subsets of $X$ is called a **basis** of the topology whenever (1) each set of $B$ is open and (2) each open set $W$ can be represented as a union of basis sets.

The concept of basis is (implicitly) used for example, whenever one describes the open sets in $\mathbb{R}$ as unions of open intervals. To recognise that a family $B$ of open subsets is a basis of $X$, one needs only to check that to each open set $W \subseteq X$ and $p \in W$ there exists a set $B \in \mathbf{B}$ with $p \in B \subseteq W$. The description of the open sets in $\mathbb{R}^n$ with the help of $\varepsilon$-balls or open cubes is assumed known.

__Definition__. (a) A mapping $f : X \to Y$ between topological spaces $X, Y$ is called **continuous** whenever, for each point $x \in X$ and each neighbourhood $W$ of $f(x)$ in $Y$, there exists a neighbourhood $U$ of $x$ in $X$ with $f(U) \subseteq W$.

(b) A mapping $f : X \to Y$ between topological spaces is called a **homeomorphism** whenever $f$ is bijective and both $f$ and $f^{-1}$ are continuous.

Instead of "homeomorphism" one can also say "topological isomorphism", as this conveys the essential meaning*. Some further fundamental concepts of topology will be recalled later. Although many books restrict a neighbourhood (briefly: nbd) to be an open set, others (including this one) say that $U$ is a **nbd** of $x \in X$ iff there exists an open $V$ with $x \in V \subseteq U$. Thus $V$ is an 'open' nbd of $x$; if $U$ were (say) compact, it would be a 'compact' nbd, etc. (The German word is "Umgebung", hence the use of $U$ for a nbd.)

---

* One must, in this context, also be very careful not to ignore the letter e in "homeomorphism", because algebra deals also in homomorphisms.

<u>Exercise 1</u>. (Revision).

(a) Recall that a family $T$ of subsets of $X$ is called a *topology*, provided (i) $\phi$ and $X$ are in $T$, (ii) the union of any number and the intersection of any *finite* number, of members of $T$, is again in $T$. The members of $T$ are the *open* sets of the topology. If $Y \subseteq X$, the **induced topology** on $Y$ consists of the family $T_y$, where

$$T_y = \{G \cap Y | G \in T\}$$

Show that $F_Y$ is a topology on $Y$ and that the identity mapping $id_Y:(Y,T_Y) \to (X,T)$ is continuous, where $id_Y(y) = y$ for each $y \in Y$.

(b) Show that if $f:(X,T) \to (Z,S)$ is a mapping between topological spaces then $f$ is continuous iff, for each open $G \in S$, $f^{-1}G$ is open in $X$ (i.e. $G \in S \Rightarrow f^{-1}G \in T$). Sets of the form $Z-G$ are called **closed** in $Z$. Show that $f$ is continuous iff for each closed $F$ in $Z$, $f^{-1}F$ is closed in $X$.

(c) Let $(X,T)$ be a topological space, and let $D,C$ denote the discrete and the coarsest topologies on $X$. Show that

$$id_X:(X,D) \to (X,T) \quad \text{and} \quad id_X:(X,T) \to (X,C)$$

are continuous, but neither is a homeomorphism if $D \neq T \neq C$.

(d) if $X$ is a metric space with metric $\rho$, a set $G$ is declared to be open iff for each $x \in G$, some $\varepsilon$-nbd $U_\epsilon(x)$ lies in $G$, where

$$U_\epsilon(x) = \{y \in X | \rho(x,y) < \epsilon\} \qquad (\epsilon > 0) .$$

Show that these open sets together form a topology on $X$.

If now $f:X \to Y$ is a mapping between metric spaces, show that $f$ is continuous iff for every convergent sequence $(x_n)$ in $X$ with $\lim(x_n) = x$, the sequence $(fx_n)$ converges in $Y$ to $fx$. (Thus "f <u>commutes with</u> lim".)

Show that $f$ can be a homeomorphism and $(X,\rho)$ a complete metric space, without the metric in $Y$ being complete.

(e) Show that the n-sphere $S^n$ is a closed subset of $\mathbb{R}^{n+1}$. (Consider the mapping $f:\mathbb{R}^{n+1} \to \mathbb{R}$ given by $(x_1,\dots,x_n) \to x_1{}^2 + \dots + x_n{}^2)$ and look at $f^{-1}(1)$.)

The interweaving of algebraic and topological concepts is best studied, at first, in the case of topological groups. Of these far-reaching theories we can only look at the early definitions and examples. As with ordered groups, the interplay of the algebraic and topological structures is displayed through compatibility conditions.

<u>Definition</u>. Let $G$ be a set such that $(G, *)$ is a group and $(G, \text{top})$ is a topological space. We call $(G, *, \text{top})$ a **topological group** whenever the following compatibility conditions are fulfilled.

(TG 1) *The operation* $*$ *is continuous.* In more detail: given a nbd $W$ of the point $x * y$ there exist nbds $U$ of $x$ and $V$ of $y$ such that for all $u \in U$ and $v \in V$, we have $u * v \in W$ (briefly $U * V \subseteq W$).

(TG 2) *The inverse mapping* $x \to x^{-1}$ *in* $G$ *is continuous*, i.e. given a nbd $W$ of the point $x^{-1}$, there is a nbd $U$ of the point $x$, with $u^{-1} \in W$ for all $u \in U$.

*EXAMPLES*. 1. The additive group $(\mathbb{R}, +, \text{top})$ of the real numbers with the usual topology is a topological group. To verify (TG 1) one regards $f : (x, y) \to x + y$ as a function $f : \mathbb{R}^2 \to \mathbb{R}$, and being linear, $f$ is continuous. For (TG 2) one immediately has the continuity of $x \to -x$.

2. The multiplicative group $(\mathbb{R}^\times, \cdot, \text{top})$ of the real numbers, with the usual topology, is a topological group. To verify (TG 1), one regards $f : (x, y) \to xy$ as a function $f : \mathbb{R}^2 \to \mathbb{R}$, and this function is continuous. Also $x \to x^{-1}$ is continuous on $\mathbb{R}^\times = \mathbb{R} \setminus \{0\}$.

3. The groups $(\mathbb{Q}, +, \text{top})$ and $(\mathbb{Q}^\times, \cdot, \text{top})$ of rational numbers, with the induced topology of $\mathbb{R}$, are topological groups.

4. The additive groups $(\mathbb{R}, +, \text{top})$, with the usual topology of $\mathbb{R}^n$, are topological groups. To these belong especially the additive groups $(\mathbb{C}, +, \cdot)$ of the complex numbers and $(\mathbf{H}, +, \text{top})$ of the quaternions.

5. (Multiplicative group $(\mathbb{C}^\times, \cdot, \text{top})$ of the complex numbers.) For the continuity of multiplication one can fall back on the real numbers. We write $x + iy = \binom{x}{y}$ etc and have

$$\begin{bmatrix} x \\ y \end{bmatrix} \cdot \begin{bmatrix} v \\ w \end{bmatrix} = \begin{bmatrix} xv - yw \\ xw + yv \end{bmatrix} .$$

Because of the continuity of the functions $(a, b) \to ab$, $a \to -a$, and $(a, b) \to a + b$ the resulting function obtained by combining them is continuous; hence so is the function

$$\left[\begin{bmatrix} x \\ y \end{bmatrix}, \begin{bmatrix} v \\ w \end{bmatrix}\right] \to xv - yw$$

and similarly for $xw + yv$.

One can argue correspondingly to show the continuity of the inverse mapping

$$\begin{bmatrix} x \\ y \end{bmatrix} \to \frac{1}{x^2 + y^2} \begin{bmatrix} x \\ -y \end{bmatrix} .$$

6. (Multiplicative group ($\mathbb{S}$, $\cdot$, top) of the unit circle in $\mathbb{C}$.) Here the topology is that induced by $\mathbb{C}$, so the restrictions of the product and inverse mappings, to elements of $\mathbb{S}$, are each continuous.

7. (Multiplicative group ($\mathbb{H}^\times$, $\cdot$, top) of the quaternions.) Here we take the usual topology of the four dimensional space $\mathbb{R}^4 = \mathbb{H}$. The conditions (TG 1) and (TG 2) can be verified as for the multiplicative group of the complex numbers, by applying the formulae for the product and inverse, given in Section 5A.

8. (Multiplicative group ($\mathbb{S}^3$, $\cdot$, top) of quaternions of modulus 1.) Here we use the induced topology of $\mathbb{R}^4$, so the restrictions to the 3-sphere $\mathbb{S}^3$, of the product and inverse mappings of $\mathbb{H}^\times = \mathbb{R}^4 \setminus \{0\}$, are continuous.

9. (Diverse groups of matrices.) We consider only the group $GL(n, \mathbb{R})$ of the $n \times n$ matrices $A$ with $\det A \neq 0$. One considers the matrix $A = (a_{ik})$ as the vector $(a_{11},\ldots,a_{1n}, a_{21},\ldots,a_{nn}) \in \mathbb{R}^{n \times n}$, and then the topology of $\mathbb{R}^{n \times n}$ can be used (e.g. to investigate convergence 'co-ordinate-wise'). For the product and inverse mappings, one argues again as for the multiplicative group ($\mathbb{C}^\times$, $\cdot$, top) of the complex numbers. In particular, one uses for the inverse $A^{-1}$ of the matrix $A$, the formula $A^{-1} = B/(\det A)$ with $B$ the adjoint of $A$, which results from Cramer's rule.

Interesting (topological) subgroups of the group $GL(n, \mathbb{R})$ are e.g. the group $SL(n, \mathbb{R})$ of matrices with determinant 1, or the group $O(n, \mathbb{R})$ of orthogonal matrices.

Translations. Let (H, $\cdot$, top) be a topological group. Then for a fixed $a \in H$ the mapping $\alpha : H \to H$, defined by $x \to xa$, is bijective and continuous because the multiplication in H is continuous. The inverse mapping $\alpha^{-1} : y \to ya^{-1}$ belonging to it is also continuous, so $\alpha$ is a homeomorphism. This type of mapping is also called a (right) translation. Under a homeomorphism, nbds of a point $x$ are mapped onto nbds of its image point $\alpha x$. In the group H, the nbds of the neutral element e are mapped by $\alpha$ in this way onto the nbds of a. In other words the nbds of the element $a \in H$ result from those of the neutral element e. Put in yet another way: a *nbd basis of e determines the entire topology of* H. Therefore, in order to give a topology to H, one needs only to specify the nbds (or a basis of nbds) of e. The upshot of these relationships is recorded in the following statement.

LEMMA. *Let* $\psi$ : (G, ·, top) → (H, ·, top) *be an algebraic homomorphism* $\psi$ : (F, ·) → (H, ·) *and continuous at* $e \in G$. *Then* $\psi$ *is continuous on the whole of* G.

*PROOF*. Let $\psi x = y \in H$ and Y be a nbd of y in H. Then $V = y^{-1} Y$ is a nbd of the neutral element e' of H. Because of the continuity of $\psi$ at e, and since $\psi e = e'$, there is a nbd U of e with $\psi U \subseteq V$. One now sets $X = xU$ and checks that X is a nbd of x with $\psi X \subseteq Y$. □

The most important interesting properties of topological spaces derive from the validity of separation axioms, connectedness and compactness. Separation axioms tell how points and/or sets can be separated from each other by suitable open sets.

Definition. A topological space X is called a $T_2$–space or **Hausdorff** space, whever given two distinct points $x, y \in X$, there exist nbds W of X and V of y with $W \cap V = 0$.

The spaces $\mathbb{R}^n$ with the usual topology are all Hausdorff. Non – Hausdorff spaces are rather rarely met, and will not be considered in this chapter. (Historical note: F. Hausdorff wrote the text *Mengelehre* [29] which was extremely influential in the early years of the development of Topology; so the spaces are named after him.)

Exercise 2.

(a) if X is a topological space, then a sequence $(x_n)$ in X converges to x, iff for each nbd U of x, almost all the points $x_n$ lie in U. (More precisely, there exists $n \in \mathbb{N}$ such that $x_{n+k} \in U$ for all $k \in \mathbb{N}$). Prove that, if X is also a $T_2$–space, then limits are unique, i.e. $(x_n)$ cannot converge to a and b with $a \neq b$.

(b) Let Y be a subspace of X with the induced topology. Prove: Y is a $T_2$–space if X is.

(c) Show that every metric space is a $T_2$–space, and hence that $\mathbb{R}^n$ and $\$^n$ are Hausdorff.

Definition. A topological space X is called **connected**, whenever it cannot be split into the union of two non–empty disjoint open sets.

*EXAMPLES*. The topological groups ($\mathbb{R}^n$, +, top) are connected. By contrast, the group ($\mathbb{Q}$, +, top) is not connected. The groups (\$, +, top), ($\mathbb{C}^\times$, ·, top) are connected. The group ($\mathbb{R}^\times$, ·, top) is not connected.

Exercise 3.

(a) A topological space X is **path–connected** provided for each $x,y \in X$ there is a continuous mapping f of the unit interval into X with $f(x) = 0$, $f(y) = 1$. Show that then X is connected. (The converse is false.)

(b) Use (a) to prove the statements about the Examples above.

Definition. A topological space X is called **compact** whenever it is Hausdorff and has the Heine–Borel covering property, which says:

*Every open covering of* X *possesses a finite sub–covering.*

(An **open covering** is a family F of open sets G such that each $x \in X$ lies in some $G \in F$; a finite **sub–covering** of F is a set $G_1, G_2, \ldots, G_n$ of members of F such that these together form a covering of F.).

We recall that a subset A of $\mathbb{R}^n$ is compact if and only if A is closed and bounded.

*EXAMPLES.* The sphere $\mathbb{S}^n$ in $\mathbb{R}^{n+1}$ is compact for all n. Therefore, the topological groups ($\mathbb{S}$, ·, top) and ($\mathbb{S}^3$, ·, top) are compact. The spaces $\mathbb{R}^n$ themselves are not compact.

Exercise 4.

(a) Prove these statements, using the theorem recalled above.

(b) Suppose X is compact. Show that if F is a closed subset, then F is compact, and that the converse holds if X is also Hausdorff.

(c) Let $f: X \to Y$ be a continuous mapping between topological spaces. If X is compact and f is a surjection, show that Y is compact. If f is a bijection, show that f is a homeomorphism provided Y is Hausdorff.

The very useful concept of compactness is still relevant for $\mathbb{R}^n$ but in a changed form:

Definition. A topological space X is called **locally compact** if every point of X possesses a compact nbd.

*NOTE.* In Section 1E it was shown that for the real numbers the connected property [CA] and local compactness [HB] are each equivalent to completeness. These two properties are therefore nothing other than certain aspects of completeness which have been carried over to more general situations.

*EXAMPLES.* 1. The spaces ($\mathbb{R}^n$, +, top) are locally compact, and hence so are many topological groups which derive their topology as subspaces of $\mathbb{R}^n$.

2. A compact, but not connected, group is the group $O(2, \mathbb{R})$ of all orthogonal $2 \times 2$ matrices. As is well known, this group consists of rotations about the origin O, with matrices of the form

$$A = \begin{bmatrix} \cos\alpha, & -\sin\alpha \\ \sin\alpha, & \cos\alpha \end{bmatrix}.$$

together with reflections in the lines through O, with matrices of the form

$$B = \begin{bmatrix} \cos\beta, & -\sin\beta \\ \sin\beta, & \cos\beta \end{bmatrix}.$$

Via the correspondence

$$A \to \begin{bmatrix} \sin\alpha \\ \cos\alpha \end{bmatrix}$$

the subgroup of rotation matrices is obviously (topologically) isomorphic to the unit circle group ($, ·, top). The set of line reflections consists (as a coset) of the rotations under the mapping

$$A \to A \begin{bmatrix} 1 & 0 \\ 0 & -1 \end{bmatrix};$$

this means that (viewed topologically) the set of line reflections forms a second circle $ . By the (continuous) determinant mapping, the rotations are mapped on $+1$ and the reflections on $-1$. The circles $ and $ are therefore not connected to each other.

The group O(2, R)

The coset of the reflections

The subgroup of the rotations

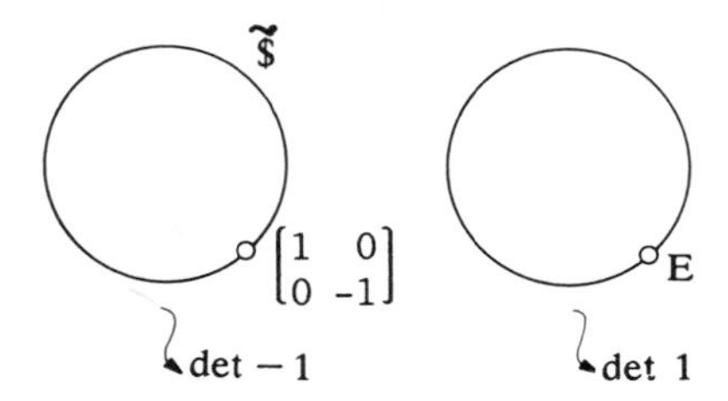


Fig. 71

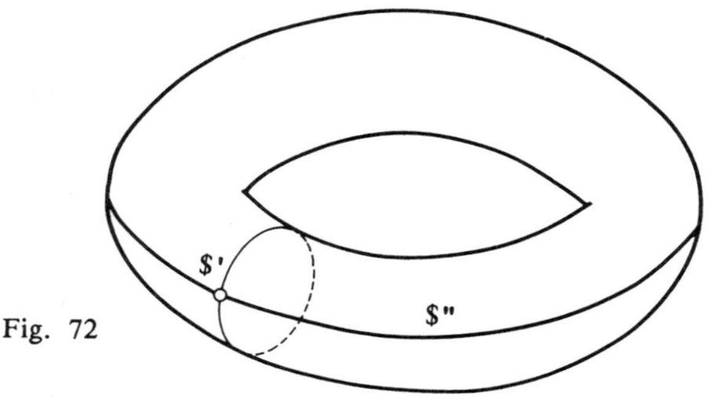


Fig. 72

3. A further compact group is formed from the torus $^1 × $^1. It has the two "axes" $' and $" (each a copy of $^1), and the group operation is calculated component-wise.

This paragraph ought not to finish without our proving at least one theorem. As we have seen, the spheres $\mathbb{S}^1$ and $\mathbb{S}^3$ can be turned into groups by defining multiplications in appropriate ways. In this respect, how fares $\mathbb{S}^2$, the usual spherical surface in $\mathbb{R}^3$?

THEOREM. *No operation can be found, such that* ($\mathbb{S}^2$, $\cdot$, top) *becomes a topological group. (Here, we naturally mean: with the usual topology for* $\mathbb{S}^2$.)

*PROOF*. We assume that ($\mathbb{S}^2$, $\cdot$, top) is a topological group, and produce a contradiction to the "Hairy Ball Theorem", or rather its 2-dimensional 'hedgehog' version (see Section 5H and the references Holmann [35] and Milnor [50]).

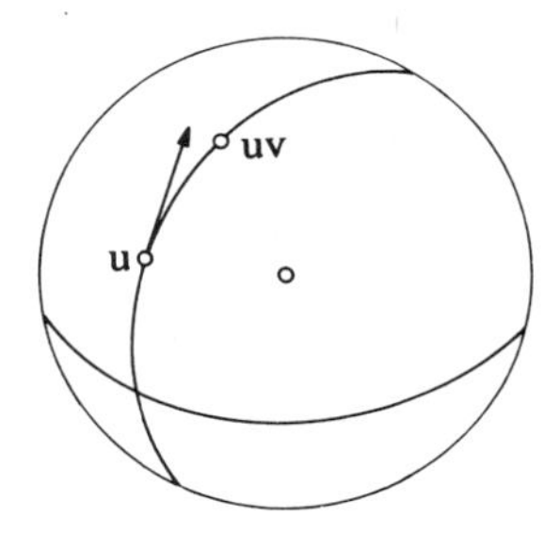


Fig. 73.

Suppose that we can find $v \in \mathbb{S}^2$ (near the neutral element $e$) such that for each $u \in \mathbb{S}^2$ the point $uv$ lies near $u$. Then (because $uv$ is not the antipodal* point of $u$) there is a unique shortest arc $A$ of a great circle from $u$ to $uv$. Because ($\mathbb{S}$, $\cdot$, top) is supposed to be a topological group, $uv$ depends continuously on $u$, and hence so does the direction of the arc $A$. If at the point $u$, we attach a tangential unit vector $t_u$ to this arc (in the direction from $u$ to $uv$) then we will obtain a continuous field of tangential unit vectors on $\mathbb{S}^2$. Such a field cannot exist, by the 'hedgehog' theorem; contradiction.

The following considerations serve to prove the existence of an element of $v \in \mathbb{S}^2$ with the required property. It comes from a typical compactness argument which we will be able to use quite often in this chapter.

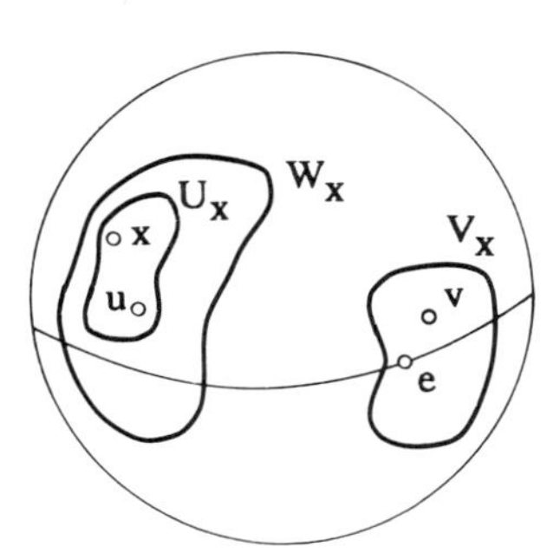


Fig. 74.

* The **antipodal** point of $u$ is $-u$ in $\mathbb{R}^3$. It lies at the end of the diameter through $u$.

Let $e$ be the identity element of the group. For each point $x \in S^2$ we choose an open nbd $W_x$ of $x$ which does not extend outside the (open) equatorial cap that has $x$ as its pole. Since $ex = x$ and the multiplication is continuous, there are open nbds $U_x$ of $x$ and $V_x$ of $e$ such that for all $u \in U_x$ and $v \in V_x$ we have $uv \in Wx$. Here we may also assume that $U_x \subseteq W_x$. The sphere $S^2$ is compact and therefore *finitely many* of the nbds $U_x$, say $U_{x(1)}, \ldots, V_{x(n)}$ – suffice as a covering of $S^2$. The intersection $V = V_{x(1)} \cap \ldots \cap V_{x(n)}$ of these *finitely many corresponding* nbds of $e$ is again a nbd of $e$. From this nbd $V$ we choose an element $v \neq e$.

Then $uv \neq u$ because $(S^2, \cdot, \text{top})$ is supposed to be a group. Moreover, the mapping $u \to uv$ is continuous because the group is topological. By the choice of $W_x$, $U_x$ and $V_x$, the points $u$ and $uv$ lie on an open equatorial cap; hence these points are certainly not antipodal, and therefore the shortest circular arc from $u$ to $uv$ is uniquely defined. This shows that we have found $v$ with the properties we required above, and the proof is complete. □

## 8B TOPOLOGICAL FIELDS.

Definition. If $(K, +, \cdot)$ is a skew-field and $(K, \text{top})$ a topological space, then $(K, +, \cdot, \text{top})$ is called a **topological skew-field** whenever the following conditions are fulfilled.

(TK 1) The topological space $(K, \text{top})$ is Hausdorff and therefore satisfies the separation axiom $T_2$.

(TK 2) $(K, +, \text{top})$ is a topological group.

(TK 3) Multiplication $(x,y) \to xy$ is continuous.

(TK 4) The inverse mapping $x \to x^{-1}$ (when $x \neq 0$) is continuous.

*COMMENTS*. (a) Instead of the conditions (TK 3, 4) one can simply demand that $(K^\times, \cdot, \text{top})$ is a topological group because multiplication must also be continuous when $x = 0$ or $y = 0$.

(b) It has become customary to say "topological field" even when one should properly say "topological skew-field" (whenever the commutative law of multiplication is not fulfilled). We use the concept from now on in this sense, and therefore also describe the quaternions as a topological field.

(c) In a trivial way, we can obtain a topological field from any field $(K, +, \cdot)$ by choosing the discrete topology (each set is open) on $K$. With the coarsest topology (only $\emptyset$ and $K$ are open) one does not obtain a topological field, because then the distinct field elements $0$ and $1$ have no disjoint nbds, and $T_2$ is not satisfied.

(d) One can refer to Bourbaki [8], Ch.III Section 6.7, to see that for a field with a topology that fulfills the conditions (TK 1 – 4), the condition $T_2$ can only be violated in the case of the coarsest topology. Thus the requirement that (K, top) should be Hausdorff is simply superfluous whenever one ignores the uninteresting marginal case of the coarsest topology.

*EXAMPLES.*
1. $(\mathbb{Q}, +, \cdot, \text{top})$, $(\mathbb{R}, +, \cdot, \text{top})$, $(\mathbb{C}, +, \cdot, \text{top})$ and $(\mathbf{H}, +, \cdot, \text{top})$ are all topological fields. In the cases $\mathbb{C}$ and $\mathbf{H}$, the usual topology in $\mathbb{R}^2$ or $\mathbb{R}^4$ is to be taken. The continuity proofs for the field–operations can be verified as we did in the previous Section for the topological group structure. (For a proof directly in terms of the modulus in $\mathbb{R}^n$, see Salzmann [65] I, 13.2.)

2. The linearly ordered field of the non–standard numbers discussed in Chapter 7 becomes a topological field whenever one chooses, as a basis for the topology, the open interval $\langle a, b \rangle$ with $a < b$. We spare ourselves the proof of all the necessary conditions.

3. Further important examples are the so–called p–adic number fields, and certain function fields, be we shall not go into these here.

<u>Definition</u>. A topological field $(K, +, \cdot, \text{top})$ is called *connected (compact, locally compact)* whenever the topological space (K, top) has these properties.

From now on we will mostly write only K for simplicity instead of $(K, +, \cdot, \text{top})$, because it is clear what is meant.

The next theorems belong already to the preparation for the proof of Pontrjagin's theorem. In his proof, there is much to do with sequences, and we therefore recall a few topological concepts and theorems concerning them. In particular, if a sequence $(a_k)$ in A converges to $\ell$, then $\ell$ is called an **accumulation point** of A provided the set $\{a_1, a_2, \ldots, a_k, \ldots\}$ is infinite; if this set is finite, then $\ell$ would equal $a_k$ for all but a finite number of k, so $\ell$ would lie in A (e.g. as an isolated point).

In a compact topological space an infinite subset has at least one accumulation point. The topological space X is called **sequentially compact** whenever in X each sequence $(a_k)$ with $k \in \mathbb{N}$ possesses a convergent subsequence. With Pontrjagin, we say that a sequence $(b_k)$ without accumulation points is divergent. In topological spaces with a *countable* basis for the open sets, the concepts "compact" and "sequentially compact" are equivalent. In particular this holds in the space $\mathbb{R}^n$ where the (open) balls with rational radii and rational centres form a countable basis.

For later reference, we number the following results.

LEMMA 1. *Let* K *be a locally compact non–discrete topological field. The following properties hold:*

*(a) There is in* K *a sequence converging to* 0, *with* $a_k \neq 0$ *for all* $k \in \mathbb{N}$.

*(b) If the open nbd* W *of* 0 *is contained in the compact set* Z, *then the sets* $a_k W$ *form a nbd–basis of* 0. *(We then also call the sets* $a_k Z$ *a nbd basis of* 0, *because they satisfy the corresponding condition, without also being open).*

*(c) If* $(a_k)$ *is a sequence converging to* 0 *with* $a_k \neq 0$ *for all* $k \in \mathbb{N}$, *and* $b_k = a_k^{-1}$, *then the sequence* $(b_k)$ *has no accumulation point in* K, *i.e. it diverges.*

*PROOF*. K is locally compact, therefore there is a compact nbd Z of 0. In this compact nbd there lies another, open, nbd W of 0. We first prove:

1. Intermediate Assertion: W is not finite. For, since K is Hausdorff, each point is closed in K. If the open set W were finite, then each point would also be open in K, as the complement of a closed set (formed from the remaining finitely many points of W, and the complement of W). In particular, 0 would then be open and bounded. Then every point $c \in K$ will be both open and closed, because the nbds of 0 define the topology of K. Thus the topology of K would be discrete: contradiction.

2. Because W is not finite, there is an infinite sequence $(c_k)$ in W of mutually distinct points $c_k$. Because $Z \supseteq W$ is compact, this sequence has an accumulation point in Z. Let this be p. Then the sequence $d_k = c_k - p$ has the accumulation point 0 and also the $d_k$ are mutually distinct. In particular at most one $d_j$ can be 0, and if we omit this, then all $d_k \neq 0$.

3. Next we prove that the sets $d_j Z$ form a nbd basis for 0. We must show : to each (open) nbd y of 0 there is some j with $d_j Z \subseteq Y$. We find j in the following way: if $z \in Z$ then $0z = 0$, so by continuity of multiplication, there exist nbds $U_z$ of 0 and $V_z$ of z with $U_z \cdot V_z \subseteq Y$. Because Z is compact, it can be covered by finitely many of the $V_z$, say $Z \subseteq V_{z(1)} \cup \ldots \cup V_{z(n)}$. For the corresponding $U_{z(i)}$, their intersection $U = U_{z(1)} \cap \ldots \cap U_{z(n)}$ is an open nbd of 0 and for each $d \in U$ one has $dz \in Y$ for all $z \in Z$. Since 0 is an accumulation point of the sequence $(d_k)$, there is a j with $d_j \in U$. For this $d_j$, we have $d_j Z \subseteq UZ \subseteq Y$ as required. Hence also $d_j W \subseteq Y$.

4. From the sets $b_k Z$ we create new sets $M_k$ by defining

$$M_k = d_1 Z \cap \;.\;.\;.\; \cap d_k Z$$

The $M_k$'s are again compact nbds of 0 and it follows from the definition that

$$M_1 \supseteq M_2 \supseteq M_3 \supseteq \ldots$$

Also the $M_k$'s form a nbd – basis of 0 (since $M_k \subseteq d_k Z \subseteq Y$). Because 0 is an accumulation point of $(d_j)$ and $d_j \neq 0$, infinitely many points of the sequence must lie in each $M_k$. We can therefore select points $a_1 \in M_1$, $a_2 \in M_2, \ldots, a_k \in M_k$ from the sequence $(d_j)$ which are all mutually distinct. This sequence $(a_k)$ converges to 0. By the same argument as with 3, we see that the sets $a_j Z$ also form a nbd–basis of 0. Therefore (a) and (b) of the Lemma are proved.

5. To prove (c) let $a_k \to 0$ and $a_k \neq 0$ for all $k \in \mathbb{N}$. We set $b_j = a_j^{-1}$. Suppose that $b_j$ has an accumulation point $b$. Because $a_k \to 0$, then by continuity of multiplication, 0 would be an accumulation point of the sequence $(a_k b_k) = (1)$, which is impossible because $1 \neq 0$ and limits are unique in any Hausdorff space like K.

This establishes Lemma 1. □

We pause for a moment on the path to the theorem of Pontrjagin and cast a look over the landscape. With topological groups, we have got to know several interesting compact examples such as $\mathbb{S}^1$, $\mathbb{S}^3$, and the torus $\mathbb{S}^1 \times \mathbb{S}^1$; but with the help of our Lemma we can show that nothing like this can happen with the topological fields.

THEOREM 2. *Every* <u>*compact*</u> *topological field is discrete and finite.*

<u>*PROOF*</u>. If the compact field K is discrete, then it must also be finite, because among the discrete spaces, only the finite ones are compact.

Conversely suppose that K is not discrete. Then we will quickly obtain a contradiction. Since K is compact, it is locally compact and we may apply our Lemma, to obtain from part (a) a null–sequence $(a_k)$ with $a_j \neq 0$ for all $j \in \mathbb{N}$. If $b_j = a_j^{-1}$, then the sequence $(b_k)$ (regarded as an infinite set of points) must have an accumulation point b in the compact field K. This is impossible from part (c) of Lemma 1. □

The next theorem guarantees us a countable basis in any locally compact (non–discrete) field. This is important because later on we shall use sequences frequently and we can use the fact that then "compact" and "sequentially–compact" are equivalent concepts.

THEOREM 3. *Let* K *be a locally compact non–discrete topological field. Then the topology of* K *has a countable basis.*

<u>*PROOF*</u>. 1. By Lemma 1, we have in K a sequence $(a_k)$ converging to 0 with $a_j \neq 0$ for all $j \in \mathbb{N}$. Let W again be an open nbd of zero contained in the compact ball Z. We have seen that the nbds $a_j W \subseteq a_j Z$ "collapse on 0". If we now set $b_j = a_j^{-1}$ and consider $b_j W$, then these sets "spread out" over the whole of K as we see in the following way. For each $x \in K$, $0x = 0$. Since multiplication is continuous, there exist (open) nbds Y of 0 and U of x with

$YU \subseteq W$. The nbd $Y$ contains a point $a_n$ and we have $a_n x \in W$, or alternatively, $x \subseteq b_n W$. We can therefore express $K$ as the countable union

$$K = \bigcup_{j \in \mathbb{N}} b_j W$$

of the open sets $b_j W$, each homeomorphic to $W$. If we can now prove that $W$ has a countable basis, then that will hold for the whole of $K$, because "countable times countable" is countable.

2. For brevity we now write $W_n = a_n W$. With $n$ fixed, the open sets $W_n + y$ with $y \in Z$ form an open covering of $Z$. Because $Z$ is compact we can find a finite sub-covering

$$K = \bigcup \left\{ W_n + y_{n,k} \middle| k = 1, \ldots, m_n \right\} .$$

It therefore remains to show that the countably many sets $W_n + y_{n,k}$ (with $n \in \mathbb{N}$ and $k = 1,\ldots,m_n$) form a basis for the open sets of $W$. To see this, we must find for each and every open subset $U$ of $W$ with $x \in U$, a set $W_n + y_{n,k}$ with $x \in W_n + y_{n,k} \subseteq U$.

We choose an $x \in U$ and consider the open set $U \setminus x$. It is a nbd of $0$ since $0 = x - x \in U \setminus x$.

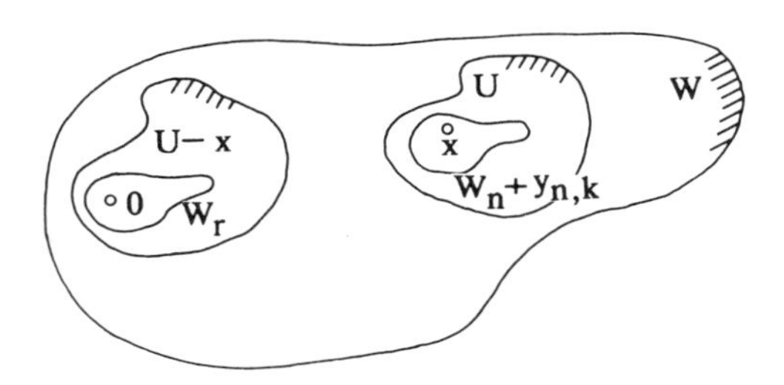


Fig. 75

By the continuity of subtraction (i.e. composition of additive inverse and addition in $(K, +)$) there is a nbd $V$ of $0$

$$V - V = \{a-b \mid a \in V \text{ and } b \in V\}$$

is contained in $U \setminus x$. Because the $W_n$'s form a nbd-basis of $U$, there exists an $n$ with $W_n \subseteq V$ and hence

$$W_n - W_n \subseteq V - V \subseteq U \setminus x .$$

Since $U \subseteq W \subseteq Z$, then $x$ lies in one of the sets $W_n + y_{n,k}$. Therefore $x - y_{n,k} \in W_n$ and

$$W_n - (x - y_{n,k}) \subseteq W_n - W_n \subseteq U \setminus x .$$

Now add $x$, to obtain

$$x \in W_n + y_{n,k} \subseteq U$$

as required. □

## 8C PONTRJAGIN'S THEOREM.

The characterization of a mathematical object through suitable properties is always only possible to within isomorphisms. For our purposes, we must define the concept of isomorphic topological fields.

Definition. Let $(K, +, \cdot, \text{top})$ and $(L, +, \cdot\ \text{top})$ be topological fields. A bijective mapping $\psi : K \to L$ is called an **isomorphism** of the topological fields, whenever the following conditions hold:

(i) $\psi : (K, +, \cdot, \text{top}) \to (L, +, \cdot, \text{top})$ is an (algebraic) isomorphism.

(ii) $\psi : (K, \text{top}) \to (L, \text{top})$ is a homeomorphism. (That is, both $\psi$ and the inverse mapping $\psi^{-1}$ are continuous.)

THEOREM. (Pontrjagin 1932). *Let* $K$ *be a connected locally compact topological field. Then* $K$ *is isomorphic to one of the topological fields* $\mathbb{R}$, $\mathbb{C}$ *or* $\mathbb{H}$.

The proof is divided into several steps, each of which we first quickly sketch out beforehand so that the main thread does not get lost in the detail.

*Step 1.* The field $K$ will be split into three mutually disjoint sets $A$, $B$, and $C$ which in the case $K = \mathbb{C}$ look like this: $A$ is the interior of the unit circle, $B$ the unit circle, and $C$, everything outside. One can define $A$ as the set of those elements $A$ for which the sequence $(a^n)$ of powers converges towards zero. The derivation of the necessary properties of $A$, $B$ and $C$ in the further proof requires lengthy, but elementary, arguments. In this step the connectivity of $K$ is not used.

*Step 2.* Using the connectivity, one proves that the characteristic of $K$ is zero, which implies that $\mathbb{Q}$ lies in $K$ and gets the usual topology. Then the topological closure $\mathbb{R}$ of $\mathbb{Q}$ must lie in $K$ and, as $\mathbb{R}$ is a sub-field of $K$, one has $K$ as a division algebra over $\mathbb{R}$.

*Step 3.* A consequence of Step 1 is the compactness of the closure $\overline{A} \subseteq A \cup B$. Using a suitable finite covering of $\overline{A}$ one obtains the finiteness of the dimension of $K$ over $\mathbb{R}$. Then the theorem of Frobenius can be invoked, to derive an *algebraic* isomorphism $\psi : (K, +, \cdot) \to (L, +, \cdot)$ where $L$ is one of the fields $\mathbb{R}$, $\mathbb{C}$ or $\mathbb{H}$.

*Step 4.* By Lemma 1, we can choose in $K$ a null-sequence $(a_k)$ with $a_k \neq 0$ for all $k \in \mathbb{N}$. Then the sets $a_kA$ form a nbd-basis of $0$. From this we prove continuity of $\psi$ and $\psi^{-1}$, and thence obtain $\psi$ as a *topological* isomorphism.

We now come to the execution of all the details.

*Step 1*. Let $K$ be a topological field. Given an element $x \in K$ we form the sequence $(x, x^2, x^3, \ldots) = (x^n)$ with $n \in \mathbb{N}$. We say generally, that a sequence $(c_n)$ diverges whenever it has no accumulation point. At the same time we can abbreviate the following definition, by writing $x^n \to 0$ instead of "the sequence $(x^n)$ converges to $0$". Let

$$A = \{x \in K \mid (x^n) \to 0\}$$

$$B = \{x \in K \mid (x^n) \text{ no sub-sequence of } (x^n) \text{ either converges to } 0, \text{ or diverges}\}$$

$$C = \{x \in K \mid (x^n) \text{ diverges}\}$$

From this definition $A$, $B$ and $C$ are certainly mutually disjoint. For example, If $K = \mathbb{R}$ then we would have

$$A = \langle -1, 1\rangle, \qquad B = \{-1, 1\}, \qquad C = \langle -\infty, -1\rangle \cup \langle 1, \infty\rangle$$

With $\mathbb{C}$ and $\mathbb{H}$, $A$ is the inside of the unit ball, $B$ its boundary surface ($\mathbb{S}^1$ or $\mathbb{S}^3$) and $C$ the exterior. If we were to form the sets $A$, $B$ and $C$ in $\mathbb{Q}$ with the usual topology, then we would have the rational numbers in the $\mathbb{R}$-intervals described above.

LEMMA 4. *Let* $K$ *be locally compact and not discrete. Then the sets* $A$, $B$ *and* $C$ *satisfy*:

(1) $A$ *is open (and a nbd of* $0$)

(2) $C^{-1} = A \setminus \{0\}$ *and* $C$ *is also open.*

(3) $K = A \cup B \cup C$.

(4) $A \cup B$ *is compact.*

(5) *The closure* $\bar{A}$ of $A$ *is compact.*

*PROOF*. Since $K$ is Hausdorff, we may choose a compact nbd $Z$ of $0$ which does not contain the unit element $1$ of $K$, and an open nbd $W$ of $0$ in $Z$. Further, let $F = Z \cup \{1\}$. This set is also compact. Using the set $F$ we shall obtain the following useful characterization of the elements of $A$. The sets $W$, $Z$, $F$ are fixed throughout the whole proof.

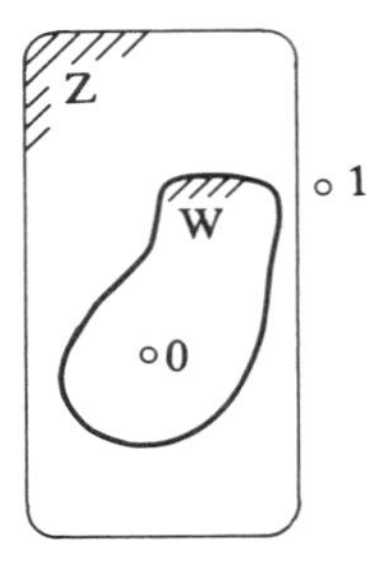


Fig. 76

<u>First Intermediate Assertion</u>. $x \in A$ iff $(\exists k : x^k F \subseteq W)$.

*COMMENT.* K is possibly not commutative, so we must distinguish between the statements $yF \subseteq W$ and $Fy \subseteq W$. Because however, the condition that $x \in A$ is "symmetrically" formulated as '$x^n \to 0$', we shall get from the Intermediate Assertion, that $yF \subseteq W \Leftrightarrow Fy \in W$. (One can naturally also follow through the whole proof with $Fx^k \subseteq W$.)

<u>*PROOF*</u> *of the Intermediate Assertion.* We first assume that $x \in A$, so $x^n \to 0$. From part (b) of Lemma 1, the sets $x^k F \subseteq W$ form a nbd basis of 0. Thus there is a k with $x^k F \subseteq W$.

Conversely suppose x is such that there exists $k \in \mathbb{N}$ with $x^k F \subseteq W$. We use induction to prove:

$$\forall n \in \mathbb{N} : x^{nk} F \subseteq W .$$

The start of the induction is looked after by our assumption that $x^k F \subseteq W$. Now suppose $x^{nk} F \subseteq W$. Multiplying by $x^k$, we get

$$x^{(n+1)k} F \subseteq x^k W \subseteq x^k F \subseteq W$$

as we wanted to show. Since $1 \in F$, we now have $x^{nk} \in W$ for all n. In the compact set Z, the sequence $x^k, x^{2k}, \ldots, x^{nk}, \ldots$ must have an accumulation point p, and we now show that $p = 0$.

For, since p is an accumulation point of the sequence, each nbd of p contains arbitrarily large powers $x^{rk}$ and $x^{sk}$ with $r < s$. Because of the continuity of division, we can so choose r and s that $x^{sk} x^{-rk} = x^{(s-r)k}$ lies in the neighbourbood of $pp^{-1} = 1$. This is even possible for arbitrarily large $s > r$. But then 1 would be an accumulation point of the set $x^{jk}$ with $j = s - r$, which contradicts the compactness of Z because $x^{jk} \in Z$ and $1 \notin Z$. Hence $p = 0$.

The sequence $(x^{nk})$ can therefore have no limit point other than zero. Because it must have at least one accumulation point in the compact set Z, the entire sequence $x^k, x^{2k}, \ldots, x^{nk}, \ldots$ converges towards zero (i.e. is a null-sequence). If we multiply this sequence by x then by continuity of multiplication the sequence $x^{k+1}, x^{2k+1}, \ldots$ is also a null-sequence. Proceeding similarly with $x^2, x^3, \ldots, x^{k-1}$, we obtain k null-sequences from which it follows by "mixing" them that $x, x^2, \ldots x^{k-1}, x^k, x^{k+1}, \ldots$ is also a null-sequence.

With this, the Intermediate Assertion is proved. □

<u>Second Intermediate Assertion</u>. A *is open and a nbd of* 0. We must show: for each $x \in A$ there is an entire nbd V of x with $V \subseteq A$. We use here the notation $X^j = \{x^j \mid x \in X\}$. By the First intermediate assertion, there is a $k \in \mathbb{N}$ with $x^k F \subseteq W$. From that it follows, by continuity of multiplication, that to each $z \in F$ there are (open) nbds $U_z$ of z and $X_z$ of x such that $(X_z)^k U_z \subseteq W$.

These (open) nbds $U_z$ form a covering of the compact set F, from which we choose a finite sub-covering $U_{z(1)},\dots,U_{z(m)}$. Then we form the (open) nbd $V = X_{z(1)} \cap \dots \cap X_{z(m)}$ of X; thus $V^k F \subseteq W$, so $V \subseteq A$ and A is open. Certainly, 0 belongs to A because $0F \subseteq W$. Therefore A is a nbd of zero, as claimed. □

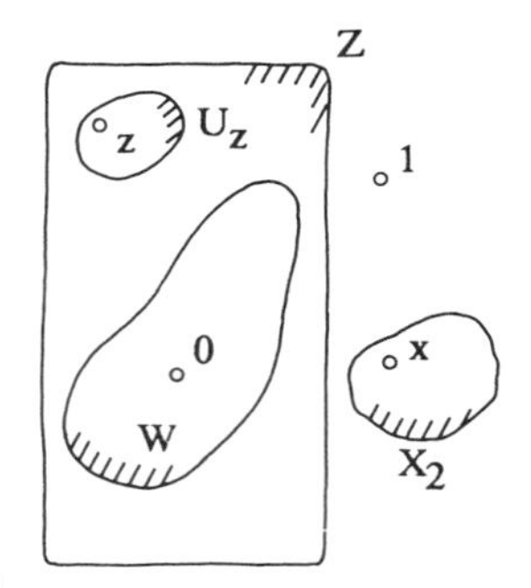


Fig. 77

From the openness of A, it follows in particular that there is an element $s \neq 0$ in A which, by the First intermediate assertion, satisfies $s^k F \subseteq W$. Because also $s^k \neq 0$, we let $t = s^k$, and so have $t \neq 0$ in A with $tF \subseteq W$. We could also have constructed such an element from the null-sequence $(a_k)$, using Lemma 1.

Third Intermediate Assertion. *If the sequence* $(a_k)$ *is divergent and* $a_k \neq 0$ *for all* $k \in \mathbb{N}$, *then the sequence* $(a_k^{-1})$ *converges to* 0.

For, assume that the sequence $(a_k^{-1})$ has an accumulation point $p \neq 0$. Then a subsequence of $(a_k a_k^{-1})$ must diverge because of the divergence of $(a_k)$, and this is impossible. Hence, to prove that $a_k^{-1} \to 0$ it suffices now to show that $(a_k^{-1})$ can have no divergent subsequence. If there were such a divergent subsequence in $(a_k^{-1})$, then we can restrict our considerations to this subsequence. We therefore simply assume that both $(a_k)$ and $(a_k^{-1})$ are divergent and look for a contradiction. To that end we use an element $t \neq 0$ with

$$tF \subseteq W \quad \text{and} \quad Ft \subseteq W ,$$

the existence of which is ensured by a remark at the end of the proof of the Second intermediate assertion. Because $tF \subseteq W$ then $t^k \to 0$, while the sequence $(t^{-k})$ diverges. If now a is any element of K, then with sufficiently large negative n the element $at^n$ does not belong to the compact set $Z \supseteq W$, while it does belong to W for sufficiently large positive n. There is therefore a whole number $r = r(a)$ such that $at^n$ does not belong to Z for $n < r$, while it lies in Z for $n = r$. If $at^r = zt$ were in Zt then we would have $at^{r-1} \in Z$ which is simply false. Therefore

$$at^{r(a)} \in Z \setminus Zt .$$

Correspondingly we obtain an element $s = s(a) \in \mathbb{Z}$ with

$$t^{s(a)}a \in Z \setminus tZ .$$

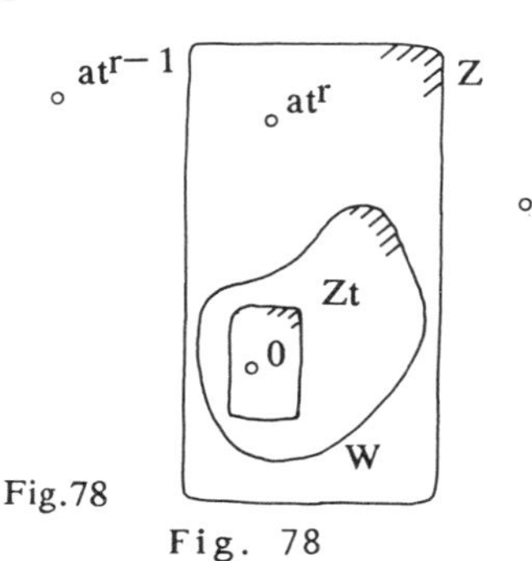


Fig.78

Fig. 78

With our divergent sequence $(a_k)$ we consider the associated sequence $r_k = r(a_k) \in \mathbb{Z}$. If the $r_k$ were bounded, say $\leq R$, then we would have $a_k t^R \in Z$ for every $k$; so the sequence $(a_k t^R)$ would lie completely in the compact set $Z$ and would have an accumulation point there, which would also give an accumulation point for $(a_k)$. Therefore the $r_k$ increase unboundedly. The same holds for the numbers $s_k = s(a_k^{-1})$ because of the divergence of $(a_k^{-1})$. We set

$$b_k = a_k t^{r_k} \quad \text{and} \quad c_k = t^{s_k} a_k^{-1} .$$

As each $b_k \in Z$ there is an accumulation point $b$ of $(b_k)$ in $Z$. From $(b_k)$ we can then choose a sub-sequence, convergent to $b$ (since $K$ has a a countable basis) and assume for simplicity that $b_k \to b$. Now $b \neq 0$ because all $b_k$ lie in $Z \setminus Zt \subseteq Z \setminus Wt$ and $W_t$ is an open nbd of $0$. Similarly we may assume that the sequence $(c_k)$ converges to $c \neq 0$. But

$$c_k b_k = t^{r_k + s_k} ,$$

and because of the continuity of multiplication, we get

$$c_k b_k \to cb \neq 0$$

whereas

$$t^{r_k s_k} \to 0$$

because $t^n \to 0$ and $r_k$, $s_k$ are unbounded.

This yields the required contradiction, and the Third intermediate assertion is proved. □

<u>Fourth Intermediate Assertion</u>. $C = (A \setminus \{0\})^{-1}$ *and* $C$ *is open.* For, by the Third intermediate assertion, if $c \in C$ then $c^{-1} \in A$, so $C^{-1} \subseteq A \setminus \{0\}$. From Lemma 1(c), if $a \in A \setminus \{0\}$ then $a^{-1} \in C$, whence $(A \setminus \{0\})^{-1} \subseteq C$ iff $A \setminus \{0\} \subseteq C^{-1}$. The set $C$ is the image of the open set $A \setminus \{0\}$ under the homeomorphism $K \setminus \{0\} \to K \setminus \{0\}$ given by $x \to x^{-1}$; therefore $C$ is open. □

<u>Fifth Intermediate Assertion</u>. $K = A \cup B \cup C$.
Let $x$ be an element of $K$. If in the sequence, there were a subsequence converging to $0$ then there exists $k \in \mathbb{N}$ such that $x^k F \subseteq W$. Then by the First intermediate assertion $x \in A$. If $(x^n)$ contains a divergence subsequence, then there is a convergent subsequence in $(x^{-n})$, by transition to inverses. So $x^{-1} \in A$ and $x \in C$. If $(x^n)$ has no subsequence which is either divergent or convergent to zero, then $x$ belongs by definition to $B$. Because $A$ and $C$ are open, $B = K \setminus (A \cup C)$ is closed. □

Sixth Intermediate Assertion. $A \cup B$ *is compact.*
For, assume that in $A \cup B$ we have a sequence $(a_k)$ without an accumulation point. Thus $(a_k)$ is a divergent sequence. (Finitely many zeros can be thrown out, so we may assume $a_k \neq 0$). Then $(a_k^{-1})$ converges to zero, so there exists $j$ with $a_j^{-1} \in A$, which would imply that $a_j \in C$ – an impossibility. Therefore, each sequence in $A \cup B$ has an accumulation point in $A \cup B$. The compactness of $A \cup B$ now follows, because $K$ has a countable basis. □

*NOTE*. Since $A \subseteq A \cup B$, we have $\overline{A} \subseteq \overline{A \cup B} = A \cup B$. Thus $\overline{A}$ is a closed subset of the compact set $A \cup B$ and as such is itself compact. From the examples above, one naturally supposes that $\overline{A} = A \cup B$ but we cannot yet assert this.

With this, Lemma 4 is proved. □

*Step 2*. In Step 1 we have done no more than (essentially) prepare our tools. With the additional assumption that $K$ is connected, these tools will allow us to prove that $K$ has characteristic zero. Thus $K$ induces the same topology on its prime field $\mathbb{Q}$ as does $\mathbb{R}$. Without the hypothesis of connectivity, one cannot prove this statement because the so-called "p-adic" fields $\mathbb{Q}_p$ are locally compact and have $\mathbb{Q}$ as prime field. They are however, totally disconnected and induce on $\mathbb{Q}$ a topology different from the usual one.

A connected topological field cannot be discrete. In a discrete field one has, for example, both the disjoint non-empty open sets $\{1\}$ and $K \setminus \{1\}$.) Therefore the theorems so far proved for locally compact, non-discrete fields hold also for locally compact connected fields.

THEOREM 5. *A locally compact connected topological field* $K$ *has characteristic* $0$.

*PROOF*. For clarity in this proof, we denote the neutral element of multiplication in $K$ by $e$ (instead of 1). We must show that for all $m \in \mathbb{N}$, the field element.

$$me = e + \ldots + e \quad \text{(with } m \text{ summands)}$$

is different from zero. We can achieve this if, e.g. we can find a suitable field element $w \neq 0$ and deduce the existence of an element $x$ with $w = mx$ $(= mex)$. It is this existence statement that we shall obtain from the hypothesis that $K$ is connected.

For the proof, we must add not only elements but also subsets of $K$. We write

$$X + Y = \{x + y \mid x \in X,\ y \in Y\}.$$

This addition has the following properties.

If $X$ is any set, and $Y$ open, then $X + Y$ is the union of open sets $x + Y$ with $x \in X$; therefore $X + Y$ is open. If $X$ and $Y$ are compact then $X + Y$ is also compact as one can easily verify by considering a sequence $(z_k) = (x_k + y_k)$ in $X + Y$.

From Step 1 above we take the (open) set $A$ and its (compact) closure $\bar{A}$. Let $A_n$ be given by

$$A_n = A + \ldots + A \quad \text{(m summands)} .$$

Correspondingly, $\bar{A}_n$ is defined (i.e. $\bar{A}_n = (\bar{A})_n$) Because $A$ is open, $A_n$ is also open, and because $\bar{A}$ is compact, so also is $\bar{A}_n$. The closure $\overline{A_n}$ of the open set $A_n$ is equal to $\bar{A}_n$ as the following argument shows: since $A \subseteq \bar{A}$ then $A_n \subseteq \bar{A}_n$, and on taking the closures, $\bar{A}_n \subseteq \overline{A_n}$. If $z \in \bar{A}_n$, then we can express $z$ as a limit of some sequence $z_i = z_{1i} + \ldots + z_{ni}$ with $z_{ji} \in A$ and so obtain $z \in \overline{A_n}$, whence $\bar{A}_n \subseteq \overline{A_n}$. Altogether then, we have $\bar{A}_n = \overline{A_n}$, as asserted.

Next we consider the boundary of $A_n$:

$$\beta_n = \bar{A}_n \setminus A_n .$$

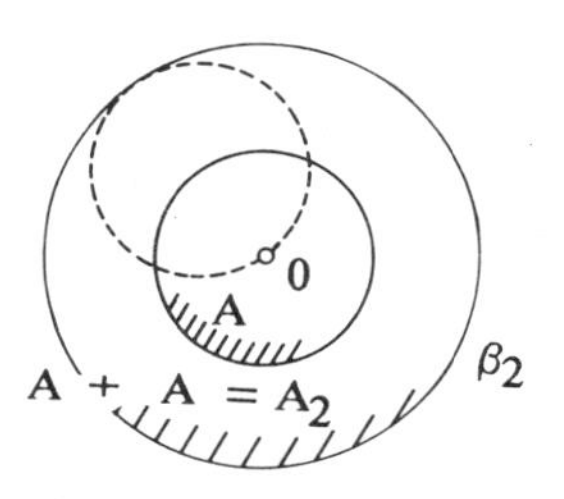


Fig.79

Is $\beta_n$ empty? The field $K$ is certainly not compact (see Theorem 2) so $K \neq \bar{A}_n$ which means $K \setminus \bar{A}_n \neq 0$. If $\beta_n$ were empty then $\bar{A}_n = A_n$ would be open and we would have a splitting of $K$ into two non-empty open subsets $A_n$ and $K \setminus A_n$. That cannot happen because $K$ is connected.

The boundary $\beta_n$ is therefore non-empty. Thus there exists an element $z \in \beta_n$, a fact that will help us later.
Let $2 \leq m \in \mathbb{N}$. We shall construct an element $w_m$ with the property

$$(*) \qquad mw_m \in K \setminus \bar{A}_{m-1}, \quad (\text{so } mw_m \neq 0) ,$$

and we first choose an (open) nbd $U$ of $0$ with

$$U + \ldots + U = U_m \subseteq A \quad \text{(m summands)}$$

(This is possible because addition is continuous.)
We cover the compact set $\bar{A}$ with the open sets $a + U$ where $a \in \bar{A}$; and then we choose a finite covering $a_1 + U, \ldots, a_k + U$. Now let $z \in \beta_n$, (as forewarned above) and $n = km$. Then

$$z = x_1 + x_2 + \ldots + x_n \quad \text{with} \quad x_j \in \bar{A} \quad \text{for } j = 1, \ldots, n .$$

Because the $n = km$ elements $x_j$ lie in the union of the $k$ subsets $a_i + U$, then one of these, say $a_1 + U$, contains at least $m$ of the $x_j$. Let these be $x_1,\ldots,x_m$ (say). Then we have

$$x = x_1 + \ldots + x_m \in \bar{A}_m$$

$$y = x_{m+1} + \ldots + x_n \in \bar{A}_{n-m}$$

$$z = x + y \in \beta_n,$$

so $z \in \bar{A}_n$ and $z \notin A_n$. Suppose now that one of the summands, say $x$, lay in the open set $A_m$. Then because of the above remark on the sum of subsets, $z$ must lie in the open set $A_m + \bar{A}_{n-m}$ and not on its boundary $\beta_n$. That is not correct, so $x \in \beta_m$. Further we have

$$x_i = a_1 + u_i \text{ with } u_i \in U \text{ for } i = 1,\ldots,m .$$

We write $u_1 + \ldots + u_m = u \in U_m \subseteq A$, and therefore

$$x = ma_1 + u .$$

Since $m \geq 2$ while $x \in \beta_m$ and $u \in A$, then $ma_1$ cannot lie in $\bar{A}_{m-1}$; therefore we have an element $w_m = a_1 \in K$ with $mw_m \neq 0$, which is what we were looking for. □

We have just shown that, starting with the hypothesis "connected", we can derive the algebraic consequence "characteristic zero". We still need however, a statement about the topology that $K$ induces on $\mathbb{Q}$. The arguments leading to this will now be given, and we shall assemble the results at the end, in the form of Theorem 6.

1. The set $A$ that was defined in Step 1 is an open nbd of zero in $K$, which (following Pontrjagin) we call the **principal** nbd of $K$. It is contained in the compact set $\bar{A}$ (recall: $\bar{A} \subseteq A \cup B$). From Lemma 1 we have a sequence $a_k \to 0$, with $a_k \neq 0$ for all $k \in \mathbb{N}$, such that the sets $a_k \bar{A}$ form a nbd basis of $0$ in $K$. Also the open sets $a_k A$ form a nbd basis of $0$. (This holds for each sequence of the given type, and all such nbd bases are mutually equivalent). In the topological group $(K, +, \text{top})$ the topology is determined as soon as we know a nbd basis of the zero element. Therefore the topology of $K$ is ultimately fixed by the principal nbd $A$ of zero.

2. The same considerations apply to the topology of $\mathbb{Q} \subseteq K$, if we consider instead of $A$ the principal nbd $A \cap \mathbb{Q}$ of the zero in $\mathbb{Q}$. The rational numbers lie in $K$ as those elements of the form $(me)(ne)^{-1}$ (with $m \in \mathbb{Z}$, $n \in \mathbb{N}$; $e$ is the neutral element of $K$). We write $m/n \cdot e$ for $(me)(ne)^{-1}$, and claim:

$$A \cap \mathbb{Q} = \left\{ m/n \cdot e \;\middle|\; |m| < n \right\} .$$

If this claim be granted, then $A \cap \mathbb{Q}$ is nothing other than the usual rational open interval $\langle -1, 1 \rangle$. (Had we been working with the non-connected locally compact p-dimensional field $\mathbb{Q}_p$, then $A \cap \mathbb{Q}$ would look completely different!).

For the proof of our claim, we consider only the case $m > 0$ so $m < n$; the case $m < 0$, (that is $-1 < m/n < 0$) is treated similarly. As in Step 1, we use the disjoint partition $K = A \cup B \cup C$. Because $A \setminus \{0\} = C^{-1}$, our claim is equivalent to:

$$\text{if} \quad m > n \quad \text{then} \quad m/n \cdot e \in C\ .$$

This we now assume to be false so $m/n \cdot e \in A \cup B$. The element $m/n \cdot e \in \mathbb{Q}$ commutes with every element of $K$, because the centre of $K$ (see Chapter 5 B) is a sub-field of $K$ which must contain the sub-field $\mathbb{Q}$. Now for any $q \in A \cup B$ that commutes with all elements of $A$, we shall show that

$$aq \in A \quad \text{for all} \quad a \in A\ ;$$

for, because $a^n \to 0$, the sets $a^n \cdot (A \cup B)$ form a nbd basis of $0$, from which (with the commutativity) via

$$(aq)^n = q^n a^n \quad \text{and} \quad (aq)^n \in a^n \cdot (A \cup B)$$

it finally follows that $(aq)^n \to 0$. From $Aq \subseteq A$ we have $\bar{A}_q \subseteq \bar{A}$ for the closures. With the element $w_m \in \bar{A}$ from Theorem 5, we have $w_m \cdot m/n \cdot e \in \bar{A}$ or with the notation $\bar{A}_n$ used there:

$$mw_m \in n\bar{A} = \bar{A}_n .$$

But by condition (*) from Theorem 5, we have

$$mw_m \in K \setminus \bar{A}_{m-1}\ ,$$

and since $n < m$ we have a contradiction. Therefore $m/n \cdot e$ lies in $C$ when $m > n$ which is what we had to show.

3. Because the principal nbd $A \cap \mathbb{Q}$ is the rational interval, then $K$ induces in $\mathbb{Q}$ the usual topology. We can therefore deal with the rational numbers in $K$ in the usual way e.g. we can now simply choose $a_n = 1/3^n$ for the zero sequence $(a_n)$ etc. Our next task is to show that not only $\mathbb{Q}$ but also $\mathbb{R}$ is contained in $K$. We have $\mathbb{Q} \subseteq \mathbb{R}$ and $\mathbb{Q} \subseteq K$, and each of the topological fields $\mathbb{R}$, $K$ induces the usual topology on $\mathbb{Q}$. In $K$, $\mathbb{Q}$ has closure $\bar{\mathbb{Q}} = L$, which at first sight is merely a closed set in $K$. We maintain however, that $L$ is a field topologically isomorphic to $\mathbb{R}$. To prove this, we must show the existence of a topological isomorphism $\varphi : \mathbb{R} \to L$ (as in the diagram in which the other arrows all denote inclusions).

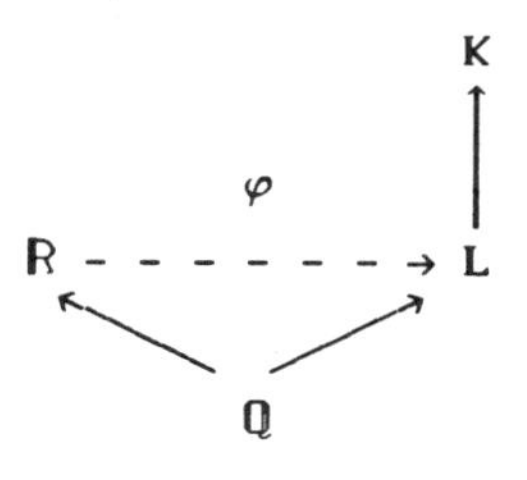

Let $r \in \mathbb{R}$ and $(r_i)$ a sequence of rational numbers converging to $r$. Being convergent, $(r_i)$ is a Cauchy sequence; and as such, this also converges in the locally compact field $K$. (Proof as exercise!) Tentatively, we now assign to the real number $r$, the limit of the sequence $(r_i)$ in $K$. and now we try using a different sequence. Thus if $(s_i)$ is another sequence of rational numbers converging to $r$ in $\mathbb{R}$ then $(r_i - s_i)$ is a null sequence, and therefore $\lim r_i = \lim s_i$ also in $K$. (Details are left again as an exercise). Without ambiguity, therefore, we can define

$$\varphi r = \lim r_i \quad \text{in} \quad K \quad .$$

The limit of the sequence $(r_i)$ lies in the closure $L = \overline{\mathbb{Q}}$ in $K$. Therefore $\mathbb{R}$ is mapped by $\varphi$ into $L$. The enthusiastic reader could now prove for himself the following assertions, or else read the proof given in Pontrjagin [59], Section 25 F.

1. $\varphi$ is bijective,

2. $\varphi$ is an algebraic isomorphism,

3. $\varphi$ is a homeomorphism.

We will now simply identify the real numbers with their images in $K$, and so regard $\mathbb{R}$ as a subfield of $K$.

4. Not only the rationals, but also all the real numbers lie in the centre of $K$. To see this, we represent each real number $r$ (in $K$) as the limit of a sequence $(r_n)$ of rational numbers $r_n$. Each of the rational numbers $r_n$ satisfies:

$$r_n x = x r_n \quad \text{for all} \quad x \in K \quad .$$

By continuity of the multiplication in $K$, we can take limits, and obtain $rx = xr$ for all $x \in K$.

We may now assemble our results, into one statement:

<u>THEOREM 6</u>. *The real numbers lie as a topological subfield in* $K$, *and they commute with every element of* $K$.

<u>Step 3</u>. Because the field $K$ contains $\mathbb{R}$ then $K$ is a division algebra over $\mathbb{R}$, and in particular, a vector space over $\mathbb{R}$. (See Section 5F.) Recall that the definition of a division algebra requires commutativity and this we have proved in Theorem 6. We now want to use the Theorem of Frobenius from Section 5F, so we must now look at the dimension of $K$ as a vector space over $\mathbb{R}$.

<u>THEOREM 7</u>. *The dimension of* $K$ *over* $\mathbb{R}$ *is finite.*

<u>*PROOF*</u>. Choose $u \in \mathbb{R}$ with $|u| < 1$, – say $u = 1/3$. In any case $u \in A$ in $K$. We cover $\overline{A}$ with the open sets $v + uA$ where $v \in \overline{A}$, and choose from these a finite sub-covering $v_1 + uA, \ldots, v_m + uA$.

*CLAIM*. The elements $v_1,\dots,v_m$ generate the vector space K of R, which is to say that for each $y \in K$ there exist $y_1,\dots,y_m \in R$ with

$$y = y_1 v_1 + \dots + y_m v_m .$$

If y does not lie in A, there is nevertheless some $k \in \mathbb{N}$ with $u^k y \in A$. If we can express $z = u^k y$ as a linear combination of $v_1,\dots,v_m$, then the same holds for $y = u^{-k} z$. It suffices therefore to consider $z \in A$. Now z lies in one of the sets $v_i + uA$, and therefore

$$z = v_1 + u z_1 \quad \text{with} \quad z_1 \in A .$$

For what we have to do, it is convenient to use all the $v_j$ but with coefficients $\varepsilon_{oi} = 1$ and $\varepsilon_{oj} = 0$ for $j \neq i$. Thus we now write z as

$$z = e_{01} v_1 + \dots + \varepsilon_{om} v_m + u z_1$$

with $z_1 \in A$ and $\varepsilon_{oj} \in \{0, 1\}$.

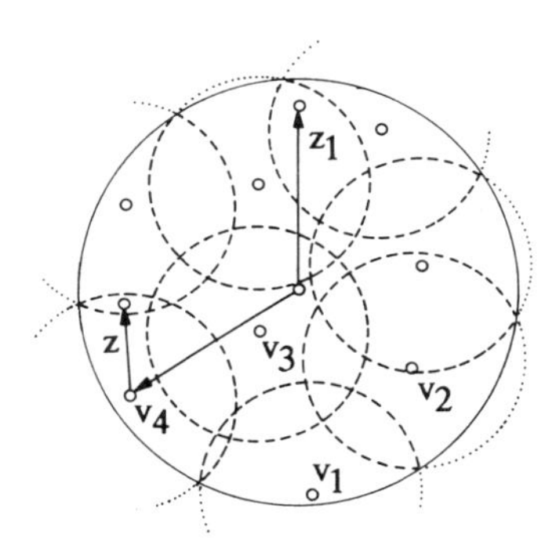


Fig. 80

Applying the same procedure to $z_1$, we get

$$z_1 = \varepsilon_{11} v_1 + \dots + \varepsilon_{1m} v_m + u z_2$$

with $z_2 \in A$ and $\varepsilon_{ij} \in \{0, 1\}$. By iteration we have

$$\begin{aligned} z = {} & \varepsilon_{o1} v_1 + \dots + \varepsilon_{om} v_m \\ & + \varepsilon_{11} u v_1 + \dots + \varepsilon_{1m} u v_m \\ & \dots\dots \\ & + \varepsilon_{k1} u^k v_1 + \dots + \varepsilon_{km} u^k v_m + u^{k+1} z_{k+1} \quad \text{with} \quad z_{k+1} \in A \end{aligned}$$

with $z_{k+1} \in A$. Rearrangement gives

$$\begin{aligned} z = {} & (\varepsilon_{o1} + u\varepsilon_{11} + \dots + u^k \varepsilon_{k1})\, v_1 + \\ & \dots\dots\dots \\ & + (\varepsilon_{om} + u\varepsilon_{1m} + \dots + u^k \varepsilon_{km})\, v_m + u^{k+1} z_{k+1} . \end{aligned}$$

$$y_j = \sum_{i=1}^{\infty} (\varepsilon_{ij} u^i) .$$

The remainders $u^j z_j$ lie in the sets $u^j A$ which form a nbd basis of 0. Thus if

$$y_j = \sum_{i=1}^{\infty} (\varepsilon_{ij} u^i) ,$$

the representation we require is

$$z = y_1 v_1 + \ldots + y_m v_m .$$

By Frobenius's theorem, K is now algebraically isomorphic to one of the fields $\mathbb{R}$, $\mathbb{C}$ or H.

Step 4. Let $\varphi : K \to D$ be an algebraic isomorphism of K onto one of the given fields $D = \mathbb{R}, \mathbb{C}$ or H, with the inverse isomorphism $\psi : D \in K$. It remains to show that $\varphi$ and $\psi$ are also continuous. The case $D = \mathbb{R}$ is already dealt with because K induces the usual topology on $\mathbb{R}$ and therefore $\mathbb{R}$ and K are homeomorphic. We now treat the case $D = H$, and the case $D = \mathbb{C}$ is analogous. Here H carries the usual topology of $\mathbb{R}^4$. To keep the distinction clear we write $(H, +, \cdot, O)$ and $(K, +, \cdot, \text{top})$, but we do not distinguish between the quaternion units 1, i, j, k in H and K in any particular way.) The real line with the usual topology is the "first axis" of K. Since the multiplications with i, j, k are homeomorphisms $x \to xi$ etc., then the "$n^{th}$ axis" in K is also homeomorphic to $\mathbb{R}$, when $n = 2, 3, 4$. The algebraic isomorphism $\psi$ maps linear combinations in H to linear combinations in K, so

$$\psi : (x_o, x_1, x_2, x_3)^{\dagger} \to x_o + x_1 i + x_2 i + x_2 j + x_3 k .$$

This mapping is continuous because it is a combination of the continuous functions $x_o \to x_o$, $x_1 \to x_1 i$ etc, together with the continuous addition in K.

With the inverse mapping, matters become a bit more complicated. (We follow here Führer [21], 11.22, but compare also van der Waerden [71] II, Section 101). Because the topologies of H and K are determined by their respective principal nbds, we first only consider these. The principal nbd of H is the interior G of the unit ball of $\mathbb{R}^4$, whose boundary is the 3–sphere $S^3$. In K we have A and $\bar{A}$.

We consider the image $\psi G$ in K and assert that we can find an open nbd W of 0 from (K, top) which is completely contained in $\psi G$. If this be granted for the moment, then for the inverse mapping $\varphi$, it means that there is an open nbd W of 0 in K such that $\varphi W$ lies in the open nbd G of 0 in H. This will establish the continuity of $\varphi$ as follows. For, the sets $t^n G$, $t^n A$ (with $t = 1/2$ say) form nbd bases in H, K respectively. Therefore W contains a set of the form $t^r A$. Thus if $\varphi W \subseteq G$ then $\varphi(t^{n+r} A) = t^n \varphi(t^r A) \subseteq t^n \varphi(W) \subseteq t^n G$. Hence $\varphi$ maps the nbd $t^{n+r}A$ into the nbd $t^n G$, so $\varphi$ is continuous at the point 0. This suffices however, because with topological groups we know that continuity at the point 0 implies continuity everywhere.

It remains therefore, only to prove the above assertion concerning $\psi G$ in W. The continuous image S of the compact set $\$^3$ is compact in the $T_2$-space K; in particular then, it is closed. The complementary set $K \setminus S$ is an open set which contains 0. We now construct a symmetrical nbd W of 0 in the following way. Multiplication in K is continuous and $0 \cdot v = 0$ for each $v \in K$. Therefore, given the nbd $U = K \setminus S$ of 0 in K, there is a nbd $\langle -\varepsilon, \varepsilon \rangle$ of 0 in $\mathbb{R}$ (with positive $\varepsilon \in \mathbb{R}$) and a nbd V of 0 in K such that

$$\langle -\varepsilon, \varepsilon \rangle \cdot V \subseteq U .$$

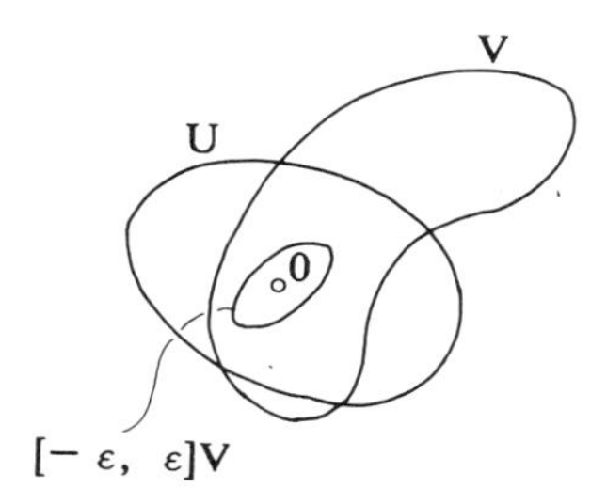


Fig.81

The set $W = \langle -\varepsilon, \varepsilon \rangle V$ is then an open nbd of 0 because it is the union of open sets $\alpha \cdot V$ with $\alpha \in \langle -\varepsilon, \varepsilon \rangle$, each of which contains 0.

If now there were some $w \in W$ with $w \notin G$, then $\varphi w \notin G$ and we could find an $\alpha \leqq 1$ with

$$\alpha \varphi w \in \$^3 ;$$

then we would have

$$\psi(\alpha \varphi w) = \alpha w \in \psi \$^3 = S$$

which is not possible, because $W \subseteq K \setminus S$. Therefore $W \subseteq \varphi G$ and the proof of Pontrjagin's theorem is now complete. □

*Concluding Note*: We have now arrived at the end of our study of the number systems. At the beginning we had a central concept of Analysis, namely the completeness of $\mathbb{R}$, which we have since met again and again in variations. With the complex numbers and quaternions, it occurred mainly in algebraic questions. In this last chapter, topology came up in its own right. Much more could still be said, but for the present this is enough.

# Appendix: Notation and terminology

This note need not be read before starting the text proper, but readers may find it a helpful reminder if they have forgotten certain material. The notation and language used in the book is fairly standard but the reader may find it convenient to see it recalled, or used to express things already known. The author uses the language of sets and functions in an informal way. He frequently defines sets in the manner:–

$$X = \{x \in A \,|\, x \text{ has some property}\} .$$

Sometimes also, he uses the form 'X $=_{def}$ ...' to indicate that X is to be a name for (or defined by) the material following "def". For example, if Y is another set then

$$X \backslash Y =_{def} \{x \in X \,|\, x \notin Y\}$$

defines the set of all elements of X that do not lie in Y (even if $Y \not\subseteq X$). If $a \leqslant b$ in the set $\mathbb{R}$ of real numbers, the closed, open, and half-open intervals with end-points a,b are denoted respectively by

$$\langle\!\langle a,b\rangle\!\rangle = \{x \in \mathbb{R} \mid a \leqslant x \leqslant b\}, \quad \langle a,b\rangle = \{x \in \mathbb{R} \mid a < x < b\}$$

$$\langle\!\langle a,b\rangle = \{x \in \mathbb{R} \mid a \leqslant x < b\}, \quad \langle a,b\rangle\!\rangle = \{x \in \mathbb{R} \mid a < x \leqslant b\}$$

Also $\mathbb{N}$ denotes the set of whole numbers, $\mathbb{N} = \{1,2,\ldots\}$ and zero is *excluded*. The empty set is denoted by $\emptyset$ . The **power** *set* $\mathcal{P}X$ of a set X is the family of all subsets of X; thus for example $\mathcal{P}\emptyset$ is the singleton $\{\emptyset\}$. The **direct** (or Cartesian) product of the sets A, B is the set of all ordered pairs (a,b) given by:

$$A \times B =_{def} \{(a,b) \mid a \in A \text{ and } b \in B\} .$$

Products of a finite number n of factors are defined in the obvious way, using induction on n. In the text, products with an infinity of factors will be defined when needed.

Sometimes it is necessary to consider a family F of sets $X_\alpha$ where $\alpha$ runs through some other set J. Then F is 'indexed' by J, and we can form the union of all the $X_\alpha$'s — written, respectively:

$$\bigcup \{X_\alpha \mid \alpha \in J\} = \{x \mid \exists\alpha \in J : x \in X_\alpha\} .$$

If J is finite, say, $\{1, 2,\ldots,n\} \subseteq \mathbb{N}$ , then this set is just the familiar

$$X_1 \cup X_2 \cup \ldots \cup X_n,$$

Similarly we can form the intersection

$$\bigcap \{X_\alpha \mid \alpha \in J\} = \{x \mid \forall\alpha \in J : x \in X_\alpha\} .$$

and again if $J = \{1,2,\ldots,\mathbb{N}\}$ this intersection is the familiar

$$X_1 \cap X_2 \cap \ldots \cap X_n .$$

If $f:A \to B$ is a mapping, then the value f(a) is often abbreviated to fa. The *graph* of f is the subset of $A \times B$ given by

$$G_f = \{(a,b) \in A \times B \mid b = f(a)\}.$$

In some theoretical work, one often refers to f while meaning $G_f$ (and in strict, logical work — but not in this text — one *defines* a "mapping" to be a subset with the properties of $G_f$).

In writing various properties, frequent use is made of the quantifiers $\forall$, $\exists$ ('meaning for all', and 'there exist') as well as the logical symbols $\Rightarrow$ ('implies') and $\Leftrightarrow$ ('if and only if', which we often write as 'iff'). We indicate here their use by recalling some terminology concerning a mapping $f:A \to B$. Thus

$$f \text{ is injective} \Leftrightarrow : a_1 \neq a_2 \text{ in } A \Rightarrow f(a_1) \neq f(a_2)$$

where the colon indicates that what follows is to be read as a complete sentence S (here, "if $a_1$, $a_2$ are any two distinct elements of A, then $f(a_1)$ is distinct from $f(a_2)$"). With such usage there is still an informality, because a full statement of S would be written

$$(\forall a \in A): a_1 \neq a_2 \Rightarrow f(a_1) \neq f(a_2).$$

Nowadays many mathematicians (especially when lecturing at a blackboard), adopt an informal mixture of words and symbols as in the earlier version. Continuing with properties of f we write

$$f \text{ is surjective } \Leftrightarrow: (\forall b \in B)(\exists a \in A) \quad b = f(a),$$

and the sentence after the colon is often expressed as "every b in B is the image of some $a \in A$". Also

$$f \text{ is bijective } \Leftrightarrow: f \text{ is injective and surjective.}$$

One often says accordingly that f is an **injection**, a **surjection** or a **bijection**. If f is a bijection, an important theorem states that there is a unique mapping $B \to A$, denoted by $f^{-1}$, such that the compositions

$$f \circ g: B \to B, \quad g \circ f: A \to A \quad \text{where} \quad (f \circ g)b = f(g(b)))$$

are the identity mappings on B, A respectively, where $id_A$, for example, is given by

$$(\forall a \in A): id_A(a) = a \ .$$

If the sets A, B have additional structure, then we are interested in mappings $f:A \to B$ that "preserve the structure". In algebra, this means that A and B each have a binary operation — like addition or multiplication — such that for all $a_1, a_2 \in A$,

$$f(a_1 * a_2) = f(a_1) \cdot f(a_2)$$

where $*, \cdot$ denote the relevant operations in A,B respectively. We then call f a **homomorphism** (relative to the operation $*$); if f is also a bijection it need not happen that its inverse $f^{-1}:B \to A$ is also a homomorphism, but with the common operations it does happen, and if so then f is called an **isomorphism**.

Residue Classes. Examples occur most commonly when A, B are groups, rings, or vector spaces, and we assume that the reader has met these before, in order to recall the notion — used several times in this book — of calculating with residue classes. As an example consider arithmetic "modulo 7" in the ring $\mathbb{Z}$ of integers. Here we calculate by throwing away multiples of 7, so for example $169 = 7 \times 24 + 1$ so we write $169 \equiv 1 \pmod 7$ and 1 is called the **residue** of 169, modulo 7. But we can also work with the residues themselves, and have

$$169 = (13)^2 \equiv 6^2 \pmod 7 = 7 \times 5 + 1 \equiv 1 \pmod 7,$$

— a technique which tells us at once that, say

$$(169)^{73} \equiv 1^{73} \pmod 7 = 1$$

when direct calculation of the remainder is impossible. Similarly with any other modulus n; the set of residues itself forms a ring $\mathbb{Z}_n$ (perhaps more manageable than the ring $\mathbb{Z}$ of all integers), and in particular

$$\mathbb{Z}_7 = \{0, 1, 2, \ldots, 6\} \ .$$

We can do the same thing with polynomials, and work (say) modulo $x^2 + 3$. Then, for example

$$x^3 + 4x^2 - 5x + 1 = (x+4)(x^2+3) - 8x - 11$$

$$\equiv -8x - 11 \pmod{x^2+3}$$

because we discard multiples of $x^2+3$. In particular, mod $x^2+3$ we have

$$x^2 = 1(x^2+3) - 3 = -3$$

$$x^3 = x(x^2+3) - 3x \equiv -3x$$

$$x^4 = x^3 \cdot x \equiv -3x^2 = -3(x^2+3) + 9 \equiv 9$$

and also

$$x^4 = (x^2)^2 \equiv (-3)^2 = 9 .$$

Here, one can see that a polynomial $\sum a_r x^r$ is congruent (mod $x^2+1$) to $\sum a_r y_r$ where $y_r$ is either a power of $-3$ or $x$ times such a power; so the set of residues is the set of linear polynomials $ax+b$.

Cosets. The formal construction with any ring $R$ is to choose a subring $I$ and then take the set $J$ of all multiples $r \cdot i$ where $r \in R$ and $i \in I$. Then $J$ is a special kind of subring, called an **ideal**; and if $j \in J$ then all multiples $rj$ still lie in $J$. We now look at all $s \in R$ for which $r-s \in J$, and the set of these forms the **coset** $r+J$. Here we recall that if $R$ is an additive group, and $J$ is a subgroup, then we can form the quotient group $S = R/J$, of which the elements are the cosets $r+J$ added according to the rule

$$(a+J) + (b+J) = (a+b) + J$$

This rule is "independent of representatives", in that if we choose $a'$, $b'$ from $a+J$, $b+J$ respectively and form $a' + b'$ then its coset is still $(a+b) + J$. The mapping $h:a \to a+J$ is then a homomorphism

$$h:R \to R/J \qquad (\dagger)$$

of additive groups. But more is true; since $J$ is an ideal (†) we can multiply cosets by the rule

$$(a+J)\cdot(b+J) = (ab) + J$$

using the multiplication in $R$. Again this is independent of representatives. It then follows that $R/J$ is a ring and $h$ in (†) preserves multiplication as well as addition: in other words, $h$ is a homomorphism of rings. The ring $R/J$ is called the (quotient or factor) **ring of residue classes**, and results from simply regarding elements as "the same" if they differ by an element of $J$.

Strictly speaking, then, with arithmetic in $\mathbb{Z}$ modulo 7, the corresponding ideal $J$ consists of all integral multiples of 7, and then $\mathbb{Z}/J$ is bijective with $\mathbb{Z}_7$ as described above. Similarly, with the polynomials mod $x^2+2$.

Conversely, given a homomorphism $g:R \to T$ of rings, the inverse image $I = g^{-1}(0_T)$ of the zero of $T$ is an ideal in $R$, called the **kernel** of $g$. Thus $I$ is a subring with the characteristic property of ideals that

$$(\forall r \in R)(\forall i \in I) \quad ri \in I$$

i.e. if $i \in I$ and $i$ is multiplied by any $r$ in the larger ring, then $ri$ also lies in $I$.

The notion of a kernel applies to any homomorphism $g:A \to B$ of (possibly non-commutative) groups. Here, if $e_B$ is the identity element of $B$, then $g$ has a kernel $N$, where

$$N = \text{kernel}(g) =_{def} g^{-1}(e_B)$$

It can be shown that $N$ is not only a subgroup of $A$ but it satisfies a property akin to that for an ideal in a ring:

$$(\forall x \in A)(\forall x \in N) \quad x^{-1}ax \in N\,. \qquad (\dagger\dagger)$$

Any subgroup satisfying this property is called a **normal divisor** (or normal subgroup). If $A$ is commutative then $(\dagger\dagger)$ holds for *any* subgroup.

Each normal divisor $K$ is in turn the kernel of a surjective homomorphism $h:A \to A/K$, where the elements of the quotient group $A/K$ are subsets of $A$, still called **cosets**, of the form

$$aK = \{ax \mid x \in K\},$$

and $h(a) = aK$. Cosets are multiplied according to the rule

$$aK * bK = abK$$

(which can be verified to be independent of representatives, in the above sense). If the group operation is written as $+$, then $ax$ becomes $a+x$, and $aK$ becomes $a+K$ as for the rings above.

Similar constructions apply to vector spaces, their subspaces, quotient spaces, and homomorphisms.

Finally, a ring may have an order relation (as in Chapter 1). The resulting properties are said to be *order–theoretic*, just as properties that refer only to group structure are *group–theoretic* and those that ignore all other structure are *set–theoretic*.

Metric Spaces. A standard example of a vector space is the space $\mathbb{R}^n$ ordered n–tuples $x = (x_1, x_2,\ldots,x_n)$, where each $x_i \in \mathbb{R}$, and addition and scalar multiplication are defined by the rules

$$x+y = (x_1+y_1,\ x_2+y_2,\ldots,x_n+y_n)$$

$$\lambda\cdot x = (\lambda x_1,\ldots,\lambda x_n) \qquad \lambda \in \mathbb{R}$$

Additionally, $\mathbb{R}^n$ has topological structures as follows. The Euclidean distance, $d(x, y)$ between $x$ and $y$ is given by

$$d(x,y) = [(x_1-y_1)^2+\ldots+(x_n-y_n)^2]^{\frac{1}{2}}$$

which expresses Pythagoras's theorem when $n=2$ or 3. When $n=1$, $\mathbb{R}^1$ is just $\mathbb{R}$ with distance $d(x,y) = |x-y|$. This distance satisfies the three conditions

$$d(x,y) = d(y,x) \geq 0$$

$$d(x,y) = 0 \quad \text{iff} \quad x = y$$

$$d(x,y) \leq d(x, z) + d(z, y) \qquad \text{(triangle inequality)}$$

for all $x, y, z \in \mathbb{R}^n$. These conditions are those required of any given function $\rho(x, y)$ of two variables with $x, y$ in a set $X$, for $\rho$ to be declared a **metric**; the pair $(X, \rho)$ is then a **metric space**, and if $\varepsilon > 0$, the $\varepsilon$-ball around $x$ is the set $\{y \in X | \rho(x, y) < \varepsilon\}$. With these balls one defines limits, continuity, etc. just as with $\mathbb{R}^n (=\mathbb{R})$, to obtain the standard *topology* (i.e., the open subsets of $\mathbb{R}^n$) in the sense explained on p.202, other metrics can give this same topology.

If now $(Y,\sigma)$ is a metric space then the *product metric* on $X \times Y$ is defined by

$$\tau\Big[(x_1, y_1), (x_2, y_2)\Big] = \Big[\rho^2(x_1, x_2) + \sigma^2(y_1, y_2)\Big]^{\frac{1}{2}} ;$$

it is an easy exercise to verify that $\tau$ satisfies the above rules for a metric, given that $\rho, \sigma$ do. (Indeed, if for the moment the Euclidean metric of $\mathbb{R}^n$ is denoted by $d_n$, then clearly $d_{n+1}$ is the product metric of $d_n$ and $d_1$.) The resulting open sets of $X \times Y$ then form the *product topology*, and this depends only on the topologies of X and Y.

If $(X, \rho)$, $(Y, \sigma)$ are metric spaces, and $f : X \to Y$ is a mapping that 'preserves distance', i.e. for all $x_1, x_2 \in X$

$$\sigma(fx_1, fx_2) = \rho(x_1, x_2)$$

then $f$ si called an **isometry**, (and is obviously an injection). In particular, if $X, Y$ are also vector spaces of finite dimension, then $f$ will also be a surjection if it is linear. Hence the linear isometries of $\mathbb{R}^n$ form a *group* (the orthogonal group) with respect to composition.

Text-books give the basic theory of these algebraic and topological structures, but often from a very abstract point of view. Less abstract treatments can be found in the books of Anderson [1], Griffiths and Hilton [23], or Higgins [31].

On matters of Analysis, (convergence of sequences, continuity, bounds of continuous functions, the intermediate value theorem, etc) there are many texts that can be consulted, but Scott and Timms [67] is written by geometers and accords well with Artmann's spirit. In most areas of mathematics, readers will need to try different texts, to find the one most appropriate to their taste.

# Comments on the literature

In many places in the text, indications have been given, concerning sources or works that take the material further. Before we list all the References, it is convenient here to mention some works that are especially important for the book.

a) Introductory:

A. Oberschelp [52]: *Construction of the number system.* Very formally written, with a complete discussion of the constructions for the various domains of numbers.

S. Lang [42]. *Algebraic Structures.* Concerned with the foundations of algebra, but in the German edition number systems are often mentioned in the book's examples.

P. Halmos [25]. *Naïve Set–Theory.* Basic ideas of Set Theory, in a simple form, very well written, and easily accessible.

Behnke÷Fladt÷Süss [6]. *Outline of Mathematics* I. A collection of articles, of which we mention particularly the article "Construction of the number system" by G. Pickert and L. Görke [57], as well as "Complex numbers and Quaternions" by G. Pickert and H.G. Steiner [58].

(b) More comprehensive special treatises:

W. Felscher [17]: *Naïve Sets and abstract numbers*, I, II, III. (Altogether, some 750 pages). Very strong on foundational questions, approached set–theoretically. Detailed historical commentary, with general mathematical explanations that go beyond the particular subjects of discussion.

H. Salzmann [65]. *Number–domains* I, II. (Altogether some 800 pages.) Here, the accent lies more on the topology and algebraic structures that are important for foundations of Geometry. This work has greatly influenced the author's own lectures.

H. Gericke [22]. *History of the concept of number.*

# References


1. Anderson, I.T. *Introduction to Field Theory*, Oliver & Boyd, Edinburgh (1964).

2. Armitage, J V. & Griffiths, H B. *A Companion to Advanced Mathematics*, Cambridge (1969).

3. Artmann, B. *Einführung in die neue Algebra*, Vandenhoek and Ruprecht, Göttingen (1980).

4. Becker, O. *Grundlagen der Mathematik in geschichtlicher Entwicklung*, Verlag Karl Alber.

5. Beckmann, F. "Neue Gesichtspunkte zum 5. Buch Euklids", *Arch.f. History of Exact Sci.4* (1967/68) pp.1 – 144.

6. Behnke– Fladt– Süss. *Grundzüge der Mathematik I*, Vandenhoeck & Ruprecht, Göttingen (1958).

7. Behrend, F A. "A Contribution to the Theory of Magnitudes and the Foundations of Analysis" *Math. Zeitschrift 63* (1956) pp. 345 – 362.

8. Bourbaki, N. *General Topology,* Parts I, II, Addison - Wesley.

9. Brenner, J L & Lyndon R C. "Proof of the Fundamental Theorem of Algebra". *Amer. Math. Monthly 88* (1981) pp. 253 – 256.

10. Courant, R & Robbins, H. *What is Mathematics?* Oxford (1941)

11. Dieudonné J. *Linear Algebra and Geometry* Kershaw/London (1969)

12. Dugac, P. "Elements d'analyse de K. Weiersbrass", *Arch Hist. Exact Sciences 10* (1973) pp. 41–176.

13. Eder, G. *Atomphysik, Quantenmechanik II,* BI Mannheim (1978),

14. Eisenberg, M D & Guy, R. "A Proof of the Hairy Ball Theorem", *Amer. Math. Monthly 86* (1979) pp. 572 - 574.

15. Endl, K. & Luh, W. *Analysis* I, II, III. Akademische Verlagsgesellschaft, Frankfurt (1972– 4).

16. Erdös, P, Gillman, L & Henriksen M. "An Isomorphism Theorem for Real– Closed Fields" *Ann. of Math 61* (1955) pp.542 - 554

17. Felscher, W. *Naïve Mengen und abstracte Zahlen*, (3 Vols).BI Mannheim (1978/9)

18. Felix, J. "Widerspruchsfreiheit und Unabhängigkeit", *Jahrbuch uber Überblicke Math.* (1977), ed Fuchssteiner, BI Zürich.

19. Fowler, D. "Book II of Euclid's Elements and a pre– Eudoxan theory of Ratio", *Arch Hist. Exact Sciences 22* (1980) pp. 5 - 36.

20. Franz, W. *Topologie I*, Sammlung Göschen, Berlin/New York (1973).

21. Führer, L. *Allgemeine Topologie mit Anwendungen*, Vieweg, Braunschweig (1977).

22. Gericke, H. *Geschichte des Zahlbegriffs*, BI Mannheim (1970).

23. Griffiths, H B , & Hilton, P J. *Classical Mathematics: a Contemporary Interpretation*, Springer, New York (1980).

24. Griffiths, H. B , & Howson A G. *Mathematics: Society and Curricula*, Cambridge (1974).

25. Halmos, P. *Naïve Set Theory*, Van Nostrand (1960).

26. Hamel, G. "Eine Basis aller Zahlen und die unstetigen Lösungen der Funktionalgleichung f(x + y) = f(x) + f(y)", *Math. Annalen 60* (1905) pp 459 - 462.

27. Hardy G H, & Wright, E M. *Introduction to the Theory of Numbers* (4th Edition), Oxford (1960)

28. Hatcher W. S. "Calculus is Algebra", *Amer. Math. Monthly 89* (1982) pp.362 - 370.

29. Hausdorff, F. *Mengenlehre*, Dover Reprints.

30. Henle - Kleinberg, *Infinitesimal Calculus*, MIT Press, Cambridge (Mass) (1979)

31. Higgins, P J. *A First Course in Abstract Algebra*, Van Nostrand Reinhold (1975)

32. Hilbert, D. *Foundations of Geometry* (Transl. E.J. Townsend) Open Court, Chicago (1889).

33. Holland, G. "Ein Vorschlag zur Einführung der reellen Zahlen als Dezimalbrüche", *Math. Phys. Semesterberichte 18* (1971) pp 87 - 110.

34. Holland, G. "Dezimalbrüche und reele Zahlen", *Der Mathematikunterricht*. Heft 3/1973 pp 5 - 26.

35. Holmann, H. "Merkwürdige Eigenschaften der Kugeloberfläche", *Math. Phys. Sem. Ber.* X (1964) pp. 99 - 108.

36. Joyce, J. *Ullysses*, Bodley Head (1937).

37. Kamke, E. *Mengenlehre*, Sammlung Göschen. De Gruyter, Berlin (1955)

38. Kaplansky, I. *Set Theory and Metric Spaces*, Allyn and Bacon, Boston (1972).

39. Keisler, J. *Foundations of Infinitesimal Calculus*, Prindle, Weber & Schmidt, Boston, (1976).

40. Kirsch, A. "Die Funktionalgleichung $f(x + x') = f(x) + f(x')$ als Thema für den Oberstufenunterricht", *Math. Phys. Sem. Ber. 24* (1977) pp.172 - 187.

41. Kowalsky H J. *Einführung in die Lineare Algebra*, Göschens Lehrbücherei, Vol. 27, De Gruyter, Berlin (1963).

42. Lang. S. *Algebraic Structures*, Adison Wesley, Reading (Mass.) (1967).

43. Langwitz, D. *Infinitesimal Kalkül*, BI Mannheim (1978).

44. Leinfelder, H. "Zum Fundamentalsatz der Algebra", *Didaktik der Mathematik 3* (1981).

45. Lenz, H. "Grundlagen der Elementarmathematik", Chap. VII of VEB *Deutscher Verlag der Wissenschaften*, Berlin (1967).

46. Lightstone, A. H. "Infinitesimals", *Amer. Math. Monthly 79* (1972) pp. 242-251.

46. Maak, W. *Differential-und Integralrechnung*, Göttingen (1960).

48. Mamona, J. *Students' Interpretations of Some Concepts of Mathematical Analysis.* (Ph.D. Thesis), Southampton (1987).

49. Milnor, J. "Some consequences of a Theorem of Bott", *Annals of Math. 68* (1958) pp. 444-449.

50. Milnor, J. "Analytic Proofs of the "Hairy Ball Theorem" and the Brouwer Fixed Point Theorem", *Amer. Math. Monthly 85* (1978) pp.521-524.

51. Mirsky, L. *An Introduction to Linear Algebra*, Oxford (1955).

52. Oberschelp, A. *Aufbau des Zahlensystems*, Vandenhoeck & Ruprecht, Göttingen (1972).

53. Otte, M. (Ed.) *Mathematiker über die Mathematik*, Heidelberg, Springer (1974).

54. Palais, R S. "The Classification of Real Division Algebras", *Amer. Math. Monthly 75* (1968) pp.366-368.

55. Pickert, G. "Wissenschaftlichen Grundlagen des Funktionsbegriffs", *Der Mathematikunterricht 15* pp.40-98.

56. Pickert, G. *Projective Ebenen*, Springer, Berlin (1975).

57. Pickert, G. & Görke, L. "Aufbau des Systems der reelen Zahlen" in Behnke - Fladt - Suss [ 6 ].

58. Pickert, G. & Steiner, H. G. "Komplexe Zahlen und Quaternionen", Ibid.

59. Pontrjagin, L. *Topological Groups*, Princeton (1946)

60. Priess-Crampe, S. *Angeordnete Strukturen: Gruppen, Körper, projective Ebenen,* Springer, Heidelberg (1983).

61. Rautenberg, W. "Ein kurzer und direkter Weg von den natürlichen Zahlen zu den reellen Zahlen mit anschliessender Begründung der Bruchrechnung", *Mathematik in der Schule*, (1966).

62. Rautenberg, W. *Reelle Zahlen in elementarer Darstellung*, Klett, Stuttgart. (1979).

63. Richter, M. *Ideale Punkte, Monaden und Nichtstandard Methoden*, Vieweg, Braunschweig (1982).

64. Rieger, C J. "Introducing the Real Numbers via Continued Fractions" *Abhandlungen der Braunschw. Wiss. Gesellschaft*, 33 (1982).

65. Salzmann, H. *Zahlbrereiche* (2 Vols). Lecture notes, Tübingen (1971/2).

66. Schafmeister, O. & Wiebe, H. *Grundzüge der Algebra*, Teubner, Stuttgart (1978).

67. Scott, B. & Timms, *Mathematical Analysis*, Cambridge (1966).

68. Stark, H. *An introduction to Number Theory*, MIT Press, Cambridge/Mass. (1978).

69. Steiner, H G. "Equivalente Fassungen des Vollständigkeitsaxioms für die Theorie der reellen Zahlen", *Math. Phys. Sem. Ber 13.* (1966) pp.180 - 201.

70. Thomas, E. "Vector fields on manifolds", *Bull. Amer. Math. Soc.* 75 (1969) pp.643 - 683.

71. van der Waerden, B L. *Science Awakening,* Noordhoff, Groningen (1954).

72. van der Waerden, B L. "Hamilton's Discovery of Quaternions", *Math. Magazine* 49 (1976) pp.227 – 236.

73. van der Waerden, B L. *Modern Algebra* (English Translation by T.J. Benac) 2 Vols, Ungar (1950).

# Index

# List of symbols

(This list refers only to those symbols that have the same meaning throughout the book, see also the Appendix.)

Index of Symbols

## Mathematics and its Applications

*Series Editor:* G. M. BELL, Professor of Mathematics, King's College London (KQC), University of London

|  |  |
|---|---|
| Faux, I.D. & Pratt, M.J. | **Computational Geometry for Design and Manufacture** |
| Firby, P.A. & Gardiner, C.F. | **Surface Topology** |
| Gardiner, C.F. | **Modern Algebra** |
| Gardiner, C.F. | **Algebraic Structures: with Applications** |
| Gasson, P.C. | **Geometry of Spatial Forms** |
| Goodbody, A.M. | **Cartesian Tensors** |
| Goult, R.J. | **Applied Linear Algebra** |
| Graham, A. | **Kronecker Products and Matrix Calculus: with Applications** |
| Graham, A. | **Matrix Theory and Applications for Engineers and Mathematicians** |
| Graham, A. | **Nonnegative Matrices and Applicable Topics in Linear Algebra** |
| Griffel, D.H. | **Applied Functional Analysis** |
| Griffel, D.H. | **Linear Algebra** |
| Guest, P. B. | **The Laplace Transform and Applications** |
| Hanyga, A. | **Mathematical Theory of Non-linear Elasticity** |
| Harris, D.J. | **Mathematics for Business, Management and Economics** |
| Hart, D. & Croft, A. | **Modelling with Projectiles** |
| Hoskins, R.F. | **Generalised Functions** |
| Hoskins, R.F. | **Standard and Non-standard Analysis** |
| Hunter, S.C. | **Mechanics of Continuous Media, 2nd (Revised) Edition** |
| Huntley, I. & Johnson, R.M. | **Linear and Nonlinear Differential Equations** |
| Jaswon, M.A. & Rose, M.A. | **Crystal Symmetry: The Theory of Colour Crystallography** |
| Johnson, R.M. | **Theory and Applications of Linear Differential and Difference Equations** |
| Johnson, R.M. | **Calculus: Theory and Applications in Technology and the Physical and Life Sciences** |
| Jones, R.H. & Steele, N.C. | **Mathematics of Communication** |
| Jordan, D. | **Geometric Topology** |
| Kelly, J.C. | **Abstract Algebra** |
| Kim, K.H. & Roush, F.W. | **Applied Abstract Algebra** |
| Kim, K.H. & Roush, F.W. | **Team Theory** |
| Kosinski, W. | **Field Singularities and Wave Analysis in Continuum Mechanics** |
| Krishnamurthy, V. | **Combinatorics: Theory and Applications** |
| Lindfield, G. & Penny, J.E.T. | **Microcomputers in Numerical Analysis** |
| Livesley, K. | **Engineering Mathematics** |
| Lord, E.A. & Wilson, C.B. | **The Mathematical Description of Shape and Form** |
| Malik, M., Riznichenko, G.Y. & Rubin, A.B. | **Biological Electron Transport Processes and their Computer Simulation** |
| Massey, B.S. | **Measures in Science and Engineering** |
| Meek, B.L. & Fairthorne, S. | **Using Computers** |
| Menell, A. & Bazin, M. | **Mathematics for the Biosciences** |
| Mikolas, M. | **Real Functions and Orthogonal Series** |
| Moore, R. | **Computational Functional Analysis** |
| Murphy, J.A., Ridout, D. & McShane, B. | **Numerical Analysis, Algorithms and Computation** |
| Nonweiler, T.R.F. | **Computational Mathematics: An Introduction to Numerical Approximation** |
| Ogden, R.W. | **Non-linear Elastic Deformations** |
| Oldknow, A. | **Microcomputers in Geometry** |
| Oldknow, A. & Smith, D. | **Learning Mathematics with Micros** |
| O'Neill, M.E. & Chorlton, F. | **Ideal and Incompressible Fluid Dynamics** |
| O'Neill, M.E. & Chorlton, F. | **Viscous and Compressible Fluid Dynamics** |
| Page, S. G. | **Mathematics: A Second Start** |
| Prior, D. & Moscardini, A.O. | **Model Formulation Analysis** |
| Rankin, R.A. | **Modular Forms** |
| Scorer, R.S. | **Environmental Aerodynamics** |
| Smith, D.K. | **Network Optimisation Practice: A Computational Guide** |
| Shivamoggi, B.K. | **Stability of Parallel Gas Flows** |
| Stirling, D.S.G. | **Mathematical Analysis** |
| Sweet, M.V. | **Algebra, Geometry and Trigonometry in Science, Engineering and Mathematics** |
| Temperley, H.N.V. | **Graph Theory and Applications** |
| Thom, R. | **Mathematical Models of Morphogenesis** |
| Townend, M. S. | **Mathematics in Sport** |
| Townend, M.S. & Pountney, D.C. | **Computer-aided Engineering Mathematics** |
| Twizell, E.H. | **Computational Methods for Partial Differential Equations** |
| Twizell, E.H. | **Numerical Methods, with Applications in the Biomedical Sciences** |
| Vince, A. and Morris, C. | **Mathematics for Computer Studies** |
| Walton, K., Marshall, J., Gorecki, H. & Korytowski, A. | **Control Theory for Time Delay Systems** |
| Warren, M.D. | **Flow Modelling in Industrial Processes** |
| Wheeler, R.F. | **Rethinking Mathematical Concepts** |
| Willmore, T.J. | **Total Curvature in Riemannian Geometry** |
| Willmore, T.J. & Hitchin, N. | **Global Riemannian Geometry** |

**Numerical Analysis, Statistics and Operational Research**

*Editor:* B. W. CONOLLY, Professor of Mathematics (Operational Research), Queen Mary College, University of London

| | |
|---|---|
| Beaumont, G.P. | **Introductory Applied Probability** |
| Beaumont, G.P. | **Probability and Random Variables** |
| Conolly, B.W. | **Techniques in Operational Research: Vol. 1, Queueing Systems** |
| Conolly, B.W. | **Techniques in Operational Research: Vol. 2, Models, Search, Randomization** |
| Conolly, B.W. | **Lecture Notes in Queueing Systems** |
| Conolly, B.W. & Pierce, J.G. | **Information Mechanics: Transformation of Information in Management, Command, Control and Communication** |
| French, S. | **Sequencing and Scheduling: Mathematics of the Job Shop** |
| French, S. | **Decision Theory: An Introduction to the Mathematics of Rationality** |
| Griffiths, P. & Hill, I.D. | **Applied Statistics Algorithms** |
| Hartley, R. | **Linear and Non-linear Programming** |
| Jolliffe, F.R. | **Survey Design and Analysis** |
| Jones, A.J. | **Game Theory** |
| Kapadia, R. & Andersson, G. | **Statistics Explained: Basic Concepts and Methods** |
| Moscardini, A.O. & Robson, E.H. | **Mathematical Modelling for Information Technology** |
| Moshier, S. | **Mathematical Functions for Computers** |
| Oliveira-Pinto, F. | **Simulation Concepts in Mathematical Modelling** |
| Ratschek, J. & Rokne, J. | **New Computer Methods for Global Optimization** |
| Schendel, U. | **Introduction to Numerical Methods for Parallel Computers** |
| Schendel, U. | **Sparse Matrices** |
| Sehmi, N.S. | **Large Order Structural Eigenanalysis Techniques: Algorithms for Finite Element Systems** |
| Späth, H. | **Mathematical Software for Linear Regression** |
| Spedicato, E. and Abaffy, J. | **ABS Projection Algorithms** |
| Stoodley, K.D.C. | **Applied and Computational Statistics: A First Course** |
| Stoodley, K.D.C., Lewis, T. & Stainton, C.L.S. | **Applied Statistical Techniques** |
| Thomas, L.C. | **Games, Theory and Applications** |
| Whitehead, J.R. | **The Design and Analysis of Sequential Clinical Trials** |